Molecular Insights into the Developmental Origins of Health and Disease

Guest Editors

Paramjit S. Tappia
Bram Ramjiawan

MDPI

Basel • Beijing • Wuhan • Barcelona • Belgrade • Novi Sad • Cluj • Manchester

Guest Editors

Paramjit S. Tappia
Asper Clinical Research Institute
St. Boniface Hospital
Winnipeg, MB
Canada

Bram Ramjiawan
Asper Clinical Research Institute
St. Boniface Hospital
Winnipeg, MB
Canada

Editorial Office
MDPI AG
Grosspeteranlage 5
4052 Basel, Switzerland

This is a reprint of the Special Issue, published open access by the journal *International Journal of Molecular Sciences* (ISSN 1422-0067), freely accessible at: https://www.mdpi.com/journal/ijms/special_issues/2G0G06C8T2.

For citation purposes, cite each article independently as indicated on the article page online and as indicated below:

Lastname, A.A.; Lastname, B.B. Article Title. *Journal Name* **Year**, *Volume Number*, Page Range.

ISBN 978-3-7258-7364-7 (Hbk)
ISBN 978-3-7258-7365-4 (PDF)
https://doi.org/10.3390/books978-3-7258-7365-4

Contents

About the Editors

Paramjit S. Tappia

Paramjit S. Tappia has over 30 years of research experience, with a particular interest in nutrition, biomarkers, and subcellular and molecular mechanisms of human disease. Dr. Tappia received his B.Sc. (Honors) in Pharmacology from the University of Sunderland, U.K., in 1985 and his Ph.D. in Biochemistry from the University of Wolverhampton, U.K., in 1992. He received postdoctoral training at the Institute of Human Nutrition, University of Southampton, U.K., from 1992 to 1995. In 1996, he joined the Institute of Cardiovascular Sciences, St. Boniface Hospital Albrechtsen Research Centre, Winnipeg, Canada, as a Research Associate. He is currently a Principal Investigator and Clinical Scientist at the Asper Clinical Research Institute, St. Boniface Hospital. To date, he has published 144 full-length papers in peer-reviewed journals with an H-index of 39 and 43 book chapters. He has edited 11 books and currently serves as Associate Editor of *Molecular and Cellular Biochemistry* and is on the Editorial Boards for four other international journals. He also serves as Editor-in-Chief of *CV Network*, which is the official news bulletin of the International Academy of Cardiovascular Sciences. Dr. Tappia has served on the scientific advisory boards as well as on organizing committees for several international meetings and has successfully trained several undergraduate and graduate students, technicians, and post-doctoral fellows.

Bram Ramjiawan

Bram Ramjiawan is the Director of Research Innovation and Regulatory Affairs and Director of Research, Asper Clinical Research Institute at the St. Boniface Hospital and Albrechtsen Research Centre in Winnipeg, Manitoba, Canada. He is responsible for the oversight of clinical research and to oversee and ensure that all clinical, regulatory and business issues are handled as required by national and international agencies. He also conducts fee-for-service work for various (industry and non-for-profit) entities. His own research pursuit involves precise detection methods (early tumor detection and vitamin D). Prior to joining the hospital, Dr. Ramjiawan was with the Government of Canada (National Research Council) as an Industrial Technology Advisor who specialized in life sciences and biomedical technologies. He is an Adjunct Professor in the Department of Pharmacology and Therapeutics in the Max Rady College of Medicine, Faculty of Health Sciences, University of Manitoba.

Preface

Drawing on the landmark work of David Barker and the Developmental Origins of Health and Disease, Drs. Paramjit Tappia and Bram Ramjiawan introduce a collection exploring how early-life environments shape lifelong health. The Reprint will appeal to researchers, clinicians, public health practitioners, and students interested in developmental and preventive medicine.

Paramjit S. Tappia and Bram Ramjiawan
Guest Editors

International Journal of
Molecular Sciences

MDPI

Editorial

Special Issue "Molecular Insights into the Developmental Origins of Health and Disease"

Paramjit S. Tappia [1,*] and Bram Ramjiawan [1,2]

1 Asper Clinical Research Institute, St. Boniface Hospital, Winnipeg, MB R2H 2A6, Canada; bramjiawan@sbrc.ca
2 Department of Pharmacology and Therapeutics, Max Rady College of Medicine, Rady Faculty of Health Sciences, University of Manitoba, Winnipeg, MB R3E 0T6, Canada
* Correspondence: ptappia@sbrc.ca

There is now a wealth of epidemiological, clinical, and experimental evidence that have concluded that the risk of developing chronic, noncommunicable diseases in adulthood may be influenced by molecular and genetic aspects [1–5], as well as lifestyle and environmental experiences during early life [6–13]. Fetal development and infancy are characterized by the rapid growth, development, and maturation of organ systems. It has now become increasingly evident that several pathophysiological conditions, including diabetes and cardiovascular disease, that occur in adolescence and adulthood, may have their origins during fetal or postnatal development. While maternal nutrition (poor quality and quantity) is the most examined aspect in terms of influencing the development of fetal organ systems, paternal stressors have also emerged as critical developmental and molecular elements that can increase offspring's predisposition to adverse health outcomes in later life.

Dr. David Barker first published findings proposing a direct link between prenatal nutrition and late-onset coronary heart disease in *The Lancet* [14]. This was the first paper to propose the core idea of the Barker hypothesis. From that time, the Developmental Origins of Health and Disease (DOHaD) framework has advanced significantly in recent years, driven by new mechanistic insights, improved analytical tools, and an expanded understanding of how early-life environments shape long-term health trajectories. These developments are strengthening the evidence base for early prevention and reinforcing the need for integrated life-course approaches to public health.

Rapid growth in epigenomic, transcriptomic, and metabolomic technologies has enabled precise mapping of how early exposures regulate gene expression and organ development. High-resolution epigenetic analyses now identify stable molecular signatures associated with maternal diet, metabolic health, stress, and environmental toxicants [15–24]. Single-cell approaches have further clarified how these exposures influence the developmental programming of key tissues, particularly through the placenta, which is now recognized as an active mediator of environmental signals.

Recent work highlights the central role of the maternal–infant microbiome axis in shaping immune function and metabolic homeostasis. Alterations in maternal microbial ecology, linked to nutrition, antibiotic use, or stress, have measurable effects on neonatal microbiome establishment and downstream health outcomes [25–30]. Parallel advances have also established that paternal factors contribute meaningfully to early programming. Epigenetic and small-RNA profiles in sperm reflect paternal nutritional and environmental exposures, challenging the maternal-centric paradigm that long dominated the field [31,32].

DOHaD research is moving beyond association toward targeted intervention. Trials addressing maternal nutrition, gestational diabetes, physical activity, and stress reduction demonstrate measurable benefits in offspring metabolic and neurodevelopmental outcomes. These findings underscore early development as a critical window of plasticity, where modifying environmental and social conditions can produce lasting health effects.

The field has increasingly acknowledged the influence of social determinants including inequity, historical trauma, and environmental disadvantage on early programming. Research conducted in partnership with marginalized communities has been particularly important in illustrating how intergenerational stress and structural conditions become biologically embedded [33–35]. These perspectives emphasize the necessity of culturally grounded, community-led approaches in DOHaD research and policy translation.

With this background and perspectives, this Special Issue provides a glimpse into the influential role of internal and external factors on fetal genetics, biochemistry and physiology. Eight outstanding papers, from experts in the field, provide a broad range of contributions detailing advances in the field of developmental biology and the reinforcement of a life-course model of health that prioritizes preconception care, maternal well-being, and early childhood development as essential components of chronic disease prevention in later life.

We earlier analyzed the phospholipid profile of the developing heart of rats exposed to low-protein (LP) diet in pregnancy [36]. It was found that maternal LP diet can induce changes in the phospholipid profile and fatty acid content of the developing heart, which may have implications for metabolism of the neonatal heart. After 25 years, the topic of cell membrane lipid composition and its influence on cellular structural integrity and functionality has been reviewed from a developmental and pathological perspective by Ali and Szabó [37]. In this review, the authors summarize the diversity of membrane lipids and their constituent fatty acids in healthy organisms and the essential roles they play in cellular function and highlight the functional significance of membrane-associated fatty acids and their influence on key cellular physiological responses.

Early life exposures to environmental/maternal chemicals can exert negative effects on the developing fetus [38]. There are three articles that demonstrate this. In the review by Tain and Hsu [39], various environmental chemicals to which pregnant mothers are commonly exposed can disrupt fetal programming, leading to a wide range of cardiovascular–kidney–metabolic phenotypes. The authors present that the aryl hydrocarbon receptor (AHR) plays a key role as a ligand-activated transcription factor in sensing these environmental chemicals that can increase the predisposition to cardiovascular diseases, hypertension, diabetes, obesity, kidney disease, and non-alcoholic fatty liver disease. The authors emphasize the importance of circumventing toxic chemical exposure during pregnancy and extend the understanding of AHR signaling, which may potentially lessen the global burden of CKM syndrome.

In the original research paper by Strunz et al. [40], the impact of polybrominated diphenyl ethers (PBDEs), commonly used as synthetic flame retardants, and they are present in a variety of products, including electronics, polyurethane foams, textiles, and building materials, on weight gain and insulin insensitivity in offspring has been examined. In a mouse model, these researchers demonstrated that maternal exposure to BDE-47 results in weight gain in female offspring together with an impaired glucose and insulin tolerance in both female and male mice. It was also found that this chemical toxin increased adipogenesis as well as neuronal dysregulation of energy homeostasis, attributed to impaired leptin signaling. The authors concluded that these mechanistic aspects may contribute to the observed weight gain as well as impaired insulin and glucose tolerance.

ETV6::RUNX1-positive pediatric acute lymphoblastic leukemia is frequently considered to have a prenatal origin. In an interventional cohort study (1221 pregnancies), the impact of maternal corticosteroid use during pregnancy and its association with this form of leukemia at birth are examined. In this study by Benítez et al. [41], it was observed that corticosteroid use for lung maturation during pregnancy was significantly associated with ETV6::RUNX1 in 39 neonates, particularly if applied before 26 weeks of gestation or if betamethasone was used. Thus, it was concluded that prenatal exposure to corticosteroids within a critical time window could increase the risk of developing ETV6::RUNX1+ preleukemic clones and potentially leukemia after birth. In addition, it was stated that modulation of ETV6::RUNX1 preleukemia may potentially prevent this subtype of childhood leukemia.

The following two articles shed some light on the developmental mechanisms that can influence brain function and lung morphogenesis. KCC2 (Potassium–Chloride Cotransporter 2) and NKCC1 (Sodium–Potassium–Chloride Cotransporter 1) are crucial for maintaining chloride ion balance inside and outside neurons, playing complementary roles in the regulation of GABAergic (gamma-aminobutyric acid) inhibition and chloride homeostasis in the nervous system. Brain-derived neurotrophic factor (BDNF) influences the functioning of these co-transporters and normal brain function. Their dysregulation is implicated in neurological disorders. In the study by Hamze et al. [42], it was reported that proBDNF delays the GABA shift polarity, thereby maintaining neurons in an immature state, which could be linked to the behavioral deficits in electroporated rats. These actions carry significant implications for cognitive processes and neural circuitry, providing insights into the intricate interplay between neurotrophic factors and neuronal functions. The authors concluded that these findings advance the understanding of neurodevelopmental processes that could potentially lead to the development of targeted therapies for brain disorders.

It is known that pulmonary branches are formed during the early stages of embryonic lung development through an intricate process known as lung branching morphogenesis that is influenced by retinoic acid (RA). Fernandes-Silva et al. [43] explored the role of RA as a metabolic modulator in an ex vivo model of lung explant culture of embryonic chicken lungs. It was revealed that RA signaling stimulation redirects glucose towards pyruvate and succinate production rather than to alanine or lactate. Of note, the inhibition of RA signal transduction reduced lung branching, which resulted in a cystic-like phenotype while promoting mitochondrial function. Accordingly, the authors suggested that RA is a regulator of tissue proliferation and that it plays an important role in determining lung metabolism during branching morphogenesis. Taken together, such information adds to the understanding of lung development and cystic-related lung disorders.

By using a rat model where maternal CKD is induced with adenine, Tain et al. [44] have demonstrated that the offspring develop high blood pressure in adulthood. These researchers also tested whether citrulline, a non-essential amino acid that can boost the production of nitric oxide (NO) and exhibits antioxidant properties, could prevent this outcome. Citrulline effectively normalized blood pressure in the offspring of CKD mothers. Its protective effects were associated with improved NO signaling, reduced renal (pro)renin receptor expression, and induced beneficial shifts in gut microbiota. Overall, perinatal citrulline supplementation enhanced NO availability and prevented hypertension programmed by maternal CKD. The study underscores the importance of maternal nutrition for fetal development but also shows how specific nutrients could be used in conditions where there is an adverse environmental experience that could predispose the offspring to chronic disease in later life.

In the last paper of this Special Issue by Lee et al. [45], an interesting topic of external pathogens, liquid biopsy, and biomarkers that could be predictive of later systemic disease is highlighted. Tick bites and related illnesses are increasing, but current diagnostic tests detect only well-known tick-borne pathogens and cannot explain poorly defined conditions like Australia's debilitating symptom complexes attributed to ticks (DSCATT). This study investigated whether blood samples could capture molecular signals originating from tick-bitten skin, offering a less invasive way to monitor local biological responses. By comparing multi-omics profiles from skin biopsies at the bite site with matched peripheral blood samples, researchers found that the pathways involved in extracellular matrix organization and platelet degranulation were activated in the skin within 72 h of a tick bite. Importantly, these same signals appeared in the blood and remained elevated for up to three months. These findings are suggestive that systemic blood profiles can reflect local tissue events after a tick bite, supporting the potential of blood-based "liquid biopsies" for future diagnostic and mechanistic studies on the development of systemic illnesses that may be tick-borne pathogen-based.

In conclusion, the last decade has marked a period of remarkable progress in DOHaD research. Technological advancements, integrative multi-omics, a broader appreciation of paternal and sociocultural influences, and the emergence of intervention studies are reshaping the field. What once seemed a revolutionary idea that early life shapes lifelong health is now an actionable framework guiding clinical practice, public health strategy, and global policy. The next advancement is clear: moving from understanding mechanisms to implementing solutions. The promise of DOHaD lies not only in explaining disease origins but in enabling societies to build healthier futures by protecting and empowering the earliest stages of life.

We extend our sincere appreciation to all contributors, each highly regarded in their respective fields, for helping to create this exceptional Special Issue. We hope that both experts and readers with a broader interest in DOHaD will find the content informative and thought-provoking. In addition, we hope that it will stimulate, motivate, and inspire biomedical scientists and researchers into further inquiry in order to advance our understanding of the essential role played by diverse maternal, paternal, and environmental factors in fetal biology and predisposition to adverse health conditions. Furthermore, this Special Issue will serve as a beneficial resource for medical students and fellows as well as graduate students that have a desire to learn about DOHaD.

Author Contributions: Conceptualization, P.S.T.; resources, B.R.; writing—original draft preparation, P.S.T.; writing—review and editing, P.S.T. and B.R. All authors have read and agreed to the published version of the manuscript.

Acknowledgments: Infrastructural support was provided by the St. Boniface Hospital Albrechtsen Research Centre, Winnipeg, MB, Canada.

Conflicts of Interest: The authors declare no conflicts of interest.

References

1. Arima, Y.; Fukuoka, H. Developmental origins of health and disease theory in cardiology. *J. Cardiol.* **2020**, *76*, 14–17. [CrossRef]
2. Zhou, L.Y.; Deng, M.Q.; Zhang, Q.; Xiao, X.H. Early-life nutrition and metabolic disorders in later life: A new perspective on energy metabolism. *Chin. Med. J.* **2020**, *133*, 1961–1970. [CrossRef]
3. Nobile, S.; Di Sipio Morgia, C.; Vento, G. Perinatal origins of adult disease and opportunities for health promotion: A narrative review. *J. Pers. Med.* **2022**, *12*, 157. [CrossRef]
4. Harary, D.; Akinyemi, A.; Charron, M.J.; Fuloria, M. Fetal growth and intrauterine epigenetic programming of obesity and cardiometabolic disease. *Neoreviews* **2022**, *23*, e363–e372. [CrossRef]
5. Alabduljabbar, S.; Zaidan, S.A.; Lakshmanan, A.P.; Terranegra, A. Personalized nutrition approach in pregnancy and early life to tackle childhood and adult non-communicable diseases. *Life* **2021**, *11*, 467. [CrossRef] [PubMed]
6. Picó, C.; Reis, F.; Egas, C.; Mathias, P.; Matafome, P. Lactation as a programming window for metabolic syndrome. *Eur. J. Clin. Investig.* **2021**, *51*, e13482. [CrossRef] [PubMed]
7. Thornburg, K.L.; Valent, A.M. Maternal Malnutrition and elevated disease risk in offspring. *Nutrients* **2024**, *16*, 2614. [CrossRef]

8. Ajuogu, P.K.; Al-Aqbi, M.A.K.; Hart, R.A.; McFarlane, J.R.; Smart, N.A. A low protein maternal diet during gestation has negative effects on male fertility markers in rats—A systematic review and meta-analysis. *J. Anim. Physiol. Anim. Nutr.* **2021**, *105*, 157–166. [CrossRef]
9. Tain, Y.L.; Hsu, C.N. Amino acids during pregnancy and offspring cardiovascular-kidney-metabolic health. *Nutrients* **2024**, *16*, 1263. [CrossRef] [PubMed]
10. Michońska, I.; Łuszczki, E.; Zielińska, M.; Oleksy, Ł.; Stolarczyk, A.; Dereń, K. Nutritional programming: History, hypotheses, and the role of prenatal factors in the prevention of metabolic diseases-a narrative review. *Nutrients* **2022**, *14*, 4422. [CrossRef]
11. Smith, E.V.L.; Dyson, R.M.; Weth, F.R.; Berry, M.J.; Gray, C. Maternal fructose intake, programmed mitochondrial function and predisposition to adult disease. *Int. J. Mol. Sci.* **2022**, *23*, 12215. [CrossRef]
12. Bar, J.; Weiner, E.; Levy, M.; Gilboa, Y. The thrifty phenotype hypothesis: The association between ultrasound and Doppler studies in fetal growth restriction and the development of adult disease. *Am. J. Obstet. Gynecol.* **2021**, *3*, 100473. [CrossRef]
13. Andonotopo, W.; Bachnas, M.A.; Akbar, M.I.A.; Aziz, M.A.; Dewantiningrum, J.; Pramono, M.B.A.; Sulistyowati, S.; Stanojevic, M.; Kurjak, A. Fetal origins of adult disease: Transforming prenatal care by integrating Barker's Hypothesis with AI-driven 4D ultrasound. *J. Perinat. Med.* **2025**, *53*, 418–438. [CrossRef]
14. Barker, D.J.; Osmond, C. Infant mortality, childhood nutrition, and ischaemic heart disease in England and Wales. *Lancet* **1986**, *1*, 1077–1081. [CrossRef] [PubMed]
15. Saavedra, L.P.J.; Piovan, S.; Moreira, V.M.; Gonçalves, G.D.; Ferreira, A.R.O.; Ribeiro, M.V.G.; Peres, M.N.C.; Almeida, D.L.; Raposo, S.R.; da Silva, M.C.; et al. Epigenetic programming for obesity and noncommunicable disease: From womb to tomb. *Rev. Endocr. Metab. Disord.* **2024**, *25*, 309–324. [CrossRef] [PubMed]
16. Dieckmann, L.; Czamara, D. Epigenetics of prenatal stress in humans: The current research landscape. *Clin. Epigenetics* **2024**, *16*, 20. [CrossRef]
17. Quinn, E.B.; Hsiao, C.J.; Maisha, F.M.; Mulligan, C.J. Prenatal maternal stress is associated with site-specific and age acceleration changes in maternal and newborn DNA methylation. *Epigenetics* **2023**, *18*, 2222473. [CrossRef]
18. Saenen, N.D.; Martens, D.S.; Neven, K.Y.; Alfano, R.; Bové, H.; Janssen, B.G.; Roels, H.A.; Plusquin, M.; Vrijens, K.; Nawrot, T.S. Air pollution-induced placental alterations: An interplay of oxidative stress, epigenetics, and the aging phenotype? *Clin. Epigenetics* **2019**, *11*, 124. [CrossRef]
19. Ghazi, T.; Naidoo, P.; Naidoo, R.N.; Chuturgoon, A.A. Prenatal air pollution exposure and placental DNA methylation changes: Implications on fetal development and future disease susceptibility. *Cells* **2021**, *10*, 3025. [CrossRef] [PubMed]
20. Street, M.E.; Shulhai, A.M.; Petraroli, M.; Patianna, V.; Donini, V.; Giudice, A.; Gnocchi, M.; Masetti, M.; Montani, A.G.; Rotondo, R.; et al. The impact of environmental factors and contaminants on thyroid function and disease from fetal to adult life: Current evidence and future directions. *Front. Endocrinol.* **2024**, *15*, 1429884. [CrossRef]
21. Navas-Acien, A.; Spratlen, M.J.; Abuawad, A.; LoIacono, N.J.; Bozack, A.K.; Gamble, M.V. Early-life arsenic exposure, nutritional status, and adult diabetes risk. *Curr. Diabetes Rep.* **2019**, *19*, 147. [CrossRef]
22. Marumure, J.; Simbanegavi, T.T.; Makuvara, Z.; Karidzagundi, R.; Alufasi, R.; Goredema, M.; Gufe, C.; Chaukura, N.; Halabowski, D.; Gwenzi, W. Emerging organic contaminants in drinking water systems: Human intake, emerging health risks, and future research directions. *Chemosphere* **2024**, *356*, 141699. [CrossRef] [PubMed]
23. Mottis, G.; Kandasamey, P.; Peleg-Raibstein, D. The consequences of ultra-processed foods on brain development during prenatal, adolescent and adult stages. *Front. Public Health* **2025**, *13*, 1590083. [CrossRef] [PubMed]
24. Galvan-Martinez, D.H.; Bosquez-Mendoza, V.M.; Ruiz-Noa, Y.; Ibarra-Reynoso, L.D.R.; Barbosa-Sabanero, G.; Lazo-de-la-Vega-Monroy, M.L. Nutritional, pharmacological, and environmental programming of NAFLD in early life. *Am. J. Physiol. Gastrointest. Liver Physiol.* **2023**, *324*, G99–G114. [CrossRef]
25. Mpakosi, A.; Sokou, R.; Theodoraki, M.; Kaliouli-Antonopoulou, C. Neonatal gut mycobiome: Immunity, diversity of fungal strains, and individual and non-individual factors. *Life* **2024**, *14*, 902. [CrossRef]
26. Ahmed, U.; Fatima, F.; Farooq, H.A. Microbial dysbiosis and associated disease mechanisms in maternal and child health. *Infect. Immun.* **2025**, *93*, e0017925. [CrossRef]
27. Vandenplas, Y.; Carnielli, V.P.; Ksiazyk, J.; Luna, M.S.; Migacheva, N.; Mosselmans, J.M.; Picaud, J.C.; Possner, M.; Singhal, A.; Wabitsch, M. Factors affecting early-life intestinal microbiota development. *Nutrition* **2020**, *78*, 110812. [CrossRef] [PubMed]
28. Bolte, E.E.; Moorshead, D.; Aagaard, K.M. Maternal and early life exposures and their potential to influence development of the microbiome. *Genome Med.* **2022**, *14*, 4. [CrossRef]
29. Faienza, M.F.; Urbano, F.; Anaclerio, F.; Moscogiuri, L.A.; Konstantinidou, F.; Stuppia, L.; Gatta, V. Exploring maternal diet-epigenetic-gut microbiome crosstalk as an intervention strategy to counter early obesity programming. *Curr. Issues Mol. Biol.* **2024**, *46*, 4358–4378. [CrossRef]
30. Chandra, M. Developmental origins of non-communicable chronic diseases: Role of fetal undernutrition and gut dysbiosis in infancy. *Children* **2024**, *11*, 1387. [CrossRef]

31. Eid, N.; Morgan, H.L.; Watkins, A.J. Paternal periconception metabolic health and offspring programming. *Proc. Nutr. Soc.* **2022**, *81*, 119–125. [CrossRef] [PubMed]
32. Sánchez-Garrido, M.A.; García-Galiano, D.; Tena-Sempere, M. Early programming of reproductive health and fertility: Novel neuroendocrine mechanisms and implications in reproductive medicine. *Hum. Reprod. Update* **2022**, *28*, 346–375. [CrossRef]
33. Phillips-Beck, W.; Sinclair, S.; Campbell, R.; Star, L.; Cidro, J.; Wicklow, B.; Guillemette, L.; Morris, M.I.; McGavock, J.M. Early-life origins of disparities in chronic diseases among Indigenous youth: Pathways to recovering health disparities from intergenerational trauma. *J. Dev. Orig. Health Dis.* **2019**, *10*, 115–122. [CrossRef]
34. Thornburg, K.L.; Boone-Heinonen, J.; Valent, A.M. Social determinants of placental health and future disease risks for babies. *Obstet. Gynecol. Clin. N. Am.* **2020**, *47*, 1–15. [CrossRef] [PubMed]
35. Krishnaveni, G.V.; Srinivasan, K. Maternal nutrition and offspring stress response-implications for future development of non-communicable disease: A perspective from India. *Front. Psychiatry* **2019**, *10*, 795. [CrossRef]
36. Tappia, P.S.; Nijjar, M.S.; Mahay, A.; Aroutiounova, N.; Dhalla, N.S. Phospholipid profile of developing heart of rats exposed to low-protein diet in pregnancy. *Am. J. Physiol. Regul. Integr. Comp. Physiol.* **2005**, *289*, R1400–R1406. [CrossRef] [PubMed]
37. Ali, O.; Szabó, A. Review of eukaryote cellular membrane lipid composition, with special attention to the fatty acids. *Int. J. Mol. Sci.* **2023**, *24*, 15693. [CrossRef]
38. Chen, Y.; He, Z.; Chen, G.; Liu, M.; Wang, H. Prenatal glucocorticoids exposure and fetal adrenal developmental programming. *Toxicology* **2019**, *428*, 152308. [CrossRef]
39. Tain, Y.-L.; Hsu, C.-N. The impact of the aryl hydrocarbon receptor on antenatal chemical exposure-induced cardiovascular–kidney–metabolic programming. *Int. J. Mol. Sci.* **2024**, *25*, 4599. [CrossRef]
40. Strunz, S.; Strachan, R.; Bauer, M.; Zenclussen, A.C.; Leppert, B.; Junge, K.M.; Polte, T. Maternal exposure to low-dose BDE-47 induced weight gain and impaired insulin sensitivity in the offspring. *Int. J. Mol. Sci.* **2024**, *25*, 8620. [CrossRef]
41. Benítez, L.; Fischer, U.; Crispi, F.; Castro-Barquero, S.; Crovetto, F.; Larroya, M.; Youssef, L.; Kameri, E.; Castillo, H.; Bueno, C.; et al. Modulation of the ETV6::RUNX1 gene fusion prevalence in newborns by corticosteroid use during pregnancy. *Int. J. Mol. Sci.* **2025**, *26*, 2971. [CrossRef] [PubMed]
42. Hamze, M.; Brier, C.; Buhler, E.; Zhang, J.; Medina, I.; Porcher, C. Regulation of Neuronal Chloride Homeostasis by Pro- and Mature Brain-Derived Neurotrophic Factor (BDNF) via KCC2 Cation–Chloride Cotransporters in Rat Cortical Neurons. *Int. J. Mol. Sci.* **2024**, *25*, 6253. [CrossRef] [PubMed]
43. Fernandes-Silva, H.; Alves, M.G.; Garcez, M.R.; Correia-Pinto, J.; Oliveira, P.F.; Homem, C.C.F.; Moura, R.S. Retinoic acid-mediated control of energy metabolism is essential for lung branching morphogenesis. *Int. J. Mol. Sci.* **2024**, *25*, 5054. [CrossRef]
44. Tain, Y.-L.; Hou, C.-Y.; Chang-Chien, G.-P.; Lin, S.; Hsu, C.-N. Perinatal use of citrulline rescues hypertension in adult male offspring born to pregnant uremic rats. *Int. J. Mol. Sci.* **2024**, *25*, 1612. [CrossRef]
45. Lee, W.; Barbosa, A.D.; Lee, A.H.-Y.; Currie, A.; Martino, D.; Stenos, J.; Long, M.; Beaman, M.; Harvey, N.T.; Kresoje, N.; et al. From local to systemic: The journey of tick bite biomarkers in Australian patients. *Int. J. Mol. Sci.* **2025**, *26*, 1520. [CrossRef] [PubMed]

International Journal of
Molecular Sciences

MDPI

Review

Review of Eukaryote Cellular Membrane Lipid Composition, with Special Attention to the Fatty Acids

Omeralfaroug Ali [1] and András Szabó [1,2,*]

1 Agrobiotechnology and Precision Breeding for Food Security National Laboratory, Institute of Physiology and Animal Nutrition, Department of Animal Physiology and Health, Hungarian University of Agriculture and Life Sciences, Guba Sándor Str. 40, 7400 Kaposvár, Hungary; omeralfaroug.ali@gmail.com

2 HUN-REN-MATE Mycotoxins in the Food Chain Research Group, Hungarian University of Agriculture and Life Sciences, Guba Sándor Str. 40, 7400 Kaposvár, Hungary

* Correspondence: szabo.andras@uni-mate.hu

Abstract: Biological membranes, primarily composed of lipids, envelop each living cell. The intricate composition and organization of membrane lipids, including the variety of fatty acids they encompass, serve a dynamic role in sustaining cellular structural integrity and functionality. Typically, modifications in lipid composition coincide with consequential alterations in universally significant signaling pathways. Exploring the various fatty acids, which serve as the foundational building blocks of membrane lipids, provides crucial insights into the underlying mechanisms governing a myriad of cellular processes, such as membrane fluidity, protein trafficking, signal transduction, intercellular communication, and the etiology of certain metabolic disorders. Furthermore, comprehending how alterations in the lipid composition, especially concerning the fatty acid profile, either contribute to or prevent the onset of pathological conditions stands as a compelling area of research. Hence, this review aims to meticulously introduce the intricacies of membrane lipids and their constituent fatty acids in a healthy organism, thereby illuminating their remarkable diversity and profound influence on cellular function. Furthermore, this review aspires to highlight some potential therapeutic targets for various pathological conditions that may be ameliorated through dietary fatty acid supplements. The initial section of this review expounds on the eukaryotic biomembranes and their complex lipids. Subsequent sections provide insights into the synthesis, membrane incorporation, and distribution of fatty acids across various fractions of membrane lipids. The last section highlights the functional significance of membrane-associated fatty acids and their innate capacity to shape the various cellular physiological responses.

Keywords: membranes; phospholipids; sphingolipids; fatty acid; de novo synthesis; desaturation; oxygenation; bioactive lipids; physicochemical; very long polyunsaturated fatty acids

1. Introduction

The biological membrane, commonly referred to as the biomembrane, holds paramount importance in both prokaryotic and eukaryotic cells. Its primary function lies in the selective regulation of molecular influx and efflux across the cellular boundary. Furthermore, it plays a crucial role in modulating intercellular communication and is involved in a vast array of complex processes, encompassing cell proliferation, differentiation, secretion, migration, invasion, and phagocytosis. However, the term "biomembrane" extends beyond the plasma membrane, as eukaryotic cells feature membranes within distinct cellular organelles [1,2], such as the endoplasmic reticulum (ER), mitochondria, nucleus, and various intracellular organelles. Additional functions of biomembranes revolve around stabilizing the consistency of cellular activities within the cell and organelles, controlling the trafficking of micromolecules (including O_2, CO_2, H_2O, H^+, K^+, HCO_3^-, Mg^{2+}, Ca^{2+}, etc.) and macromolecular compounds, and providing a surface where essential biological events take place. According to Janmey and Kinnunen [3], biomembranes represent heterogeneous, asymmetrical bilayers with complex structures that contribute to the maintenance of cellular homeostasis and functionality. Hence, biomembranes' systems exhibit considerable structural and dynamic diversity, making them an enduring area of scientific exploration.

The concept of Langmuir films, initially proposed by Langmuir in 1917, represents the earliest paradigm aimed at elucidating membrane systems [4]. Numerous subsequent paradigms have been developed in an attempt to explain membrane systems. The semi-fluid dynamics of biomembranes are merely determined by their intricate structure. The so-called "fluid mosaic model", one of the most renowned models in the biological domain, is employed to illustrate the structure and function of membranes. Singer and Nicolson introduced this model in 1972 [5], describing lipids, proteins, and carbohydrates as the primary constituents of the membrane. In light of the fact that proteins do not actually dissolve in membrane lipids, this proposal has undergone several amendments. After 25 years of Singer and Nicolson's proposal, Simons and Ikonen [6] proposed the "lipid raft" model, predicated on the clustering of sterols (namely, cholesterol in animals) and sphingolipids (SLs) within membranes to form microdomains where membrane-associated proteins are distributed. It has been established that these compartmentalized microdomains limit membrane lateral mobility and actively engage in various cellular events based on their structural arrangements [7]. Following the lipid rafts model, numerous other models have been introduced, which are either focused on revising the fluid mosaic model [8] or explaining the interaction between the similar [9] or distinct molecule classes [10] within membranes. Generally, the complexity of membranes exceeds that of model membranes due to the heterogeneous distribution of building molecules and their complex interactions. The continuous advancement of technology empowers science to delve deeper into the intricate structures of membranes, implying that the cell membrane model will invariably evolve toward increasing complexity, mirroring the progression from initial notions of membrane structure.

Lipids, proteins, and carbohydrates are pivotal biomolecules within biomembranes, exhibiting heterogeneous dispersion across membranes' structures (see Figure 1). Membrane lipids, marked by diversity and possessing distinct properties either individually or in conjunction with other moieties, contribute to bilayer development and serve essential functions. Almost 50% of the membrane matrix is composed of proteins, which exist in various structures such as including integral (embedded with lipid bilayers), peripheral (associated with the membrane surface), and anchoring (not directly attached but rather bound to lipid embedded with lipid bilayers) proteins. Hydrophobic forces or ionic interactions mediate the binding of membrane proteins to membrane lipids, forming lateral domains with certain functions such as environmental communication, adhesion, trafficking, and signaling. Carbohydrates form covalent bonds with proteins or lipids within membranes, which only occur at the outward surface of the plasma membrane, yielding glycol-complexes [2,11]. The extant biotic assemblies within biomembranes are postulated to have transited from thermodynamic reactions on analogous abiotic assemblies [12]. The interaction between membrane lipids and proteins may modulate their individual qualities, thereby altering membrane conformation.

Among the constituents of biomembranes, fatty acids comprising the lipid portion have gained great focus due to their diverse functions in cellular processes. Understanding the diversity and composition of eukaryotic biomembrane lipids, especially fatty acids, is essential for elucidating the underlying mechanisms controlling cellular functions. Furthermore, it sheds light on the potential roles that particular lipids and fatty acids may play in various physiological and pathological processes, including inflammation and metabolic disorders. The current review primarily focuses on a healthy organism, intending to highlight the enormous diversity of biomembrane lipids and, as a secondary objective, characterize the biological roles of distinct fatty acids embedded into the cellular membranes. In addition, this review enhances our knowledge of fundamental cellular processes and subtly underscores the potential for therapeutic strategies centered on the lipid composition and fatty acid metabolism of biomembranes, which are likely promising foundations for further scientific inquiry.

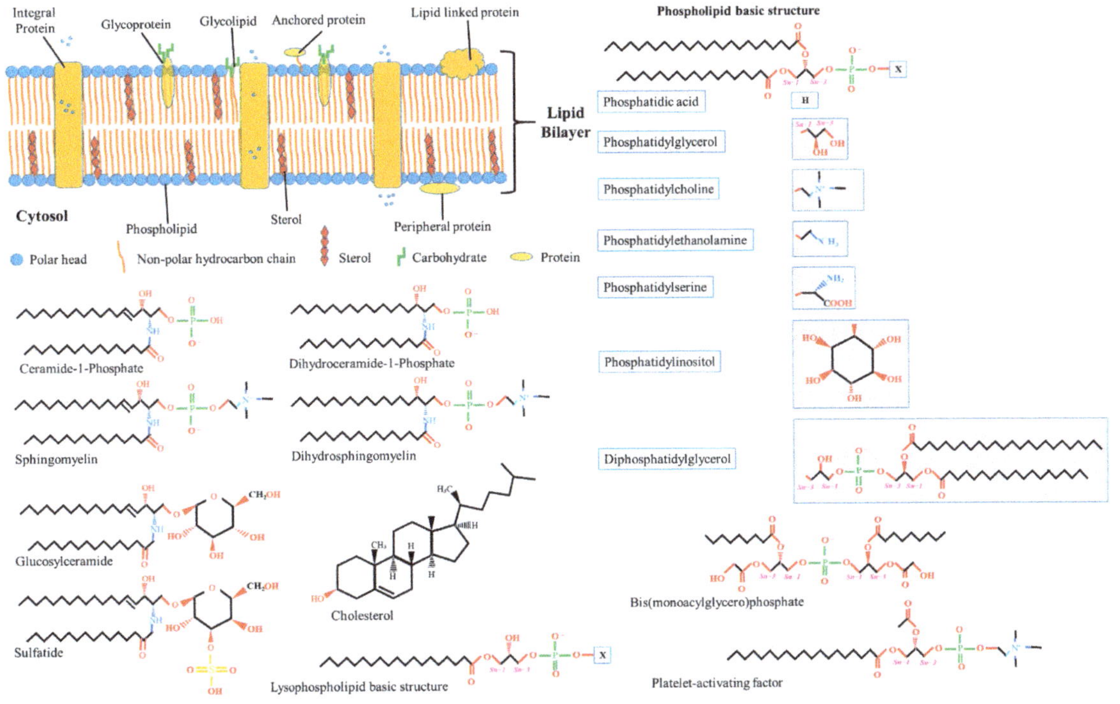

Figure 1. Schematic representation of biological compartments of the cell membrane and the molecular structure of the different lipids available in membranes. The molecular structures of different lipids have been adapted from the PubChem database (an open database for the public, available at https://pubchem.ncbi.nlm.nih.gov, accessed on 17 September 2023). Abbreviation: *Sn*, stereospecific numbering in the glycerol; ≐ and ▲, chiral carbon centers.

2. Lipid Bilayer

Lipids have gained recognition and have become a subject of considerable interest among scientists since the original publication of Chevreul's work [13], which delineated the concept of fatty acids. Lipids are widely acknowledged for their crucial role in forming cellular structures and mediating various physiological and life-sustaining processes. The concept popularity of a lipid layer's existence on the cell's surface can be traced back to Overton's reports between 1885 and 1899, although a comprehensive elucidation of the membrane structure did not emerge until 1925 [14]. It was Gorter and Grendel who, employing a Langmuir monolayer, initially identified the presence of a lipid bilayer within blood chromocytes. Their discovery revealed a distinctive 2:1 ratio between the cellular surfaces covered by lipids and the estimated total cell surface area [15]. Consequently, a lipid bilayer emerges as a supramolecular matrix comprising two leaflets of lipid molecules residing within the biomembrane. Each leaflet necessitates a specific lipid composition characterized by certain physicochemical properties to finely modulate targeted functions.

Despite enduring exposure to changing conditions of temperature, pressure, and solvents, the lipid composition of animal cell membranes remains relatively stable, indicating a relatively confined capacity for drastic alterations in response to external stimuli. Nonetheless, the layers of membranes remain far from static; elements can transfer within (lateral diffusion) and between (vertical or flip-flop diffusion) leaflets. Lipid transporter proteins, namely, flippase, floppase, and scramblase, mediate the movement of lipids across membrane layers. In contrast, the retrograde traffic is responsible for the backward movement of lipids from membranes to organelles [2,16]. The ER, mitochondria, and Golgi apparatus are responsible for biosynthesizing most of the lipid classes in biomembranes, including glycerophospholipids, cholesterols (CHOL), and SLs. Conversely, lipid hydrolysis transpires within the lysosome, specifically the intralysosomal luminal vesicles, where numerous water-soluble hydrolases are active [17–19]. Lipids are transported to lysosomes through endocytic and autophagocytic pathways. The products generated from lipid hydrolysis are either utilized within the cell or expelled via exocytosis at the plasma membrane.

Thousands of lipid structures have been identified in mammals [20], with the coexistence of hundreds within a single cell remaining a probable [21]. The chemical properties of membrane lipids are characterized by distinctive features. These include the head-group or backbone structure, hydrocarbon chain length, degree of unsaturation, the presence of chirality, ionization, chelating power, and lipid concentration. Nevertheless, lipid classification is not arbitrary and can be predicated on physical properties, chemical properties, or biosynthetic qualities [22,23]. Within mammalian cell membranes, the preponderant lipid class is glycerophospholipids, also known as phospholipids. Characterized by a hydrophilic head group lining surfaces and a hydrophobic tail interposed in between, this class constitutes the bulk of the membrane lipid matrix. Other minor lipid classes recognized within biomembranes include glycolipids and sterols, with plasma membranes distinctively characterized by a considerable abundance of sterols. A schematic representation delineating the principal lipid classes identified in biomembranes is available in Figure 1. It is well-established that the lipid composition of biomembranes exhibits variations across organelles [24,25] and tissues; it dynamically adapts within the cell in response to specific cellular activities. The distinctive biophysical state of membrane lipids and the fatty acid composition may influence membrane rigidity, serve specific functions, and reveal the cell's physiological state.

2.1. Glycerophospholipids

In 1811, the pioneering work of Vauquelin led to the identification of phosphorus in cerebral lipid extracts [26], and since then, phosphorus-containing lipids have become an intriguing field of investigation. This class of polar lipids is commonly referred to as 'glycerophospholipids' or simply "phospholipids". It is the most prevalent lipid class in mammalian membranes, accounting for 50–60 mol% of the overall membrane lipid matrix [27]. The foundational structure of phospholipids closely resembles that of diacylglycerol (DAG, featuring a glycerol backbone with two acyl (fatty acid) chains at *sn*-1 and *sn*-2 positions); it is further distinguished by the inclusion of a polar phosphorus group at the *sn*-3 position. Hence, lipids within this class exhibit amphipathic properties, which are characterized by the presence of a hydrophilic head group and two hydrophobic fatty acids.

Over the past century, a multitude of phospholipid types have been identified, with variations in lipid structure playing a profound role in the differentiation of phospholipid varieties. The bulk of phosphate groups are attached to specific molecules or moieties, determining the exact type of phospholipid and its position within the lipid bilayer. Numerous phospholipids have been identified in mammalian membranes, including phosphatidic acid (PA), phosphatidylglycerol (PG), phosphatidylcholine (PC), phosphatidylethanolamine (PE), phosphatidylserine (PS), phosphatidylinositol (PI), diphosphatidylglycerol (DPG), bis(monoacylglycero)phosphate (BMP), platelet-activating factor (PAF), and lysophospholipids (LysoP).

2.1.1. Phosphatidic Acid

PA, often referred to as phosphatidate (see Figure 1), represents the simplest phospholipid structure and tends to accumulate in membranes in relatively minor proportions, owing to the activity of lipid phosphate phosphohydrolases [28,29]. It was initially identified as a phosphorylated isomer of DAG [30]. PA, therefore, constitutes a non-bilayer lipid characterized by a phosphate group esterified at the *sn*-3-hydroxyl of the glycerol backbone and two fatty acyl chains occupying the remaining *sn*-positions. Multiple pathways contribute to PA production (see Figure 2), including the dual acylations of glycerol-3-phosphate, phospholipid hydrolysis pathway (especially involving PC), and DAG phosphorylation [31]. The synthesis of PA from DAG is a reversible process catalyzed by DAG kinase and PA phosphatases (also referred to as lipins).

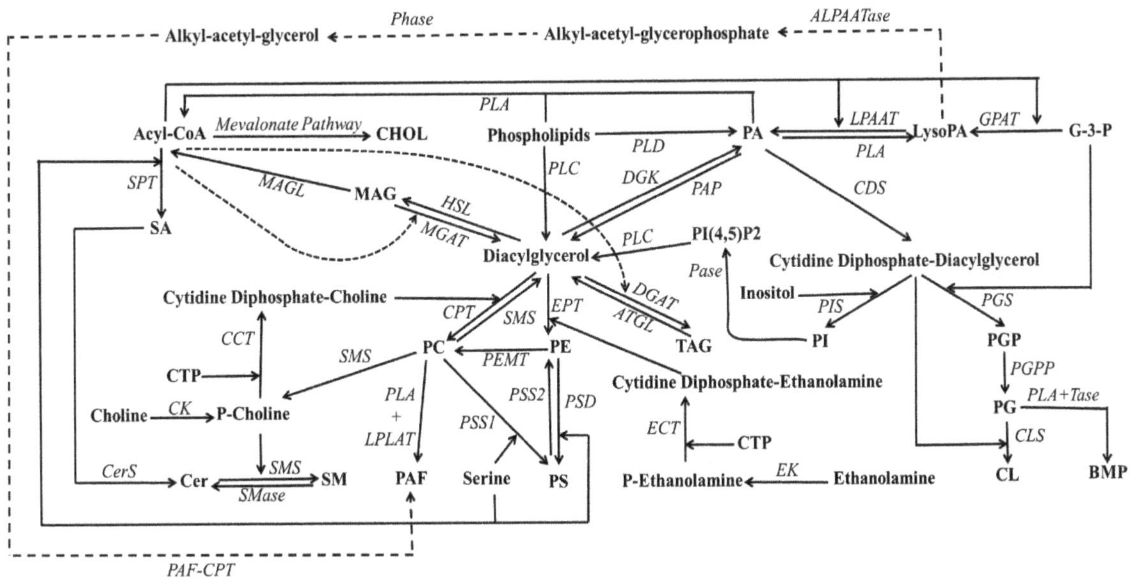

Figure 2. Schematic representation of the synthesis pathways for various phospholipids available in biomembranes. Abbreviations: ALPAATase, alkyl-acetyltransferase; ATGL, triacylglycerol lipase; CCT, cytidine 5′-triphosphate:phosphocholine cytidylyltransferase; CDP, cytidine diphosphate; CDS, CDP-DAG synthase; Cer, ceramide; CerS, ceramide synthase; CHOL, cholesterol; CK, choline kinase; CL, cardiolipin; CLS, cardiolipin synthase; CPT, CDP-choline:DAG cholinephosphotransferase; CTP, cytidine 5′-triphosphate; DG, diacylgelycerol; DGAT, DAG acyltransferase; DGK, DAG kinase; ECT, cytidine 5′-triphosphate:phosphoethanolamine cytidylyltransferase; EK, ethanolamine kinase; EPT, CDP-ethanolamine:DAG ethanolaminephosphotransferase; G-3-P, glycerol-3-phosphate; GPAT, glycerophosphate acyltransferase; HSL, hormone sensitive lipase; LysoPA, lysophosphatidic acid; LPAAT, lysophosphatidic acid acyltransferase; MAG, monoacylglycerol; MAGL, monoacylglycerol lipase, MGAT, monocylglycerol acyltransferase; P-Choline, phosphocholine; P-ethanolamine, phosphoethanolamine; PA, phosphatidic acid; PAF, platelet activating factor; PAF-CPT, platelet activating factor cholinephosphotransferase; PAP, phosphatidic acid phosphatase; Pases, phosphatases; PC, phosphatidylcholine; PE, phosphatidylethanolamine; PEMT, phosphatidylethanolamine *N*-methyltransferase; PG, phosphatidylglycerol; PGP, phosphatidylglycerophosphate; PGPP, phosphatidylglycerophosphate phosphatase; PGS, phosphatidylglycerophosphate synthase; Phase, phosphohydrolase; PI, phosphatidylinositol; PIS, phosphatidylinositol synthase; PLA, phospholipase; PLC, phospholipase C; PLD, phospholipase D; PS, phosphatidylserine; PSD, phosphatidylserine decarboxylase; PSS, phosphatidylserine synthase; SA, sphinganine; SM, sphingomyelin; SMase, sphingomyelinase; SMS, sphingomyelin synthase; SPT, serine palmitoyltransferase; TAG, triacylglycerol; Tase, transacylase.

PA is a negatively charged anionic lipid involved in cellular signal transduction and capable of reacting with divalent ions such as Ca^{2+}. Furthermore, its presence within mammalian cells is vital, as it acts as a mediator for phospholipid metabolism, a regulator for glycerolipid metabolism, neuroendocrine cell exocytosis, protein kinases, small G-proteins, and a modulator for membrane fusion and fission machinery [32–36]. Therefore, any alterations in PA levels may indicate disruptions in cellular homeostasis and the onset of metabolic and health-related consequences, as evidenced by Tanguy et al. [31], who linked the high accumulation of PA in cells to metabolic disorders.

2.1.2. Phosphatidylglycerol

When alcohol glycerol esterifies with a phosphate within a phospholipid, the resulting lipid structure is referred to as 'PG'. Benson and Maruo identified this lipid structure in 1958 [37]; it is characterized by two free hydroxyl groups. Basically, it comprises a glycerol backbone linked with two fatty acyl chains and phosphoglycerol. Within mammals, PG is synthesized in the mitochondria through multiple pathways: (1) it originates from imported PA, which undergoes a series of enzymatic reactions involving intermediates within the cytidine diphosphate-diacyl glycerol pathways in the inner mitochondrial membrane, and (2) from dephosphorylated phosphatidylglycerolphosphate catalyzed by the mitochondrial phosphatase enzyme [38].

Though PG does not constitute a substantial proportion of mammalian membranes (1–2% of membrane polar lipids), it accounts for up to 7–15% of the lipid composition in lung surfactants [39,40]. This heightened presence of PG in the lungs, where it ranks as the second most prevalent phospholipid in the lungs, underscores its crucial role in surfactant activity. Beyond the lung, the PG functionality extends to lipid–protein and lipid–lipid interactions, along with its influence on membrane rigidity. The PG molecular structure relatively resembles that of DPG and BMP, with all of them featuring more than glycerol in their structures. Furthermore, the molecular structure of PG in specific tissues has been considered to be a functional analogue of PI (having an inositol group rather than glycerol) [41]. Thus, these phospholipids may manifest similar activities, such as the inhibition of phosphatidylcholine-dependent kinase activity in swine brain [42]. Elevated levels of PG have been associated with viral infection, as PG can integrate into viral membranes during replication [43–45]. In contrast, some reports suggest that PG is involved in regulating innate immunity and suppressing viral infection [46–48], potentially including COVID-19 infection [49]. Therefore, further studies are imperative to ascertain the significant biological roles of PG in various mammalian species.

2.1.3. Phosphatidylcholine

The PC, also known as lecithin, was the first isolated phospholipid in 1850, with choline (a source of the methyl group) serving as the polar head [50]. Herein, it is very self-evident that the PC structure is not entirely endogenous, as choline is an essential nutrient for mammals. PC is a ubiquitous presence in all cell membranes, spanning prokaryotic cells (e.g., bacteria) and eukaryotic cells (i.e., cells of plants and animals). Structurally, PC exhibits two major linkage types in tissues: diacyl-PC (ester bond; most abundant in eukaryotes) and alkyl-PC (featuring an ether bond) [51]. Additionally, the less common isomer of PC is alkenyl-PC (vinyl ether bond), which is referred to as choline plasmalogens and plasmenylcholine. These lipids typically comprise two fatty acids linked to glycerol through ether and ester bonds at *sn*-1 and *sn*-2, respectively [52,53].

PC represents the most abundant phospholipid class (constituting nearly 50% of all phospholipids within bilayers), particularly in the pulmonary surfactant, where dipalmitoyl-PC predominates [54–57]. As a fundamental building block of the membrane bilayer, PC occupies the outer leaflet [58]. Remarkably, approximately 80 to 90% of the lipids in the plasma membrane's outer leaflet consist of PCs. The preponderance of PC synthesis occurs in the ER, where cytidine 5′-triphosphate (CTP):phosphocholine cytidylyltransfease (PCT) (generally known as CCT) [59,60] catalyzes the rate-limiting step in the cytidine 5′-

diphosphocholine (CDP-choline, citicoline or Kennedy) pathway [61]. This CDP moiety is not only involved in PC biosynthesis [62] but in all other phospholipids, with the exception of PA, depending on which moiety replaces choline. A distinctive pathway for PC biosynthesis exclusively takes place in the liver, where PC is generated from PE via sequential methylation [63], facilitated by the phosphatidylethanolamine *N*-methyltransferase (PEMT) that is found in the mitochondrial-associated membranes (MAM).

It has been believed that PC's relatively neutral molecular properties (having positive and negative charges but lacking net charge) and its predominance play an essential role in maintaining biomembrane integrity and functionality. Unlike other phospholipids, PC does not exhibit negative charge repulsion. PC serves as a precursor for sphingomyelin (SM) due to its choline molecule [64]. In addition, it acts as a precursor for other polar lipids, such as PA, lysophosphatidylcholines (LysoPC), PS, and PAF. PC plays a crucial part in cell signaling processes and impacts the concentration of circulating lipoproteins [56,65,66]. Furthermore, it is integral to membrane trafficking and molecule transportation. LysoPC composed of C22:6 (at the *sn*-2 position) has been demonstrated to be more effective than C22:6-free fatty acids in crossing the blood–brain barrier [67].

2.1.4. Phosphatidylethanolamine

Following PC, the second most prevalent phospholipid in mammalian tissues is PE, formerly known as "cephalin". It was the second discovered phospholipid in cerebral tissue by Thudichum in 1884 [68], constituting approximately 15–25% of the total phospholipids in mammalian cell membranes [69]. In neural tissues, PE can reach even higher levels, up to 45% [70], pointing out its essential role in this tissue. It is profoundly abundant in mitochondrial membranes and is exclusively localized in the cytosolic leaflet of the plasma membrane, in contrast to PC [58]. The structure of PE involves the esterification of the phosphatidyl group to the hydroxyl group of an amino group (namely, the ethanolamine), resulting in a small reactive head group. PE does not form a bilayer independently (on its own) but exhibits an inverted hexagonal phase. This class of lipids features various linkages, including diacyl, alkyl, and alkenyl configurations (see Figure 3). Ethanolamine plasmaologens, also known as plasmenylethanolamine, are more abundant than plasmenylcholine in many tissue types, except for the heart and smooth muscle [52].

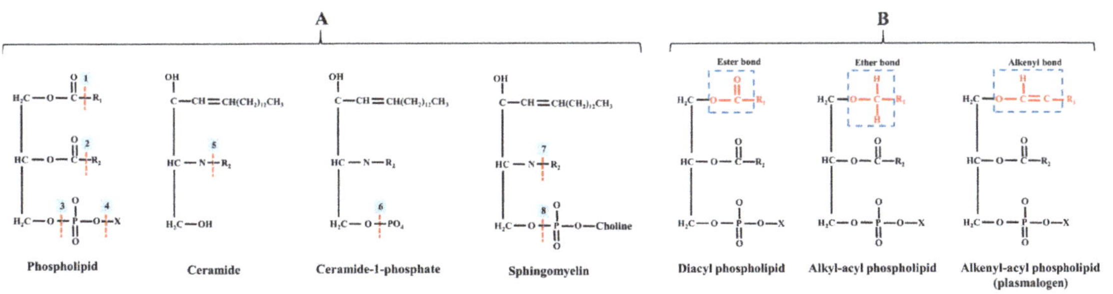

Figure 3. (**A**) Site activities of different phospholipases on membrane lipids. (**B**) Different linkage types in phospholipids. Abbreviations: 1, phospholipase A1; 2, phospholipase A2; 3, phospholipase C; 4, phospholipase D; 5, ceramidase; 6, lipid phosphate phosphatase; 7, sphingomyelin deacylase; 8, sphingomyelinase; R, fatty acid; X, head group.

In eukaryotes, the biosynthesis of PE is an outcome of multiple pathways, notably, the de novo synthesis of PE through CDP-ethanolamine [61] and the salvage pathway involving the decarboxylation of PS by phosphatidylserine decarboxylase (PSD) in the mitochondria [71]. Additional pathways involved in the remodeling of PE, which are also identified in bacteria and plants, include the following: (1) the base-exchange pathway between PE and PS [72]; (2) the degradation of sphingosine-P via sphingosine-P lyase [73];

and (3) the reacylation of Lyso-PE at MAM [74]. Notably, despite the structural resemblance of PE and PC, PE exhibits distinct chemical and biological properties. PE stands apart from PC due to its smaller head group, which manifests less affinity to water. Consequently, PE can form compact aggregation and displays a heightened thermostability [75,76]. These attributes significantly contribute to membrane rigidity, making PE an indispensable component of the membrane's architecture, permeability, and fluidity.

In terms of membrane rigidity, PE often mimics the behavior of CHOL, particularly in insects [77]. In light of these findings, the PC/PE ratio is likely to exert a substantial influence on the functionality, fluidity, stability [78], and permeability of the membrane. Furthermore, PE plays a vital role in upholding membrane integrity and participating in cellular signaling. Studies have revealed that PE is implicated in various processes, including membrane-to-membrane fusion [79], DAG generation through the involvement of phospholipase C (PLC), and the modification of membrane proteins through the mediation of reactive aldehydes [80]. PE has also been observed to induce negative curvature in biomembranes [81], which is primarily attributed to its diminutive conical head group. In addition, PE serves as a precursor for various other lipids, including *N*-acyl-phosphatidylethanolamine (NAPE), which serves as a crucial precursor during the biosynthesis of certain essential biological compounds in the brain (e.g., anandamide) [82,83].

2.1.5. Phosphatidylserine

Folch and Schneider identified serine in cephalin components in 1941 [84], marking the beginning of the discovery of PS. PS is a minor class of phospholipids in mammalian cells (2–15% of total phospholipids), which demonstrates a pronounced tendency for accumulation within the cerebral cortex [85,86]. It has also been detected in the membranes of organelles such as mitochondria and ER, where it serves as a substrate for the production of PE. Notably, the distinguishing feature of the PS structure, setting it apart from other phospholipids, is the attachment of the serine residue to the phosphatidyl group, resulting in the formation of a negatively charged head group. This characteristic renders it exceptionally reactive with divalent metals. In contrast to PC and PE, PS exclusively exists in a diacyl isomer, with *sn*-2 being markedly unsaturated [85].

In contrast to plants [87], yeasts, and prokaryotes [88], mammalian cells lack the de novo CDP-DAG biosynthetic pathway for PS biosynthesis. The biosynthesis of PS in mammalian cells transpires both in the MAM and in the cytosol of the ER and is facilitated by a calcium-dependent base exchange. This pathway is catalyzed by PS synthase-1 and -2 (PSS1 and PSS2, respectively), utilizing PC (catalyzed by PSS2) and PE (catalyzed by PSS2) as the primary precursors at both sites [89]. Subsequent to its production, a fraction of PS translocates to the plasma membrane via passive diffusion. This lipid primarily localizes to the cytosolic leaflet of the plasma membrane [90], although its migration to the outer leaflet is notable during programmed cell death [91] and cancer progression [92].

The externalization of PS on the cell's outer layer serves as a molecular signal, prompting neighboring cells, including macrophages, to engulf and phagocytose the dying cell. Beyond this role, PS plays a multifaceted biological role within cellular functions. It contributes to the recognition and communication mechanisms between cells. PS existence is crucial during PE biosynthesis, acting as a source pool [71]. Furthermore, PS has been observed to interact with SLs, resulting in elevated interdigitation under the influence of CHOL [93]. PS is also vital for the maintenance of plasma membrane integrity within mammalian cells, exerting modulation over membrane fluidity and permeability, both of which are essential for the regular function of membrane-bound proteins.

PS has been implicated in the activation of protein kinase, prothrombinase, and neuroinflammation signaling pathways, as well as being an essential element of lipid–calcium–phosphate complexes [94–97]. Consequently, PS facilitates a range of membrane-bound signaling processes, including apoptosis, activation of enzymes, immune regulation, coagulation cascade, and mineral deposition during bone regeneration.

2.1.6. Phosphatidylinositol

The earliest documented report of phosphatidylinositol (PI) traces back to the year 1930 when inositol was initially identified within a lipid extract [98]. It was not until nearly three decades later, in 1959, that Pizer and Ballou elucidated the precise structure of PI [99]. PI, an anionic phospholipid, features a distinctive inositol head group, characterized by a hexa-hydroxy-ring consisting of six carbon atoms. Within the realm of inositol-containing phospholipids (phosphoinositides), PI represents the most elementary form, with the other seven isomers constituting phosphorylated derivatives of the PI structure [100]. In eukaryotic organisms, three primary forms of phosphoinositides prevail: (1) PI, formerly recognized as monophosphoinositide, in which 1′-myo-inositol is linked to PA; (2) PI4P, where a phosphate group esterifies position 4 of inositol, formerly referred to as diphosphoinositide; and (3) PI5P, featuring a phosphate esterifies position 5 of inositol. In eukaryotes, the phosphorylation of positions 2 and 6 of PI is impeded due to steric hindrance. PI can constitute up to 10% of total phospholipids and is ubiquitously present in the cytosol of all cellular membranes and certain organelles (e.g., endoplasmic reticulum and Golgi apparatus) [101,102]. PI of eukaryotic organisms is primarily biosynthesized from PA via a de novo pathway and is catalyzed by the CDP-DAG synthase (which serves as a rate-limiting enzyme [103]) and CDP-DAG myo-inositol 3-phosphatidyltransferase [104]. These enzymes are localized in the ER, where they facilitate the formation of intermediates from PA and the attachment of myo-inositol to these intermediates, respectively. Mammalian cells possess the capability to synthesize inositol de novo from glucose-6-phosphate. Other marked three biosynthetic pathways have been identified in plants and prokaryotes, with the most recent discovery occurring a decade ago [105].

Though PI's initial discovery was nearly a century ago, our understanding of the biological functions of PI has markedly advanced over the past three decades. PI is not merely a component of bilayer lipids; it is involved in various metabolic processes [106]. Its significance extends to the brain, where it serves critical functions. In addition, it serves as the primary pool of the C20:4 n6 fatty acyl chain in animal cells, frequently occupying the *sn*-2 position [107–109]. This specific acyl chain is of paramount importance for the biosynthesis of eicosanoids, including prostaglandins [101,110]. The enzyme phospholipase A2 (PLA2) is responsible for the removal of C20:4 n6 from PI, resulting in the formation of LysoPI (see Figure 3). Consequently, an accumulation of LysoPI indicates heightened PLA2 activity, implying metabolic alterations and, potentially, the progression of cancer [111].

Furthermore, PI constitutes the major substrate of the signaling DAG in mammalian cells, a process catalyzed by the PLA2 and PLC enzymes. Thus, PI emerges as a dynamic lipid that participates in intracellular signaling, inflammation, and immune regulation. PI also contributes to the formation of glycosyl bridges that facilitate the binding of multiple proteins (known as glycosyl-phosphatidylinositol (GPI)-anchored proteins) to the cellular membrane surface [112]. PI has been shown to engage in regulating protein activities at the cellular interface. The various phosphoinositides, including PI3P, PI4P, PI5P, PI(4,5)P2 (the most abundant PI-phosphorylated structure in mammalian membranes), PI(3,4)P2, PI(3,5)P2, and PI(3,4,5)P3, while accumulating in very low concentrations (1%), significantly contribute to membrane organization. An in-depth discussion has been reviewed by Posor et al. [113]. For instance, PI(4,5)P2 functions as a cofactor for phospholipase D (PLD), an enzyme responsible for the production of PA, which serves as a signaling molecule.

2.1.7. Diphosphatidylglycerol

The DPG, also known as cardiolipin (CL), was initially isolated from bovine hearts by Pangborn in 1942 [114], and the nomenclature "cardio" reflects its association with cardiac tissues. This uncommon tetra-acylated phospholipid is exclusively confined to the inner and outer mitochondrial membranes, with the production site on the matrix side of the mitochondrial inner leaflet [115]. Basically, it can be described as PG with additional phosphatidate groups esterified to glycerol, resulting in two negative charges. The biosynthesis of CL primarily proceeds from the PA substrate [116], which is subsequently transformed into PG within the mitochondria. The conversion of PG species into CL through condensation is facilitated by cardiolipin synthase (CLs). It is postulated that the biosynthesis of CL in eukaryotic cells has evolved from prokaryotic ancestors [117].

CL plays a pivotal role in mitochondrial activity, which is evident through its substantial concentration (15–20%) among the total polar lipids of the mitochondria [118]. Thus, it dynamically contributes to the respiratory chain, interacts with adenosine diphosphate (ADP)/adenosine triphosphate (ATP) and imported complex III and IV proteins, regulates mitochondrial fission and fusion, and controls the release of apoptotic factors [119–121]. Therefore, variations in CL concentrations may be associated with mitochondrial dysfunction disorders [119].

2.1.8. Bis(monoacylglycero)phosphate

The BMP is a unique lipid involved in cellular trafficking due to its enrichment in the intracellular membranes of the late endosomes [122,123] and lysosomes [124]. Body and Gary were the first to isolate it from pig lungs in 1967 [125]. While it was initially misidentified as "bisphosphatidic acid" or "lysobisphosphatidic acid", BMP's accurate structural characterization was reported by Brotherus and Renkonen in in vitro cultured hamster kidney fibroblast cells [126]. BMP is a negatively charged structural isomer of PG, featuring an unusual *sn*-1-glycerophospho-*sn*-1′-glycerol configuration. This lipid structure is related to polyglycerophospholipids, which also encompass PG and DPG [85]. In fact, PG has been identified as the substrate for BMP production [127–129], though the precise mechanisms of their production and metabolism continue to be subjects of ongoing research. PG is a fundamental component of mitochondria and ER, and it reaches the lysosome (the BMP biosynthesis site) via autophagy. Herein, the phospholipases are less effective towards BMP, preventing the lysosomal membranes from autodigestion.

The production of BMP involves multiple reactions, including the acylation of fatty acid to glycerol's hydroxyl moiety and the esterification of phosphoric acid to ethanolamine. Despite BMP constituting a minor fraction of cellular polar lipids, comprising less than 1% of the total [130], elevated levels have been detected in rat splenic tissue [131] and alveolar macrophages [132]. Elevated BMP concentrations have been associated with lipid storage diseases and drug-induced lipidosis [132–134]. Studies on BMP have consistently increased over the past 14 years. This interest is attributed to its role in the metabolism of glycosphingolipids (GSLs) and CHOL [19,135], which, in return, influence cellular signaling, vesicle fusion, and membrane integrity.

2.1.9. Platelet-Activating Factor

The PAF is a unique bioactive ether phospholipid with a structural composition of 1-alkyl-2-acetyl-*sn*-glycero-3-phosphocholine structure, notably lacking the conventional phospholipid *sn*-1 ester bond [136]. It was initially introduced by Benveniste et al. [137] from rabbit platelets, making PAF the earliest identified phospholipid capable of inciting an inflammatory response. The biosynthesis of PAF occurs within the ER through two primary pathways: the de novo pathway from PC transferred to alkyl acetyl glycerol [138] and a biomembrane remodeling process that involves the substitution of the *sn*-2 fatty acyl chain with an acetyl group [139]. The latter pathway is catalyzed by PLA2 and acetyltransferase/transacylase.

The accumulation of PAF exhibits variations among cell types, typically representing a negligible fraction of the total phospholipids. This characteristic poses challenges in its precise quantification. The heightened accumulation rate of PAF observed in various tissues correlates with inflammatory responses [140], projecting its major involvement in the regulation of inflammatory and immune responses, as well as physiological processes such as platelet aggregation and thrombosis. In addition, PAF has been documented to exert influence over the activities of multiple physiological systems, including the cardiovascular, nervous, respiratory, excretory, and reproductive apparatuses [141–143].

However, alterations in PAF concentrations have been associated with certain diseases, syndromes, and skin cancer [144,145], albeit without serving as a direct mediator.

2.1.10. Lyso-lipids

Shifting the focus to lyso-lipids, this class is alternatively referred to as hydrolyzed lipids. This lipid class is constituted by various isomers originating from the enzymatic cleavage of acyl chains from phospholipids or SLs, which are catalyzed by phospholipase and deacylase enzymes, as illustrated in Figure 3. Thus, lyso-lipids of membranes can be categorized according to their original backbone and further classified into lysoglycerophospholipids (LysoPs) and lysosphingolipids (LsoSLs), respectively. Generally, LysoPs are amphipathic molecules carrying either an alkyl or acyl chain [146,147]. On the other hand, the LysoSLs are distinct due to the absence of the amide-linked fatty acid at the 2-amino position of the sphingoid base [148], setting them apart from their parental structure. Long ago, LysoP isomers were considered intermediates in phospholipid biosynthesis or fragments of disrupted cells. Nevertheless, they display distinct properties and functions compared to parental phospholipids. LysoP contributes to cellular homeostasis by engaging in bilayer remodeling and rigidity. Furthermore, specific LysoP molecules can function as ligands for various G-protein-coupled receptors [149], underscoring their involvement in cellular signaling.

While the current review does not emphasize this category due to its limited prevalence and identification in studies, the most abundant LysoP class is lysophosphatidylcholines/lysolecithins (LysoPC). LysoPC is generated through the hydrolysis PC, mainly catalyzed by PLA2. LysoPC possesses physical properties distinct from PC, forming micelles rather than bilayers. It is typically found in minute proportions and plays a role in the mechanism of the autoimmune response [150]. The accumulation of LysoPC within cells has been associated with metabolic irregularities [150] and phospholipid peroxidation [151,152]. Lysophosphatidic acid (LysoPA), the simplest structure within the LysoP category within mammalian membranes, is biosynthesized in most cells through the activity of lysophospholipase-D on LysoPC or via the actions of phospholipases (phospholipase A1 (PLA1) and PLA2) on PA [153]. LysoPA serves numerous functions, including the regulation of cellular differentiation, growth, proliferation, migration, and apoptosis. In the context of inflammation and cancer, it has gained significant attention, focusing on its structural features and the extent of accumulation [154,155].

2.2. Sterols

This category of membrane lipids is named according to its primary constituent, sterol. Alternatively, it can be referred to as steroid alcohol, distinguishing it from phospholipids. Sterols are characterized by a rigid, always *trans* tetracyclic hydrocarbon ring, a 3β-hydroxyl group, and a flexible side fatty acyl chain as a tail [156]. Thus, sterols possess both hydrophilic properties (represented by the hydroxyl group) and hydrophobic properties (attributed to the ring and fatty acyl chain). Notably, variations in the floppy tail of sterols account for the structural diversity observed across different biological kingdoms. Sterols are primarily found in membranes of animals (cholesterol), plants (stigmasterol or β-sitosterol), and fungi (ergosterol). It is important to note that most bacterial membranes are devoid of sterols. Among mammalian membranes, CHOL is the most commonly encountered sterol and recognized structure. Despite its widespread presence in various organisms, it is noteworthy that certain insect species lack the genes responsible for its biosynthesis [157].

Cholesterol

CHOL is a sterol isoprenoid characterized by a semi-rigid tetracyclic ring composed of three six-membered rings and one five-membered ring. It features a polar 3β-hydroxyl group and an 8-carbon chain attached to the carbon-17 position, while its side acyl chain exhibits conformational flexibility [158,159]. The polar nature of the CHOL group alone prevents CHOL from forming bilayers. However, when synthetic CHOL is combined with a PC head moiety, bilayer formation occurs [160]. Within bilayers, CHOL is asymmetrically distributed, with the majority of sterols (60–70%) located in the inner leaflet [161,162]. CHOL was initially discovered in gallstones by de La Salle in 1858, but it took another decade for researchers to identify it [163].

CHOL can be obtained from the diet or synthesized by the liver (which contributes 50% to total CHOL synthesis) and the ERs of other cells. The biosynthesis of CHOL is regulated by sterol-responsive element binding protein 2 (SREBP2)-cleavage-activating protein, which senses intracellular CHOL and modulates nuclear transcription. Importantly, cells can also import CHOL from the vascular system, where lysosomes recycle the low-density lipoprotein by transferring CHOL to the ER. The CHOL biosynthesis pathway involves a series of enzyme-catalyzed reactions generating a series of intermediate compounds. Typically, over 20 enzymes are involved in the CHOL biosynthesis pathway, using acetyl-CoA as a substrate. Though animal cholesterol is synthesized in the ER, a relatively higher proportion is found in cellular plasma membranes than in the ER [2,164]. Notably, the plasma membrane contains a significant amount of CHOL (making up to 50% of membrane lipids, primarily in the cytosolic leaflet) as compared to other subcellular organelles [162,165]. In the cytosolic leaflet, the hydroxyl group and the aliphatic chain are oriented towards the aqueous phase and the bilayer's interior, respectively.

CHOL plays an important role in modulating dynamic membrane activities [156]. Its fused ring structure (exhibiting amphiphilic properties) imparts greater membrane rigidity. Thus, CHOL levels critically influence membranes' rigidity, fluidity, and permeability [166,167]. The incorporation of CHOL into membranes reduces permeability to non-polar molecules while increasing the hydrophobic barriers to polar molecules. CHOL also has a condensing effect on hydrocarbon chains, thereby reducing the surface area occupied by lipids [168]. Additionally, CHOL participates in the formation of lipid rafts through interactions with various phospholipids, with a notably favorable interaction observed with saturated phospholipids [169]. The solubility of CHOL in membranes depends on the degree of unsaturation of the phospholipids. A high number of unsaturated double bonds has been shown to reduce CHOL solubility [170,171]. Remarkably, even among saturated phospholipids, CHOL affinity was shown to be different. CHOL's affinity to other lipid complexes relies on various factors, such as the head group tilt structure [172], hydration [173], acyl chain order [174], possible interdigitation of acyl chains [175], and the presence of hydrogen bond acceptor and donor groups [176].

CHOL serves a wide range of signaling functions through its interactions with various cellular molecules and receptors. A recent study indicated that the interaction between cholesterol and lipid-mediated innate immune memory triggers cytokine cascades as associated with COVID-19 [177]. Conversely, an imbalance in membrane-CHOL levels may likely pose severe metabolic consequences, including cancer progression [178,179]. CHOL also serves as a precursor for the biosynthesis of bile acids and steroid hormones [180,181], which mediate crucial roles in biological processes, such as carbohydrate metabolism [182–184]. Furthermore, CHOL esters play a critical role in transporting fatty acyl chains via lipoproteins in the bloodstream, and these esters are integral components of amphiphilic plasma lipoproteins [185].

2.3. Sphingolipids

SLs constitute a class of lipids that are present in cells in relatively lower proportions, typically accounting for about 10–20% of total cellular lipids. Despite their relatively lower abundance, SLs exhibit significant signaling activities [27]. The bio-functional roles of SLs can be broadly categorized into three areas: firstly, they modulate the physical properties of biomembranes; secondly, they serve as signaling molecules, acting as second messengers or secreted ligands for cell-surface receptors; and thirdly, they mediate interactions between cells and their external environment [186]. Hence, variations in the ratio of SLs can have a substantial impact on cellular activities and overall cellular survival. The initial identification of SLs can be attributed to Thudichum [68], while the elucidation of their structure, namely, the sphingosine (SO) component, was achieved by Carter [187]. Unlike phospholipids, which are glycerol-based, SLs consist of a long-chain sphingoid base as a backbone, with an amide-linked acyl chain attached instead of an oxygen ester. Notably, the hydrophobic chains (fatty acid) in sphingosine (SO) are structurally constant and non-hydrolyzable, distinguishing SLs from the variable and hydrolyzable fatty acids found in phospholipids. Numerous distinct SL structures have been identified, with structural differences primarily based on variations in backbone structure, hydrophobic chain length, and the level of unsaturation.

2.3.1. Sphingoid Bases

Among the most well-known backbone structures are sphinganine (SA) and SO bases, which serve as the primary reservoir for SL biosynthesis. In the cytosolic side of the ER, serine palmitoyltransferase (SPT) catalyzes the condensation of palmitoyl-Coenzyme A with L-serine, resulting in the formation of 3-ketosphinganine [188]. Subsequently, under the influence of 3-ketosphinganine reductase, 3-ketosphinganine is transformed into SA (as shown in Figure 4). SPT, a pyridoxal 5′-phosphate-dependent enzyme, is the rate-limiting enzyme for SA production [189]. It is worth noting that SPT is not limited to serine alone as a substrate; studies have shown that it can also employ alanine and glycine [190], leading to the production of structures known as 1-deoxysphingolipids. On the other hand, SO is biosynthesized during ceramide (Cer) production or hydrolysis, a process catalyzed by delta-4-desaturase (Δ4-desaturase, or D4D) and SPT enzymes, and ceramidase (CDase), respectively. However, it is important to highlight that free SO is not produced via the de novo pathway; rather, it is generated from the hydrolysis of Cer by CDase.

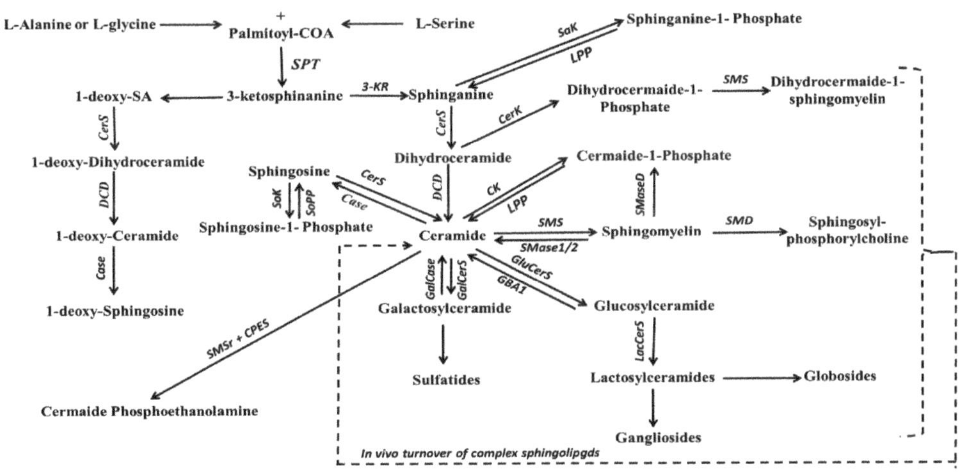

Figure 4. Schematic representation of the de novo biosynthesis pathway of major complex sphingolipids. Abbreviations: 3-KR, 3-ketosphinanine reductase; Case, ceramidase; CerK, ceramide kinase; CerS, ceramide synthase; CPES, ceramide phosphoethanolamine synthase; DCD, dihydroceramide desaturase; GalCase, galactosylceramidase; GalCerS, galactosylceramide synthase; GBA1, acid β-glucosidase; GluCerS, Glucosylceramide synthase; LacCerS, lactosylcermaide synthase; LPP, lipid phosphate phosphatase; Sak, sphinganine kinase; SMaseD; sphingomyelinaseD; SMD, sphingomyelin deacylase; SMS, sphingomyelin synthase; SMS1/2, sphingomyelin-1 or -2; SMSr, sphingomyelin synthase related proteins; Sok, sphingosine kinase; SoPP, sphingosine phosphate phosphatase; SPT, serine palmitoyltransferase.

SA and SO kinases can phosphorylate SA and SO, generating their respective 1-phosphate forms/derivatives. This pathway is reversible, and sphingoid-1-phosphate produced in this manner can undergo dephosphorylation through sphingoid-1-phosphate phosphatases. Sphingoid-1-phosphate remarkably differs in its activities from free sphingoid bases, serving not only as second messengers but also as first messengers [191]. Sphingoid-1-phosphate also serves as a substrate for phospholipid synthesis, as well as having a universal cellular survival signal [192] that is mediated by its binding to specific G protein-coupled cell surface receptors [193]. Sphingosine phosphate lyase has the capacity to cleave sphingoid-1-phosphate into phosphatidylethanolamine [194,195].

On the other hand, free sphingoid bases are essential secondary mediators, mediating various cellular processes, including growth, proliferation, DNA synthesis, and Cer biosynthesis [196]. These bases can readily traverse membranes, suggesting their potential involvement in stimulus-induced changes in membrane permeability. However, pinpointing the exact signaling functions of sphingoid bases is likely challenging due to their various signals and immense interaction with numerous cellular molecules, such as CHOL, phospholipids, and proteins [197–200]. It is necessary to highlight that dietary SLs have a proportional direct impact on their detected levels in cellular membranes and tissues [201]. In addition, a number of compounds, such as fumonisin (FUM) mycotoxins and Alternaria toxins [202], share structural similarities with free sphingoid bases, enabling them to interfere with sphingolipid metabolism and alter cellular signaling.

2.3.2. Ceramide

Cer is a non-bilayer-forming lipid characterized by its composition of a sphingosine base and a single acyl chain bonded to an amide group, thus lacking a distinct head group, illustrating its hydrophobic nature. The simple structure of Cer bears a resemblance to DAG. Cer serves as one of the simplest SLs and functions as the core building block for more complex SLs [198,203,204], which have larger molecular dimensions. Cer can be synthesized through multiple pathways: (1) de novo synthesis from SA substrate in the ER (a process catalyzed by *N*-acyl transferase/ceramide synthase (CerS)) and dihydroceramide desaturase [194]; (2) in vivo turnover of complex SL found in plasma membranes and lysosomes catalyzed by enzymes such as sphingomyelinase (SMase) or non-lysosomal glucosylceramidase) [205]; and (3) the salvage pathway in lysosomes that involves the re-acylation of SO to produce Cer [206].

The key enzyme responsible for de novo Cer generation is CerS, a family of six integral membrane proteins (CerS1–6) located in the ER of mammalian cells and regulated by corresponding six genes situated at distinct chromosomes [207]. The expression of these protein isoforms varies among different tissues [208], leading to variations in Cer proportion and acyl chain lengths. The CerS enzyme is responsible for the formation of dihydroceramide (DCer), which is an intermediate in Cer synthesis. In this step, DCer is formed by acylating a fatty acid to SA, followed by a desaturation reaction catalyzed by DCer desaturase to produce Cer. A decade ago, DCers were commonly considered to be quiescent intermediate metabolites, but recent research has unveiled their distinct functions compared to Cer [209]. Though de novo Cer production takes place in the ER [210], it has been suggested that long-chain bases are acylated in hepatic mitochondria. However, under specific events such as type 2 diabetes and FUM exposure, 1-deoxy-Cer and 1-deoxy-DCer are generated [211,212].

Cer plays a critical role in cellular signaling, regulating cell growth and apoptosis depending on the length of its acyl chain. Specific Cer species, like C16-Cer, have been proposed to be associated with apoptosis rates [213], while C18-Cer has been linked to growth arrest and a proportional decrease during cancer progression [214]. In addition, the ratio between C16 and C24:0/C24:1 Cer species has been identified as a factor in signaling induction, with C16 inducing apoptotic effects and C24:0/C24:1 exhibiting protective effects [215]. Therefore, alterations in the length of the Cer chain may potentially modify signaling, resulting in diverse metabolic effects. Recent review articles have highlighted the connections between Cer acyl chain length and cognitive functions [216] and intracellular lipid regulation [208]. Cer is also known for its ability to cluster apart from membranes, forming ceramide-rich domains with gel-phase properties. These domains are believed to act as platforms for protein–lipid interactions, selectively recruiting or excluding certain membrane components from small transit rafts. Cer-rich domains cluster these components in a stable manner, impeding their in-plane diffusion [217]. Therefore, the high hydrophobicity and complex polymorphic phase behavior of Cer [218] make Cer an important unit in lipid raft formation.

2.3.3. Complex Sphingolipids

In mammals, complex SLs are present in two isomers: phosphosphingolipids (PSLs) and GSLs. Complex SL consists of Cer bonded to complex phosphoryl or carbohydrate moieties, located either in the lumen or at the cytosolic surface of the Golgi apparatus. The transport of Cer between the ER and Golgi organelles is regulated through vesicular and non-vesicular mechanisms, which involve Cer transfer protein [219,220]. This process is primarily coupled by complex SL migrations across membrane leaflets [221] and acts as a rate-limiting factor in the production of complex SLs.

Phosphosphingolipids

In the realm of PSLs, Cer binds to a phosphate group with a polar head group, forming a polar head group mainly composed of choline, ethanolamine, or glycerol. This structural distinction sets PSLs apart from PC in that they not only act as hydrogen bond acceptors but also as hydrogen bond donors. The PSL class includes various subtypes, such as Cer-1-phosphate (Cer1P), DCer-1-phosphate (Dcer1P), Cer phosphoethanolamines (CerPE), sphingomyelins (SM), dihydrosphingomyelins, and LysoPSLs (lacking an attached fatty acyl chain). Among PSLs, SM stands out as the most studied and highlighted class of PSLs in cellular membranes. This review primarily focuses on SM, omitting detailed discussions of other PSLs. However, Cer1P is the simplest PSL with its structure involving the esterification of Cer with the phosphate group. Cer1P serves crucial roles in the regulation of eicosanoids by activating the PLA2 enzyme [198,222].

Sphingomyelin

SM, also referred to as Cer-1-PC, is primarily of animal origin and constitutes a major fraction of SLs in the plasma membrane, accounting for approximately 15% of cerebral lipids [64]. SM is essential for the transmission of nerve impulses. It was initially isolated and described by Thudichum in 1884 [68]. SM is composed of Cer linked to a phosphocholine group [223], a process catalyzed by sphingomyelin synthase (SMS) [224]. Therefore, the overall configuration of SM closely resembles that of PC. SMS is comprised of multiple isomers, including SMS1 and SMS2, each with distinct active sites, with SMS1 situated in the lumen of the Golgi apparatus and SMS2 located on the plasma membrane [225]. SMS is not solely involved in SM production; it also modulates the generation of DAG during de novo synthesis, occurring simultaneously with SM synthesis. SM can also be produced from LysoSM through fatty acid acylation or the straightforward transmission of phosphocholine to Cer [226]. However, the specific enzymes involved in the latter event have yet to be identified.

Similar to PC, SM is primarily located in the outer leaflet of membranes, but it has also been detected in the nuclear envelope membrane [227], mitochondria [228], and liver chromatin [229]. Vesicular transport is the mechanism that facilitates the migration of SM from the Golgi apparatus to the plasma membrane [230], where possible degradation by sphingomyelinase (SMase) may occur, resulting in the generation of Cer. Remarkably, SMS2 catalyzes a contrasting mechanism for SM synthesis in the plasma membrane [224]. Elevated activity of SMase in the plasma has been associated with metabolic dysfunctions and diseases [231]. However, intracellular levels of SM are not exclusively determined by SMS and SMase activities but are also influenced by the dietary uptake of SM. A review by Yang and Chen [232] delves into potential aspects of SM utilization as a dietary supplement.

SM stands apart from PC, despite sharing the same PC head group. Its distinctive characteristics result from a higher proportion of saturated acyl chains and enhanced intermolecular hydrogen bonding capabilities. Unlike PC, SM serves not only as a hydrogen bond acceptor but also as a hydrogen bond donor. Consequently, SM is capable of being involved in various cellular signaling processes, encompassing functions related to proliferation, migration, and apoptosis [233,234]. Previous studies have elucidated how SO and LysoSLs can modulate protein kinase activities [235,236]. Furthermore, both Cer and SM play a role in modulating the uptake of cholesterol esters from high-density lipoprotein (HDL) particles by the target cells [237]. SM also plays a major role in the formation of lipid rafts, engaging in interaction with CHOL to form membrane microdomains [238–240], wherein roughly 70% of the cellular total SM is concentrated [241]. This favorable interaction between SM and CHOL can be attributed to the specific attributes of SM molecules, including their elongated saturated chains and reactivity properties (hydrogen donor and acceptor).

Glycosphingolipids

This lipid class closely resembles SM due to their shared origin from Cer. It is commonly referred to as GSL as it distinguishes itself from SM by replacing the complex phosphoryl group with a carbohydrate moiety. GSLs are largely derived from glucose moiety, resulting in the formation of glucosylceramide (GlcCer). In addition, GSLs can also be synthesized from a galactose moiety under the activity of galactosyltransferase, leading to galactosylceramide (GalCer) formation. It is essential to highlight that GSLs vary in their carbohydrate acylation locations. GlcCer is primarily produced at the cytoplasmic surface of the Golgi apparatus, whereas GalCers is made on the luminal side of the ER and is subsequently transported to the Golgi apparatus for further structural modifications to generate various GSLs [242,243]. Within cellular membranes, GSLs are believed to exhibit a preference for partitioning into lipid rafts and are involved in communication with the surrounding environment.

Hundreds of complex GSL structures are currently identified in biological systems, the vast majority of which are gangliosides, which are primarily composed of sialic acid and oligosaccharides [244]. Due to the intricate nature of this lipid class, in-depth classification and discussion have been deliberately avoided. However, two of the simplest GSL structures are glucosylceramide (GlcCer) and GalCer, often referred to as 'cerebrosides', featuring either a glucose moiety or a galactose moiety, respectively. The crucial translocation of GlcCer to the luminal leaflet of the Golgi apparatus is an essential step for its subsequent conversion into LacCer, an irreversible pathway involving the addition of a galactose molecule. In addition to these, there exist other GSL complexes such as sulfatides (containing sulfate) and globosides (featuring two or more sugar moieties, typically D-glucose, D-galactose, or N-acetyl-D-galactosamine), which have been identified as GSL derivatives in cellular contexts [245]. For the sake of simplification, scientists have categorized GSLs into two main groups: (1) neutral GSLs, which are characterized by glycosyl groups devoid of acids and remaining unsubstituted, and (2) acidic or amphoteric GSLs, whose glycosyl groups contain one or more sialic acids or a sulfate or phosphate group [246,247].

GlcCer and GalCer function as precursors for numerous complex GSLs, some of which possess additional carbohydrate groups numbering as high as 20 [248]. The addition of these carbohydrate moieties takes place in the Golgi luminal leaflet following the flip-flop translocation of simple GSLs. Majorly, GSLs serve two distinct functions [249]. Firstly, they act as cell receptors to their binding ligands, thereby acting as antigens while facilitating cell adhesion. Secondly, they function as signaling modulators by interacting with other membrane constituents, particularly growth factor receptors. Thus, GSLs play an essential role in immune-cell functions, with a large number of GSL molecules serving as tumor-associated antigens [250–253].

3. Fatty Acids of Biomembranes

Within the context of biomembranes, the matrix comprises an array of complex molecules, with fatty acids serving as fundamental building blocks. Fatty acids exist typically in two forms: saturated and unsaturated monocarboxylic acids, whereas both are characterized by a terminal carboxyl (-COOH) group and a terminal methyl (-CH_3) group designated as carbon 1 (Δ) and omega (ω or n), respectively. Over the past century, numerous nomenclature systems have been proposed, including trivial, systematic, Δ^x, $n - x$, and lipid numbers [22,23]. The trivial nomenclature, though prevalent, lacks systematic patterns. In contrast, the systematic nomenclature adheres to a more regular and structured approach, based on the nomenclature of parent hydrocarbons. It involves adding the suffix "oic" to the hydrocarbon name after removing the terminal "e". This nomenclature also encompasses the identification of the position of the first double bond from the (n), with the series of fatty acids being named accordingly (e.g., n-3, n-6, n-7, and n-9 series). These distinctions among n-fatty acids lead to variations in their properties, consequently influencing the structure and function of biomembranes [254].

Concerning complex lipids, phospholipids, and SLs addressed in this review, fatty acids play a central role as their primary constituents. Therefore, it is essential to provide a concise overview of their biosynthesis, incorporation into complex molecules, and their biological functions in mammals.

3.1. *Synthesis of Fatty Acids*

Fatty acids can either be derived from the diet or biosynthesized within the cytosol and ER through an indigenous pathway known as de novo fatty acid synthesis. This synthesis is a complex process influenced by several determinants, including species, transcription genes, dietary composition, age, gender, stored lipids, and both endogenous (metabolic and interactive molecules) and exogenous (environmental) factors. A multitude of genes regulate the synthesis of fatty acids, which can vary among different species. In eukaryotic organisms, nearly 5% of the overall genes are responsible for a significant proportion of lipid structures [255]. Remarkably, the liver X factor (LXR) contributes to the regulation and synthesis of saturated, mono-, and polyunsaturated fatty acids by targeting their transcriptional genes [256]. It also indirectly influences encoding factors involved in lipogenesis, such as sterol regulatory element-binding protein 1c (SREBP1c) [257], peroxisome proliferator-activated receptor gamma (PPAR-γ) [258], and carbohydrate response element-binding protein (ChREBP) [259].

Numerous organisms can produce a wide variety of fatty acids, but only a limited number of molecular structures are synthesized in significant quantities at the natural physiological rate [260,261]. Generally, the synthesis activity of fatty acids is relatively low in normal adult cells, with the exception of certain tissues, including the brain, liver, adipose, and lungs [262,263]. The liver, known as a lipogenic organ, is predominantly responsible for the de novo pathway, although the white adipose tissue (which consists of lipogenic cells) and mammary glands in animals and humans also possess the capability to produce fatty acids through de novo lipogenesis [264–267]. Under conditions of energy equilibrium, the liver takes up a remarkable proportion (30–50%) of free fatty acids continually absorbed from the diet. These assimilated lipids are either directly incorporated into phospholipids and triglycerides (TAGs) or subjected to modifications (including elongation and/or desaturation) to produce new/modified fatty acids.

Lipogenic cells can synthesize fatty acids endogenously from non-fat molecules, such as glucose or amino acids (see Figure 5). In this process, pyruvate, a metabolite of glucose, enters the mitochondria, where it undergoes oxidative decarboxylation through the pyruvate dehydrogenase complex to form acetyl-CoA. Within the mitochondria, acetyl-CoA can also be derived from the degradation of ethanol, proteins (deamination and oxidation), and fatty acids undergoing β-oxidation. Subsequently, these produced acetyl-CoA enter the tricarboxylic acid cycle (TCA, citrate cycle, or Szent–Györgyi–Krebs cycle). This event is very crucial since mitochondrial acetyl-CoA molecules are not permeable to mitochondrial membranes. Thus, they are initially located within the mitochondria, whereby they endure a condensation reaction with oxaloacetate to form citrate, a process catalyzed by citrate synthase during TCA [268,269]. The citrate molecule is then expelled from the TCA cycle to the inner mitochondrial membrane, and subsequently to the cytosol. This citrate transporting event requires a dicarboxylate antiporter solute carrier family 25 (SLC25A1) [270].

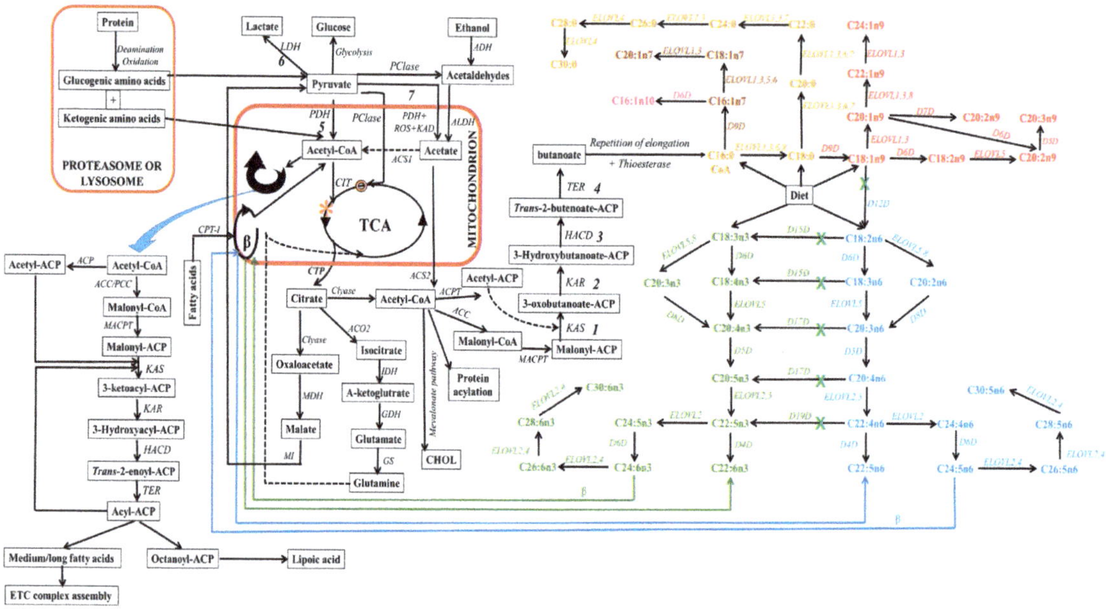

Figure 5. Schematic depiction of eukaryotic fatty acid biosynthesis and modification pathways, with emphasis on enzyme involvement (in italics). In this schematic, mitochondria are denoted by the red box, while proteasome/ribosome components are indicated by the orange box. However, light orange text delineates the elongation pathway for diverse saturated fatty acids. Within the diagram, both green and blue hues, accompanied by corresponding colored arrows, elucidating the discrete pathways for n3 and n6-fatty acid synthesis, respectively. On the other hand, text with color spectrum transitioning from pink to red designates the pathways for the synthesis of various monounsaturated fatty acids. The green "X" indicates the unattainability of this pathway in mammals, particularly higher eukaryotes, owing to the absence of a specific enzyme. ELOVL8 is a fish-specific elongase. Abbreviations: 1, condensation; 2, reduction; 3, dehydrogenation; 4, reduction; 5, aerobic conditions; 6, hypoxia or anaerobic conditions; 7, aerobic conditions; ACC, acetyl-CoA carboxylase; ACP, acyl carrier protein; ACPT, acyl carrier protein transacylase; ACS1, acetyl-CoA synthetases-1; ACS2, acetyl-CoA synthetases-2; ADH, alcohol dehydrogenase; ALDH, aldehyde dehydrogenase; ACO2, aconitase; β, beta oxidation; CHOL, cholesterol; CIT, citrate synthase; Clyase, citrate lyase; CPT-I, carnitine-palmitoyl transferase-I; CTP, citrate transporter protein, EAR, enoyl-ACP reductase; ETC, electron transport chain; GDH, glutamate dehydrogenase; GS, glutamine synthetase; HACD, β-hydroxyacyl-ACP dehydrase; IDH, isocitrate dehydrogenase; KAD, keto acid dehydrogenase; KAR, β-ketoacyl-ACP reductase; KAS, β-ketoacyl-ACP synthetase; MACPT, malonyl-CoA:ACP transacylase; MDH, malate dehydrogenase, MI, malic enzyme, PCase, pyruvate carboxylase; PCC, propionyl-CoA carboxylase; PDH, pyruvate dehydrogenase; ROS, reactive oxygen species; TCA, tricarboxylic acid cycle; TER, trans-enoyl-ACP reductase; O, oxaloacetate; *, citric acid).

In the cytosol, citrate can undergo distinct metabolic pathways to generate various metabolites (see Figure 5). For instance, ATP-citrate lyase enzymatically cleaves citrate into acetyl-CoA, which leads to the carboxylation of acetyl-CoA and the formation of malonyl-CoA. Within cellular cytoplasm, acetyl-CoA (an active form of acetate) and malonyl-CoA serve as the primary substrates that initiate the process of carbon chain elongation. It is worth noting that propionyl and short-branched acyl units for priming can also be utilized in specific cases. This occurs, for instance, when adipose tissue contains monomethyl-branched fatty acids [271]. Additionally, the elongation process during the synthesis of branched fatty acids in specific glands requires the incorporation of methylmalonyl units [272].

It is imperative to emphasize the critical role played by the acyl carrier protein (ACP), which binds to acetyl-CoA and malonyl-CoA, a process catalyzed by malonyl-CoA:ACP transacylase. This integration event facilitates cytosolic elongation in higher eukaryotes by sequentially transferring these substrates from one enzyme/enzyme domain to another throughout sequential biosynthetic cycles. This active participation of ACP is vital for fatty acid biosynthesis and the functions of fatty acid synthase (FAS), which is a multi-enzyme system regulated by the encoded FASN gene. This cytosolic de novo pathway comprises a series of reactions catalyzed by acetyl-CoA carboxylase (ACC) and FAS, which serve as rate-limiting enzymes [273–275]. These reactions include ATP-dependent carboxylation of acetyl-CoA to form malonyl-CoA, Claisen condensation to extend malonyl-ACP and form 3-oxobutanoate, ketoreduction to yield 3-hydroxybutanoate, dehydration to yield butenoate, enoyl reduction to yield butanoate, and repeating elongation reactions (see Figure 5). This process results in the elongation of carbon chains up to the length of C16 or C18 in the cytosol [276–278].

ACC, the rate-limiting enzyme in the de novo pathway, facilitates the irreversible decarboxylation of acetyl-CoA through the addition of CO_2 to produce malonyl-CoA. The resulting malonyl-CoA attaches to ACP and also serves as a two-carbon donor within a cyclic sequence of reactions facilitated by FAS, leading to the generation of a variety of fatty acid species. The end products (acyl-ACP) of cytosolic de novo synthesis are primarily palmitic acids (C16:0), with lower extents of myristic (C14:0) or stearic (C18:0) acids originating from acetyl-CoA. The determination of chain length during cytosolic de novo biosynthesis involves three enzymes: acyltransferases, ketosynthases, and thioesterases [279]. It is essential to highlight that there are two ACC isoforms: ACC1 (also known as ACCα), which is highly expressed in adipose and hepatic tissues, and ACC2 (also known as ACCβ), which is highly expressed in the heart and skeletal muscles [280]. The mechanism for the conversion of acetyl-CoA to malonyl-CoA by ACC is suggested to differ depending on the ACC type due to their different expression locations [281]. ACC1 is a cytosolic enzyme, whereas ACC2 is located in the outer mitochondrial membrane. On the other hand, FAS enzymes are exclusively cytosolic and can catalyze the formation of C16:0 from acetyl-ACP (substrate) and malonyl-ACP (2-carbon donor) [274,275]. FAS comprises numerous large-multifunctional protein domains (type I FAS) in eukaryotic and specific bacteria, while a monofunctional polypeptide domain is present in the case of major bacteria (type II FAS) [279,282–285].

In animals, the FAS pathway undergoes termination through a process involving a thioesterase, resulting in the liberation of the free fatty acid as the final product. The termination of the repeating elongation process exhibits the greatest degree of variation in fatty acid biosynthesis. Nevertheless, the conversion of the cytosolic elongated product to the CoA-ester is vital for further biosynthetic pathways that generate new fatty acid structures.

3.1.1. Elongating Fatty Acids through Non-Cytosolic Mechanisms

Generally, the incorporation of the product into lipid structures or its involvement in subsequent elongation and/or desaturation processes is contingent upon the specific requirements of the organism at a given time. Further elongation pathways are not exclusive to fatty acids derived from de novo fatty acid synthesis in the cytosol; they also act on fatty acids derived from the diet, further extending and/or desaturating them to produce longer saturated, monounsaturated, or polyunsaturated fatty acids that are vital for all biomembranes. Following the production of palmitic acid in the cytosol, further modifications of this fatty acid may occur within cell organelles. In mammals, these modifications involve elongation (chains of 18 carbons or longer) and/or desaturation (formation of monenoic/monounsaturated fatty acids).

The ER and mitochondria regulate the modification (elongation) of fatty acids [278] in order to provide sufficient specific signals and functions. Already existing saturated fatty acids are elongated by the sequential addition of two carbon atoms, resulting in the formation of new fatty acids [286]. The principal fatty acid elongation pathway at the cytosolic side of the ER involves a series of four independent reactions: (1) condensation, (2) reduction, (3) dehydration, and (4) a final reduction step [287]. Major enzymes involved in the elongation process include 3-ketoacyl-CoA synthases (elongase enzymes or ELOVLs for reaction 1), 3-ketoacyl-CoA reductase (for reaction 2), hydroxyacyl-CoA dehydratase (for reaction 3), and *trans*-2,3-enoyl-CoA reductase (for reaction 4). The ELOVL family, sometimes referred to as type III FAS, serves as the rate-limiting enzyme family in the elongation pathway. This family consists of seven subtypes in mice, rats, and humans, and their regulation is governed by ELOVL-encoded genes.

ELOVLs catalyze the condensation of acyl-CoA and malonyl-CoA, which is responsible for elongating fatty acids and determining their carbon chain length, thereby influencing the cell's fatty acid composition and signaling. ELOVLs exhibit variation in substrate specificity, tissue distribution, and regulation [288]. Based on their final products (see Figure 5), ELOVLs are classified into groups: ELVOLs that elongate saturated and monounsaturated fatty acids (ELOVL1, 3, 6, and 7), ELVOLs that produce very long-chain polyunsaturated fatty acids (ELOVL2 and 4), ELVOL5, which acts on a wide range of substrates with carbon chains ranging from 16 and 22, and ELOVL8, which acts on a wide range of substrates with carbon chains ranging from 16 to 20. ELOVL8 is a distinct subtype that has been recently discovered but is believed to be specific to fish [289]. It is widely acknowledged that these genes are primarily regulated at the transcriptional level; however, additional regulatory mechanisms may exist, including allosteric inhibition. In mammals, ELOVL4 is the sole enzyme capable of catalyzing the formation of fatty acids with more than 26 carbons [290]. These polyunsaturated fatty acids with more than 28 carbon atoms are primarily found in the retina, brain [291–294], testis [295,296], spermatozoa [297], epidermis [298], meibomian gland [299,300], and Vernix caseosa [301].

An additional pathway for elongating fatty acids occurs in non-cytosolic fatty acid synthesis, especially in the mitochondria. Both animals and yeasts possess mitochondria that contain FAS II enzymes (mtFAS II), which appear to interact with ACP-linked molecules [302,303]. It should be noted that the ACC enzyme has not been identified in the mitochondria of most mammalian species, including humans. However, a recent isoform of ACC1 has been identified in the mitochondria of mice [304]. Thus, isoforms of ACC1, and potentially mitochondrial propionyl-CoA carboxylase [305], are believed to regulate the decarboxylation of acetyl-CoA to yield malonyl-CoA within mammalian mitochondria. Nevertheless, these reactions occur at a limited rate, suggesting that imported malonate may play a role in mitochondrial fatty acid synthesis.

Mitochondrial fatty acid elongation relies on nicotinamide adenine dinucleotide phosphate (NADPH)-dependent enoyl-ACP reductase, with acetyl-ACP and acyl-ACP serving as substrates [306]. This pathway appears to be energetically unfavorable and represents a minor pathway in eukaryotes [278], primarily contributing to the generation of fatty acids used in the lipogenesis of mitochondrial membranes and cellular respiration. Unlike animals, where thioesterase-mediated termination is involved, mitochondrial termination entails channeling the mitochondrial acyl-ACP into the lipid biosynthetic pathway [307]. The primary generated product of mtFAS II activity is an octanoyl chain, which serves as a substrate for lipoic acid synthesis—a vitamin that acts as a scavenger for free radicals [308–310] and enhances energy metabolism as a cofactor [311]. Although this pathway can also generate medium and long fatty acids [312], their exact biological roles remain uncertain. However, studies by Nowinski et al. [264] and Angerer et al. [313] have suggested that these mitochondrial long-chain fatty acids are involved in the electron transport chain (ETC) complex assembly.

Furthermore, a similar pathway for fatty acid elongation is proximal fatty acid elongation, which is characterized by reversible β-oxidation. In this pathway, acetyl-CoA acts as the carbon donor, and peroxisomal *trans*-2-enoyl-CoA reductase substitutes acyl-CoA dehydrogenase to facilitate a thermodynamically favorable reaction [314]. It is important to emphasize that CoA is implicated in the fatty acid catabolism of the reversible pathway, while ACP plays a role in mitochondrial fatty acid elongation. However, the precise functions of fatty acid elongation within peroxisomes remain insufficiently characterized from a scientific standpoint. In practice, the extent of elongation is typically assessed using the elongase estimated index, which is determined by the ratio of C16:0 to C18:0.

3.1.2. Desaturation of Fatty Acids

Within the context of de novo fatty acid synthesis, a process characterized by the removal of two hydrogen atoms to create a double bond often intersects with the fatty acid elongation pathway. This synergy ensures the production of long-chain and very long-chain unsaturated fatty acids, both mono- and polyunsaturated. Notably, the enzymes responsible for fatty acid desaturation, known as fatty acid desaturases, are ubiquitous across all domains of life with the exception of archaea, where they are notably absent [315]. However, it is of significant importance to underscore that the synthesis of polyunsaturated fatty acids can also occur independently of the classical series of desaturase and elongase enzymes. Several studies, including those conducted by Smith and Tsai [316], Kaulmann and Hertweck [317], Napier [318], and Metz et al. [319], have extensively documented an alternative pathway for the biosynthesis of long-chain polyunsaturated fatty acids in both prokaryotes and lower eukaryotes. This alternative route relies on semi-fatty acid synthesis systems, specifically known as polyketide synthases (PKS). PKS employs the same four fundamental reactions as FAS. However, the PKS cycle is frequently condensed, resulting in the formation of highly modified carbon chains featuring numerous keto and hydroxy groups, along with trans-configured double bonds that exhibit various functional roles [320,321].

Among the plethora of desaturase families found in different species, researchers have categorized them into three distinct types, as described by Cerone and Smith [322]. The first family, acyl-acyl carrier protein desaturase, is exclusive to plastids of higher plants. The second family, acyl-lipid desaturases, is found in the ER membranes of higher plants and cyanobacteria. The third family is the family of acyl-CoA desaturases, which can be present in both eukaryotes and prokaryotes, and these enzymes use a cytochrome b5-like system during oxygen reactions [323]. Within the context of this review, with a primary emphasis on mammals, the discussion will be specifically on the acyl-CoA desaturase families. In mammals, a critical in vivo biosynthetic route for the production of long-chain polyunsaturated fatty acids is known as the 'Sprecher pathway' [324]. This pathway relies on two fatty acid desaturase enzymes, two ELOVLs, and a peroxisomal β-oxidation process.

The desaturase pathway encompasses diverse enzyme families, such as stearoyl-CoA desaturase (SCD) [325] and fatty acid desaturase enzymes (FADS) [326]. Each of these enzymes acts on distinct substrates. For example, FADS primarily targets polyunsaturated substrates, while SCD predominantly acts on saturated substrates. SCD, also referred to as delta-9 desaturase (Δ9-desaturase, or D9D), is an ER enzyme that catalyzes the formation of monounsaturated fatty acids, including palmitoleic (C16:1 n7) and oleic (C18:1 n9) acids, from saturated fatty acids, such as palmitic and stearic acids, respectively. These enzymes exhibit varying specificities and can work on a range of fatty acids with different carbon chain lengths, from 16 to 26 carbon chains. This variation hinges on the specific isoform present, with some species harboring multiple homologues of D9D. For instance, two homologues (D9D-1 and D9D-2) have been identified in marine copepods [327].

Basically, within the D9D reaction, a double bond is introduced at the 9th position from the carboxyl group of the fatty acid. In addition, a multitude of desaturase enzymes present in plants, bacteria, and insects perform the initial double bond introduction on saturated fatty acids at various positions, including Δ3, Δ5, Δ7, and Δ11 [328–332]. The activity of D9D is modulated by dietary fatty acid intake and is subject to up-regulation by the expressions of SCD gene isoforms [333]. Since Bloch's 1960 pioneering discovery of SCD [315], a plethora of gene isoforms have come to light. In mice, the SCD exhibits four distinct isoforms (SCD-1–4) [334], whereas in humans, only two isoforms (SCD-1 and SCD-5) have been identified [335]. These gene isoforms can vary in substrate preference, Δ position, and double bond conformation [336]. SCD-1 is the most studied isoform among them [325]. The importance of the SCD pathway is underscored by its involvement in cellular stress, lipid metabolism, inflammation, and autoimmunity. Among the isoforms, SCD-1 is the most extensively studied and is associated with various physiological processes [337–342].

Fatty acids featuring a Δ9 double bond are eligible for elongation, but it is important to note that these fatty acids may also be derived from the diet. However, D9D activity is indirectly determined by assessing the ratio between C18:0 and oleic acid. Notably, the de novo elongation of oleic acid results in the formation of longer chains of monounsaturated fatty acids. Furthermore, oleic acid, in particular, may undergo a desaturation process often followed by elongation, where the double bond is introduced before the Δ9 position. Nevertheless, unlike in plants and a limited number of animal species, mammals lack the enzymatic capability to catalyze the introduction of the second double bond in oleic acid, particularly not after the Δ9 position.

Fatty acid desaturase genes (FADs) comprise a gene family responsible for encoding a variety of desaturase enzymes. These enzymes play a significant role in catalyzing the synthesis of polyunsaturated fatty acids by introducing multiple double bonds at positions within fatty acids. Among the genes involved in fatty acid desaturation, mammals have been identified with four distinct FADs [335]: (1) FAD-1, (2) FAD-2, (3) FAD-3, and (4) FAD-6. Each of these FAD types tends to have alternative transcriptions [343–345], which can express distinct desaturase activities at different Δ-positions. Generally, FAD enzymes can be categorized into FAD-1, responsible for generating omega-3 (n3) fatty acids; FAD-2, involved in generating omega-x (nx) fatty acids; and FAD-3, which contributes to the formation of omega-6 (n6) fatty acids. While little is known about the substrates of FAD-6 and their major roles have not yet been defined, it appears to be homologous to human FAD-2 [346–348] and likely plays a role in the synthesis of polyunsaturated fatty acids.

The transcription factor FAD-6 has been described to possess Δ4-, Δ5-, and Δ8-desaturation activities, with a notable impact on docosahexaenoic acid (C22:6 n3, or DHA) synthesis from n3-docosapentaenoic acid (C22:5 n3, or DPA-n3) in golden pompano fish [349]. On the other hand, Park et al. [345] detailed the existence of nine alternative transcriptions for FAD-3, potentially arising from splicing events. Initially identified through cloning efforts by Marquardt et al. [350], FAD-3 has been traditionally classified as a non-polyunsaturated desaturase, primarily due to its limited substrates, namely, vaccenic acid (C18:1 trans-11) and sphingoid bases. In this regard, it should be noted that FAD-3 may facilitate the unanticipated Δ13-desaturation of trans-vaccenate [351]. These limited substrates give rise to the production of 11E,13Z-octadecadienoic acid [352] and 4E,14Z-sphingodienine [353], respectively.

In contrast to FAD-3, FAD-1 and FAD-2 are the principal contributors to the biosynthesis of polyunsaturated fatty acids in mammals. Nevertheless, mammals lack two crucial desaturases, namely delta-12 desaturase (Δ12-desaturase, or D12D) and delta-15 desaturase (Δ15-desaturase, or D15D) [354–356]. These enzymes are often referred to as "methyl-end desaturases" due to their ability to introduce a new double bond between an existing unsaturated bond and the methyl terminus ($-CH_3$) of the fatty acid. Thus, mammals are de novo incapable of introducing a new double bond after the Δ9 position of oleic acid. Hence, mammals must obtain polyunsaturated fatty acids from their diet, leading to the concept of essential fatty acids. These essential fatty acids, such as linoleic acid (C18:2 n6, or LA) and α-linolenic acid (C18:3 n3, or ALA), cannot be endogenously synthesized by mammals and must be sourced from dietary intake [357–359]. Nonetheless, the enzyme activities of FAD-1 (Δ5-desaturase, or D5D) and FAD-2 (Δ6-desaturase, or D6D) have been identified in mammals. These enzymes are responsible for introducing double bonds at the Δ5 and Δ6 positions, respectively [360,361]. Notably, both D6D and D6D are highly expressed in the liver, with D6D exhibiting particularly high expression levels [362].

In mammals, dietary LA, ALA, and other unsaturated fatty acids, whether from the diet or synthesized endogenously, serve as substrates for the generation of polyunsaturated fatty acids characterized by an increased number of double bonds and longer carbon chains. The enzyme D6D plays a crucial role in the initial steps of synthesizing arachidonic acid (C20:4 n6, or AA), eicosapentaenoic acid (C20:5 n3, or EPA), and DHA by catalyzing the conversion of LA and ALA into γ-linolenic acid (C18:3 n6) and stearidonic acid (C18:4 n3), respectively [362]. However, the biosynthesis of AA and EPA, in particular, involves an additional desaturase enzyme, delta-5 desaturase (D5D), which utilizes the substrates dihomo-γ-linolenic acid (C20:3 n6, or DGLA) and eicosatetraenoic acid (C20:4 n3) to yield AA [360] and EPA [363], respectively. According to Vagner and Santigosa [364], the substrate affinities of D6D appear to exhibit contrasting and debatable characteristics. Consequently, further investigations are imperative to substantiate a definitive conclusion. It is essential to emphasize that the distinctive substrate affinities of D6D play a critical role in determining the ratio of n6- to n3-polyunsaturated fatty acids (referred to as n6/n3 or n6:n3). The role of FAD-6 in determining the ratio of n6/n3-polyunsaturated fatty acids and its influence on the metabolic flux of these fatty acids have been highlighted [365,366].

Remarkably, D6D and D5D are also involved in the synthesis of n9-polyunsaturated fatty acids, specifically, Mead acid (C20:3 n9), which is produced in the absence of LA and ALA [367], when only monounsaturated fatty acids are available as substrates. Elevated levels of Mead acid are matched with the proportional depletion of n6- and n3-polyunsaturated fatty acids and serve as a biomarker for LA and ALA deficiency in diets. For instance, the ratio between trienoic and tetraenoic acids (such as Mead acid:AA) functions as a biomarker for the lack of dietary essential fatty acids [368,369]. Park et al. [370] have identified two pathways for Mead acid production, involving D6D and D5D. These pathways vary in substrates of D6D, which can either utilize oleic acid or gondoic acid (C20:1 n9) to yield C18:2 n9 and C20:2 n9, respectively. D5D catalyzes the direct conversion of C20:2 n9 into Mead acid by introducing a double bond at the Δ5 position. Furthermore, authors have also reported the novel activity of Δ7-desaturase (D7D, regulated by FAD-1), leading to the formation of C20:2 n9 from C20:1 n9.

Notably, AA, EPA, and DHA are biologically essential fatty acids with enormous contributions to membrane phospholipids. In the Sprecher pathway, the primary fatty acid synthetic pathway in mammals, the biosynthesis of DHA from EPA is not direct but rather involves a series of reactions: (1) elongation to a 24-carbon chain, (2) desaturation via D6D, and (3) peroxisomal β-oxidation for chain shortening. However, the synthesis of n6-docosapentaenoic acid (C22:5 n6, or DPA-n6) and DHA in eukaryotes also occurs through Δ4-desaturase (D4D) pathways, depending on the species. In lower eukaryotes, marine vertebrates, and humans, D4D (encoded by the FAD-2 gene), which is expressed to a lesser extent, plays an essential role in introducing a double bond at position Δ4, directly producing DPA-n6 and DHA from docosatetraenoic acid (C22:4 n6, or adrenic acid) and DPA-n3 substrates, respectively [371–374]. It is noteworthy that this reaction bears resemblance to that facilitated by FAD-6 expression, suggesting the possibility of FAD-6 up-regulating D4D activities.

A further marked expression of FAD-2 identified within mammalian cells is Δ8-desaturase (D8D). This enzyme establishes an autonomous pathway distinct from D6D, introducing an additional double bond to substrates like ALA, LA, and their elongated polyunsaturated fatty acids, resulting in the formation of very long-chain polyunsaturated fatty acids [375,376]. For instance, the emergence of D8D was observed when DGLA and eicosatetraenoic acid were derived from eicosadienoic acid (C20:2 n6) and eicosatrienoic acid (C20:3 n3), respectively [377]. In a study conducted on mouse liver, D8D expressed low activity, which is primarily associated with n3-unsaturated fatty acids, such as eicosatetraenoic acid (C20:4 n3), EPA, DPA-n3, DHA, and nisinic acid (C24:6 n3) [375].

In general, more than a hundred FAD-2-related desaturase enzymes have been identified in various animal species, although they are notably absent in mammals. For instance, the Δ17 (17-desaturase, or D17D) and Δ19 (19-desaturase, or D19D) desaturases have been identified in lower animal families (excluding mammals) and algae [378,379]. These enzymes play an essential role in the production of EPA and docosapentaenoic acid-n3 (C22:5 n3, or DPA-n3) from AA and adrenic acid, respectively. The activities of enzymes involved in lipogenesis are subject to intricate regulation by a matrix of genes and exogenous factors. Notably, polyunsaturated fatty acids have been shown to inhibit the transcription of hepatocellular genes responsible for encoding lipogenic enzymes [380,381]. It is a common practice to assess desaturase enzyme activity indirectly by determining their coefficients based on the ratio of the fatty acids generated to the substrates utilized.

Desaturase enzymes exhibit a broader scope of activities beyond their involvement with fatty acid substrates incorporated into phospholipids, as they are also active in SLs. Ordinarily, SLs are characterized by their predominantly very long saturated or monounsaturated nature. The determination of fatty acid chain length within SLs is intricately governed by the activities of the CerS type [382]. However, it is worth noting that polyunsaturated fatty acids show a slight accumulation in SLs within testes and spermatozoa [383,384] as compared to phospholipids within the same tissues. The classification of desaturases responsible for introducing double bonds into Cer structures has been presented by Nachtschatt et al. [385]. This classification delineates three distinct categories of desaturases: (1) α-hydroxylases [386], (2) D4D and C4-hydroxylases [387], and (3) D8D [388]. These desaturases play a pivotal role in diversifying the composition of SLs, particularly in terms of the introduction of double bonds, which contribute to the functional and structural heterogeneity of these important lipid molecules.

3.2. Incorporation of Fatty Acids into Lipids of Biomembranes

Fatty acid synthesis and their subsequent incorporation into biomembranes represent critical processes in the life of all organisms. The one exception to this rule is the archaea domain, which employs isoprenoids as membrane lipid side chains rather than fatty acids [389]. Understanding how fatty acids are incorporated into membrane lipids is of paramount importance. Fatty acids may become part of phospholipids either through acylation of glycerol-3-phosphate during the biosynthesis of phospholipids or through the action of lysophosphatidic acyltransferases and phospholipases that remodel the structure of pre-existing phospholipids [390,391] (which can be seen in Figure 6).

The vital nature of these processes is underscored by the coexistence of diverse fatty acid structures within complex biomembrane lipids [392]. This structural diversity arises from various factors, including the preferences of *sn*-positions for particular fatty acids, the substrate preferences of biosynthetic enzymes, and the dynamic lipid remodeling program. Notably, studies by Shindou et al. [393] and Coleman et al. [394] have elucidated the preferences of specific enzymes for distinct fatty acids. For instance, acyl-CoA synthetase long-chain family members 3 and 4 demonstrate preferences for AA and EPA, while 1-acylglycerol-3-phosphate O-acyltransferase-α prefers myristic acid, palmitic acid, and LA. Meanwhile, 1-acylglycerol-3-phosphate O-acyltransferase-β exhibits a preference for AA.

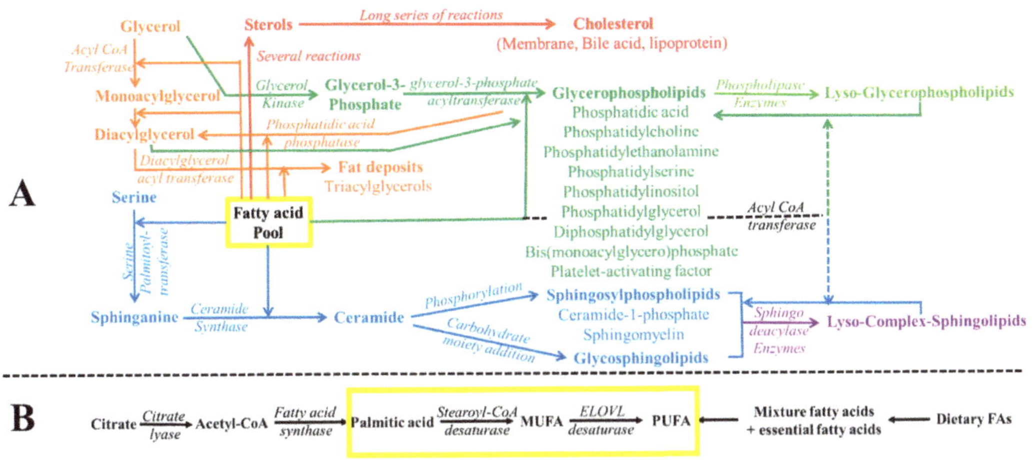

Figure 6. Schematic representation of (**A**) fatty acids incorporated into various lipids and (**B**) the origins of fatty acid pool formation (the de novo pathway and dietary sources). The depiction employs color coding to symbolize distinct metabolic pathways. The yellow box designates the fatty acid pool, signifying its integration into diverse membrane lipids. The brown shade denotes processes related to non-membrane and non-polar lipid formation. The red color represents the metabolic routes responsible for sterol production. Additionally, the green color signifies the integration of fatty acids into various phospholipids, while the blue hue corresponds to the incorporation of fatty acids into diverse sphingolipids.

In the context of lipid remodeling, the replacement of fatty acids within existing phospholipids plays a central role. For instance, the incorporation of C20 fatty acids generally takes place post-de novo synthesis, necessitating the employment of the remodeling pathway [395]. Hence, this process involves the conversion of one distinct phospholipid into another [85], highlighting the importance of the Lands cycle in this process. The Lands cycle [396,397] is a central component of the remodeling process, enabling the attainment of specific structural configurations required for the generation of particular cellular signals. In this pathway, phospholipases initiate cleavage (deacylation) of fatty acids from phospholipids, resulting in the formation of free fatty acids and lysophosphatidate. On the other hand, acyltransferases function in a contrasting manner, acylating the requisite fatty acids into lysophosphatidate to generate phospholipids. This process is CoA-dependent, which is primarily due to the extensive utilization of CoA intermediates as substrates. Other remodeling pathways include the CoA-independent transacylation pathway and the direct transacylation pathway, which are specifically suited for highly unsaturated fatty acids such as AA, EPA, and DHA. In these pathways, transacylases catalyze the migration of fatty acids between molecular species of phospholipids. Remarkably, these remodeling pathways exhibit substantial variability across different tissues [395], as comprehensively reviewed [390].

Additionally, it is noteworthy that fatty acids are predominantly incorporated into SLs by the rate-limiting enzyme CerS. However, a remodeling mechanism can also come into play, modulating specific signaling and functional roles. For example, the work of Markham et al. [398] suggests that the accumulation of very long-chain fatty acids contributes to the formation of microdomains through increased hydrophobicity, membrane leaflet interdigitation, and the transition from a fluid to a gel phase. These structural transformations have significant implications for cellular function and signaling within membranes.

The movement of membrane phospholipids between bilayers involves the activities of various transmembrane lipid transporter proteins, namely, flippase, floppase, and scramblase. These proteins play distinct roles in the translocation of lipids and exhibit differential mechanisms of action. For instance, flippase facilitates the translocation of lipids from the exo-leaflet to the inner-leaflet, contrasting with the activity of floppase, which operates in the opposite direction, moving lipids from the inner-leaflet to the exo-leaflet. Notably, both flippase and floppase depend on ATP as an energy source for their functions. Conversely, scramblase functions as an ATP-independent transporter and orchestrates the bidirectional movement of lipids across membranes in a non-specific manner, allowing for the concurrent translocation of lipids from the inner to the outer leaflet and from the outer to the inner leaflet. Comprehensive insights into these membrane fatty acid transporters have been provided by the extensive reviews conducted by Samovski et al. [399] and Glatz et al. [16]. These reviews offer a detailed discussion of the mechanisms and significance of these proteins in lipid translocation processes within biological membranes.

3.3. *Fatty Acid Composition in Biomembranes*

The fatty acid composition of biomembranes plays a critical role in maintaining cellular homeostasis and ensuring proper functionality. It is worth noting that this composition is not static but can be modified/remodeled in response to homeoviscous adaptation. This process was initially described in algae [400] and has been later extended to non-homeothermic animals. In contrast, mammals are generally considered to have stable biophysical properties and lipid order within cellular membranes, but they can respond to changes in their dietary fatty acid compositions. Furthermore, certain mammalian cells potentially exhibit a lower degree of homeoviscous adaptation in response to the membrane curvature elastic stress [401]. The fact that there is not a single standardized composition for fatty acids in biomembranes should not be viewed as a flaw but rather as an indication of the intricate and dynamic nature of biological systems. The complexity and adaptability of fatty acid composition in biomembranes underscore the potential evolutionary advantage conferred by the ability to alter lipid structures.

The available literature, which will be discussed in subsequent sections, provides extensive data regarding the proportions of specific fatty acids in various lipid complexes within biomembranes. This section merely focuses on the major fatty acids identified in phospholipids and SLs, as these fractions are the central aspects of this review. Regardless, the length and degree of saturation of fatty acyl side chains in CHOL-esters can vary between different cells. This variation is primarily influenced by dietary factors and cell metabolism, and it has a direct impact on the stability and fluidity of the cellular membrane.

3.3.1. Fatty Acid Profile of Phospholipids

Commonly, the acyl chains of phospholipids found at the *sn*-1 and *sn*-2 positions of the glycerol backbone are typically composed of a saturated fatty acid (such as C16:0 or C18:0) and an unsaturated fatty acid (with carbon chains of 18, 20, 22, or 24), respectively. Furthermore, these acyl chains exhibit variations in terms of their length, the number of double bonds, and the position of hydrogen atoms adjacent to these double bonds (whether in *cis* or *trans* configuration). It is important to note that *trans*-double bonds are relatively rare within mammalian membranes, while *cis*-double bonds are frequently abundant. In fact, the occurrence of *trans*-double bonds in mammals is far less frequent than in bacteria [402,403]. In some instances, identical acyl chains can be found at both *sn*-1 and *sn*-2 positions of glycerol [404,405]. It is worth highlighting that even when the number of carbons is the same, a slight mismatch may occur due to the *sn*-positions. This mismatch is a consequence of the *sn*-2 chain bending perpendicular to the membrane's plane [406,407].

In the case of PA, the composition of the two fatty acids can vary across different cell types and constituents. Typically, dietary fatty acids and metabolic disorders play a substantial role in determining the composition of these acyl chains. PA is primarily composed of saturated and monounsaturated fatty acids, with carbon chains ranging from 14 to 24. Concerning PG, this class exhibits variations in the length and number of unsaturation of its fatty acids, depending on the cell type and the dietary fatty acids involved. In the PG of rat lungs, palmitic acid is the most abundant fatty acid, with unsaturated fatty acids constituting a smaller portion [408]. PG is unique among phospholipids due to its higher degree of unsaturation, with unsaturated fatty acids more likely to occupy the *sn*-1 position. Research by Xie et al. [409] suggests that the degree of unsaturation in PG may have varying effects on mouse keratinocyte proliferation.

Commonly, the fatty acid composition of PCs is typically determined post-synthesis, with various events, such as deacylation and reacylation, taking place during its remodeling [396,397]. These events, part of the Lands cycle, can also impact the composition of other phospholipids due to homeostatic mechanisms or metabolic implications [410]. Generally, PC exhibits variations in its fatty acid composition across species and cell types [411–416]. Saturated fatty acids, like palmitic or stearic acids, are typically abundant at the *sn*-1 position, while C18 unsaturated chains or longer polyunsaturated fatty acids like AA and DHA are more prevalent at the *sn*-2 position.

The fatty acid composition of PE is highly dependent on the particular cell, tissue, and physiological conditions. In contexts like chicken egg, rat liver, and brain, palmitic and stearic acids tend to occupy the *sn*-1 position, while AA, oleic, and DHA are more common at the *sn*-2 position [411,412,417]. Notably, PE in the erythrocyte membrane exhibits a greater tendency for the accumulation/recruitment of polyunsaturated fatty acids compared to PC [418]. Consequently, PE in this context contains more polyunsaturated fatty acids, primarily AA and DHA, at the *sn*-2 position, despite its diacyl structure bearing similarities to that of PC.

The composition of fatty acids in PS differs depending on the tissue type. Nonetheless, lipid remodeling and selective insertion of fatty acids are common processes that occur along the biosynthetic pathway. Therefore, the fatty acid composition of newly synthesized PS differs from that of its precursors, PE and PC. When PS was isolated from rat liver and cow brain and analyzed regiospecifically [412,414], it was observed that stearic acid was more abundant at the *sn*-1 position, while the proportion of palmitic acid was lower here. These data underscore the distinct fatty acid incorporation pattern exhibited by PS in comparison to PC and PE. Furthermore, the acylation of fatty acids at the *sn*-2 position has been shown to be tissue-specific, with high proportions of AA and DHA in the livers of rats and the brains of bovines, respectively. Similar findings regarding DHA in brain regions of mice and rats have been reported by Kim et al. [419] and Hamilton et al. [420]. However, stearic acid was the most abundant fatty acid at the *sn*-1 position in these cases. Remarkably, the incorporation of these unsaturated fatty acids into PS appears to be more extensive than in the case of PC, as revealed by these studies. The fatty acid composition of PS at its *sn*-positions plays a crucial role in determining its functional properties to varying degrees [93,421].

In mammalian cells, the composition of PI is characterized by the presence of stearic acid and AA in the *sn*-1 and *sn*-2 positions, respectively, as consistently demonstrated in various studies [411,412,422,423]. These two fatty acids collectively account for a substantial portion of PI acyls, typically ranging from 50% to 80% in the brain and liver. Additionally, oleic acid is frequently identified as the third most abundant fatty acid in the rat brain, while palmitic acid and DHA exhibit lower levels of acylation in this phospholipid. However, it is worth noting that Ulmann et al. [424] reported a distinct fatty acid composition in the rat brain, where oleic acid was the most prevalent, followed by stearic and palmitic acids. This variability in PI composition across studies may be attributed to a range of factors, including dietary influences and genetic variations. In general, PI exhibits a strong preference for AA in its acyl composition. Remodeling processes can lead to the deacylation of pre-existing PI, resulting in the formation of lyso-PI via the action of phospholipases. The incorporation of AA into lyso-PI is facilitated by lysophosphatidylinositol acyltransferase 1 [425].

The structure of fatty acids within CL greatly influences its shape and properties. Saturated chains tend to produce lamellar forms of CL, whereas unsaturated chains contribute to non-lamellar structures [38]. Thus, the distinctive fatty acid composition of CL is believed to play a critical role in its diverse biological functions across different cellular and subcellular membranes. CL remodeling is an essential process for CL formation, during which LA and DHA are primarily incorporated, influencing susceptibility to oxidation [426–429]. LA has been reported as the majority of CL fatty acids, often accounting for 80–90% of the composition [430]. Although C18 polyunsaturated fatty acids are the predominant constituents of CL, dietary fatty acid intake can influence its composition. Increased dietary supplementation levels of C20 polyunsaturated fatty acids and DHA have been shown to elevate their proportions in mammalian CL, as indicated by Berger et al. [431] and Wolff and Entressangles [432].

The structure of BMP exhibits variations in chain length and the degree of fatty acid unsaturation. Oleic acid is frequently identified as the most abundant fatty acid in BMP [126,132,433]. However, in certain cell types, polyunsaturated fatty acids such as LA and/or EPA and DHA have been reported to be highly accumulative [434–437]. In alveolar macrophages, for instance, Holbrook et al. [438] noted that oleic acid predominates along with either AA or DHA.

In intact tissues (e.g., neural tissue), palmitic acid typically represents the most abundant fatty acid within the alkyl group of PAF. Stearic and oleic acids may also be present but to a relatively lesser extent. The *sn*-2 position of PAF is often esterified with the acetyl group or other short-chain fatty acids [439]. However, in rat nervous tissue, *sn*-2 is predominantly occupied by unsaturated fatty acids, primarily AA and adrenic acid (C22:4n6) [440], indicating the profound impact of *sn*-2 composition on PAF activities. These longer n6-type fatty acids serve as potent precursors for eicosanoids.

3.3.2. Fatty Acid Profile of Sphingolipids

In general, SLs exhibit a prevalent composition of very long-chain saturated and/or monounsaturated fatty acids, typically ranging from 18 to 34 carbon chains. Notably, some sphingolipid structures have been reported to contain odd-numbered fatty acid chains [201,441], suggesting a possible proportional elevation of these chains in the tissues of ruminants and coprophagous species. Despite the fact that the enzyme SPT utilizes palmitoyl-Coenzyme A to produce SA, it has the capability to utilize various other fatty-CoA substrates as well [442]. Sphingoid bases within SLs predominantly feature saturated aliphatic chains, with some instances of mono- and di-unsaturated chains. SO possesses a fixed *trans*-double bond between carbons 4 and 5. The chain length of sphingoid bases' fatty acids typically falls within the range of 14 to 32 carbons [443].

Within the context of Cer, the variability in chain length is attributed to the diversity of CerS isoforms. In most scenarios, CerS5 and CerS6 predominantly provide chains with 14–18 carbons, CerS1, CerS4, and CerS2 contribute chains with 18–24 carbons, and CerS3 supplies chains with up to 34 carbons [382]. Cer features an acyl chain linked to an amide that is frequently saturated and considerably longer compared to those found in SLs containing sugar moieties [240]. The length of the Cer fatty acid is determined by ER elongase and desaturase complexes, rather than the Golgi apparatus, which produces complex SLs. Cer and SM are primarily acylated by palmitoleic and oleic acids, in addition to long and very long-chain fatty acids. In most organisms, including mammals, Cer can contain 2-hydroxylated fatty acids [444,445], which is a characteristic favoring interaction with SMS2 over SMS1 [446].

SM structures are characterized by long fatty acids with relatively high chain length inequality, leading to their interdigitation sensitivity. The cause of interdigitation is not fully understood but may involve the regulation and/or induction of proteins in response to specific fatty acids within SM under certain membrane constituents. In comparison to PC, SM typically contains more saturated and longer acyl chains, typically ranging from 16 to 24 carbons [415,447], and may include odd-numbered fatty acids [448]. The palmitic, stearic, behenic (C22:0), lignoceric (C24:0), and nervonic (C24:1 n9) acids are among the most frequently occurring fatty acids in SM [415,447–450]. Mammalian germ cells, in particular, are abundant in very long-chain fatty acids with up to 34 carbons [297,451], some of which may exist in a 2-hydroxylated form in certain mammals [452]. Double bonds are relatively rare in SM fatty acids, and, when present, they are often located at a distance from the membrane–water interface [447].

GSLs exhibit a greater variation in the chain length of their fatty acids compared to phospholipids, with some GSLs containing fatty acids with more than 16 carbons [453]. Although it is challenging to detect long and/or unsaturated fatty acids in GSL, a few studies have reported the presence of high proportions of long and very long fatty acids. For instance, stearic acid has been found to be more dominant than palmitic acid in the GSLs of the mouse brain [454]. Furthermore, the majority of GalCer in the brain is composed of very long fatty acids [455]. In gangliosides, lignoceric acid (C24:0) is the most abundant acyl chain, constituting up to 70% of total fatty acids in myelin [456], while stearic acid is the predominant component, making up 80% of total fatty acids in the human brain [457].

4. Fatty Acids and Cellular Functionality

The different chemical structures, physicochemical properties, and physiological functions of various fatty acids exhibit remarkable diversity. These distinctions have led to various categorizations, depending on the specific scientific focus. Traditionally, fatty acids have been recognized for their dual biological functions: as structural components of biomembranes and as sources of energy. However, contemporary perspectives on fatty acids have expanded to encompass their roles as bioactive molecules that contribute significantly to overall health. Pioneering work by Burr and Burr [458] and von Euler [459] underscored the diverse biological functions of fatty acids, particularly LA and ALA, highlighting their crucial roles in cellular signaling processes. This section focuses on fatty acids' impact on membrane properties and avoids delving into their role in energy provision via β-oxidation (primarily relies on TAGs), which is a subject beyond the scope of this review.

4.1. Influence of Fatty Acids on Bilayer Properties

Fatty acids play an essential role as constituents in cellular membrane assembly, profoundly influencing the physicochemical attributes of these membranes. Biomembranes exhibit a discerning preference for incorporating specific fatty acids, particularly favoring long and very long polyunsaturated fatty acids that contribute to the formation of fluidic membranes. A study by Rodriguez-Estrada et al. [460] has associated long-chain lipid metabolites derived from LA and ALA with the preservation of membrane properties. Sensor proteins typically monitor and regulate the physicochemical properties of membranes [7,461,462]. Fatty acids exhibit variations in chain length and degree of unsaturation across different cellular contexts, catering to specialized functions. For instance, the study of Matveyenka et al. [463] has highlighted the correlation between the rate of insulin aggregation and the length and degree of the unsaturation of fatty acids. Therefore, maintaining equilibrium among various fatty acid species (saturated, monounsaturated, and polyunsaturated) within membranes holds a position of critical importance, as any deviations from this equilibrium could lead to modifications in membrane integrity and cellular metabolic signaling. Notably, Baccouch et al. [464], Hashimoto et al. [465], and Ibarguren et al. [466] have reported the effects of fatty acid composition on various aspects of membrane behavior, including fluidity/viscosity/rigidity, thickness, permeability, phase transitions, fusion, lateral pressure, flip-flop dynamics, and structural integrity.

The optimal functionality of membrane-bound enzymes, ion channels, and receptors is intrinsically linked to membrane rigidity and permeability, influencing the diffusion of biomolecules within the lipid bilayer. Incorporating higher proportions of long saturated fatty acids enhances membrane rigidity, as these fatty acids are notably stable, having higher melting points than unsaturated fatty acids [467], and tend to form close clusters [468], particularly at physiological temperatures, resulting in heightened membrane rigidity. In contrast, polyunsaturated fatty acids contribute to greater conformational flexibility in membranes, which is a trait dependent on their chain length, degree of unsaturation, and the positioning of hydrogen atoms relative to the double bond. The presence of unsaturated fatty acids introduces curves/bends (also known as "kinks") in the hydrocarbon chains, leading to the formation of less densely packed lipids and more fluidic membranes [469]. However, the influence of polyunsaturated fatty acids on membrane fluidity may vary, particularly in different bilayer states [470]. For instance, EPA and DHA have demonstrated negligible effects on fluidity in liquid-crystalline states [471]. In contrast, within different membrane models, polyunsaturated fatty acids with four or more double bonds, specifically AA, EPA, n6-DPA, and DHA, have been reported to decrease membrane thickness [464,472–474], increase the tilt angle [472], and elevate membrane fluidity [475,476]. The degree of rigidity contributed by double bonds within fatty acids is contingent on various factors, including their conformation (*cis* or *trans*), the degree of unsaturation, and their relative positioning concerning the carboxyl group. According to Roach et al. [477], the membrane properties of fatty acids associated with *cis*-unsaturated fatty acids were markedly different from those of saturated and *trans*-unsaturated fatty acids. Typically, *cis*-isomers exhibit greater polarity and possess relatively higher boiling points compared to *trans*-isomers, although not as high as those of saturated fatty acids. Notably, the position of the double bond exerts a more substantial impact on boiling points than the number of double bonds [478]. *Cis*-double bonds have been identified as expanding the spatial area occupied by the fatty acid, thereby increasing membrane fluidity [479,480] and permeability. Moreover, phospholipids containing long-chain n3-fatty acids have shaped more disordered and flexible membrane structures compared to LysoPLs containing n6-fatty acids, underscoring the significant role of n3-fatty acids in shaping membrane integrity.

Unsaturated fatty acids influence biomembrane rigidity not only through their intrinsic molecular structure but also by modifying the proportions of other membrane constituents that contribute to rigidity. For instance, Schumann et al. [481] and Stillwell [482] investigated the role of polyunsaturated fatty acids in modulating raft characteristics, including size, stability, and distribution. Notably, polyunsaturated fatty acids have a reduced affinity for cholesterol (CHOL) compared to their saturated counterparts. Consequently, an increased incorporation of unsaturated fatty acids within biomembranes can result in loosely packed lipid structures. This, in turn, leads to the displacement of raft-associated proteins and the removal of SM and CHOL from lipid rafts. These alterations result in shifts in membrane rigidity and permeability [483–492]. Therefore, the degree of unsaturation plays a crucial role in modulating the flip-flop rate and the asymmetry/distribution of membranes. Cheng et al. [493] and Armstrong et al. [494] have substantiated a positive correlation between the trans-membrane flip-flop rate and the proportion of unsaturated fatty acids. In assessing membrane rigidity, ratios such as phospholipid/CHOL, PC/PE, and PC/SM (the unsaturation index) have been routinely employed.

It is rational to posit that membranes characterized by inadequate lipid packing correspondingly exhibit elevated permeability. Indeed, a considerable body of research has established a link between polyunsaturated fatty acids and heightened membrane permeability, reflecting the rate at which molecules traverse biomembranes. It has been observed that the incorporation of long and very long polyunsaturated fatty acids, such as ALA, AA, EPA, and DHA, increases the permeability and elasticity of biomembranes [464,469,495–500]. This augmentation facilitates the translocation of ions and molecules across the membrane. Mondal et al. [469] attribute the elevated membrane elasticity to the disruption of the robust hydrogen-bond network surrounding the charged lipid head groups by the polyunsaturated fatty acids. The effects of polyunsaturated fatty acids, particularly EPA and DHA, on elasticity (and consequently permeability) can exhibit variability within the same cell [501], depending on the presence of other cellular constituents. Notably, the presence of CHOL can modulate membrane properties [495]. DHA, in particular, elevates permeability more significantly than its precursor, ALA [502], underscoring the pivotal role of the degree of unsaturation and its elevated incorporation levels in the functions of vital cells. DHA promotes heightened hydration within the head group and inter-chain regions, thereby increasing permeability. This is primarily attributed to the elevated number of double bonds. As elucidated by Mitchell and Litman [503], the presence of water within the hydrocarbon bilayer region exhibits a positive correlation with the high number of double bonds.

In addition to its effects on membrane rigidity and permeability, the incorporation of DHA profoundly impacts various aspects of biomembrane dynamics. DHA remarkably alters lipid packing, phase behavior, curvature, elasticity, interleaflet lipid flip-flop rates, lipid phase separations, membrane fusion, and vesicle formation [464,494,504–508]. According to Mitchell and Litman [509], the packing-free volume increases in the following order: 16:0-18:1PC has a lower relative abundance than 16:0-22:6PC, which, in turn, has a lower relative abundance than 22:6-C22:6PC. The potential effect of highly unsaturated fatty acids on phospholipids, which are characterized by loose packing, appears to be closely linked with the promotion of membrane elasticity, vesicle exfoliation (the formation of "blebs"), fusion, and flip-flop processes. In this regard, fatty acids actively participate in the processes of cell fusion and modulate cell phase behavior. During cell fusion, two distinct lipid bilayers merge, resulting in the formation of a continuous bilayer structure and the mixing of the internal contents of the lipid bilayers. Consequently, alterations in fusion processes have been associated with curvature stress in membranes [510,511]. The impact of the degree of unsaturation on membrane fusion has been previously demonstrated by Ahkong et al. [512], Meers et al. [513], Ehringer et al. [502], and, more recently, Li et al. [504]. The configuration of double bonds plays a critical role in determining the extent of biomembrane fusion. According to Creutz [514], AA and oleic acid are particularly effective fusogens, whereas saturated and trans-unsaturated fatty acids exhibit negligible fusogenic activity.

In terms of phase behavior, different fatty acid compositions contribute to various phase transitions, including gel-to-fluid, hexagonal, and liquid phases. Short-chain saturated fatty acids and mono- and polyunsaturated fatty acids result in lower viscosities, contributing to the formation of more fluid membranes compared to long-chain saturated fatty acids [466]. The impact of unsaturated fatty acids is particularly evident in thermal hysteresis, especially the transition between the fluid and hexagonal phases in PE, which is reportedly impeded by oleic acid, LA, and ALA [515,516]. Stearic acid and hydroxylated fatty acids induce a modest shift toward a higher melting temperature (the gel-to-fluid phase transition temperature) in bilayers containing C14:0/C14:0-PC [517]. On the other hand, PCs containing DHA exhibit higher melting points than those containing ALA and AA [518]. Despite DHA's loose packing property [507], the presence of a saturated fatty acid at the *sn*-1 position in a PC molecule may affect its packing stability by altering both intra- and intermolecular van der Waals interactions.

4.2. Relative Functional Significance of Polyunsaturated Fatty Acids

The multifaceted role of fatty acids within various membrane lipids, particularly n3-fatty acids, has been documented over the past century. Diets rich in n3-fatty acids have been extensively associated with elevating the proportions of n3-fatty acids in biomembranes, thereby contributing to the maintenance of cardiovascular, vascular, and neural health [519,520]. Moreover, these dietary choices have shown promise in ameliorating conditions such as atherosclerosis, hypercholesterolemia, and cancer [487,521,522]. Fatty acids exhibit a multitude of physicochemical properties that serve diverse purposes by modifying the characteristics of bilayer lipids, thus influencing signal transduction. Notably, the length of a fatty acid exerts a marked influence on cellular signaling and metabolic processes. For instance, SLs containing short-chain fatty acids have been observed to augment susceptibility to apoptosis [523]. Membrane lipids enriched with monounsaturated fatty acids also play specific functional roles. Cao et al. [524] have reported that palmitoleate can function as a lipid-regulating hormone, often referred to as a 'lipokine', by enhancing sensitivity to glucose and inhibiting lipogenesis and hepatic inflammation. Furthermore, the well-documented antitumor and apoptotic properties of C18-monounsaturated fatty acids in carcinoma cells [525] underscore their potential utility in anticancer medications.

The acyl chain length of SLs, particularly Cers, significantly influences TAG accumulation and the hepatic uptake of fatty acids, which is attributed to the disruption of CD36/FAT expression [526]. This discovery underscores the role of CerS2 in catalyzing the generation of very long-chained Cers. In a cardiac context, long-saturated and polyunsaturated fatty acids have been shown to up-regulate voltage-dependent calcium release in cardiac myocytes [527], implicating their involvement in cardiac damage. Sassa and Kihara (2014) have presented a comprehensive review detailing the metabolism of very long-chain fatty acids and their contributions to the health and pathophysiology of various tissues, including the skin, meibum, retina, testis, and brain. The extensive body of available literature underscores the remarkable significance of polyunsaturated fatty acids, which have been the subject of substantial research due to their diverse bio-functional roles across various cell types.

The essentiality of LA and ALA in mammals transcends their role as diet-derived fatty acids; they are also fundamental precursors for the synthesis of long and very long polyunsaturated fatty acids. However, it is imperative to note that not all absorbed dietary LA and ALA are available for elongation and desaturation processes, as a fraction of these fatty acids are utilized for generating the energy source ATP during the β-oxidation process. Therefore, a prolonged deficiency in LA and/or ALA can lead to severe consequences, often manifesting as clinical symptoms [528–532]. LA, specifically, serves as a critical substrate for the biosynthesis of arachidonic acid (AA) and adrenic acid, both of which play crucial roles in early brain development [533–535]. Furthermore, LA is indispensable for the formation of n-hydroxyceramides, which covalently bond with epidermal proteins, thereby curtailing water loss and bolstering the skin's barrier function [536]. In addition, it has been demonstrated that LA exhibits antibiotic-like properties, manifesting as an antibacterial effect that inhibits microbial adhesion to cells, a characteristic shared by numerous polyunsaturated fatty acids [537–540].

In general, polyunsaturated fatty acids exert substantial influence over the epidermis and its barrier properties. Notably, dietary supplementation of γ-linolenic acid has demonstrated anti-inflammatory properties [541] and has proven effective in enhancing skin characteristics in a dry skin model by reinforcing the skin's barrier function and limiting dehydration [542]. Similar observations have been made with the supplementation of EPA and DHA [543], where an increase in the production of specific Cer families with anti-inflammatory properties was evident. It is worth noting that the effects of these fatty acids varied across distinct skin regions, including the epidermis, dermis, and hypodermis [543]. It is of particular interest that, among n6-fatty acids, γ-linolenic acid and DGLA have gained recognition for their anti-inflammatory attributes, similar to those of EPA and DHA (n3-fatty acids). Notably, γ-linolenic acid is found in inflammatory cells at relatively modest concentrations, and increasing its dietary intake does not necessarily lead to a proportional increase in its intracellular levels [544,545]. Given the efficient conversion of γ-linolenic acid to DGLA in mammals, it is conceivable that DGLA-derived lipid mediators play a role in mediating the anti-inflammatory effects associated with γ-linolenic [546].

The essentiality of ALA initially became apparent through observations of its ability to alleviate symptoms related to LA deficiency [547]. Its significance grew further when it was established that ALA serves as a precursor for EPA and DHA [548,549]. These C20 and C22 n3-polyunsaturated fatty acids are known to constitute a significant portion of the membrane lipids in critical tissues such as the brain [550], retina [551], and testis [552], reflecting their involvement in neurotransmission, visual excitation, and spermium maturation. The implications of n3-polyunsaturated fatty acids, especially DHA, on these tissues have been extensively documented in numerous studies [467,553–566]. These studies have proposed numerous biological functions for n3-polyunsaturated fatty acids, including the modulation of membrane proteins, gene expression, neurogenesis, enhancement of microcirculation, learning processes, and cellular protection. Notably, in neural tissue, the selectivity of PS declines under DHA deficiency [420]. The role of n3-fatty acids incorporated into PS in improving memory [567] and protecting against age-related lipid metabolic disorders, especially in the presence of DHA-enriched PC [567], is well acknowledged. For instance, DHA inhibits the production of amyloid-beta (Aβ) peptides associated with cognitive impairments, thereby mitigating amyloidogenesis, oxidative stress, and apoptosis [520]. The overall impact of polyunsaturated fatty acids on oxidative stress remains a subject of debate, as Shefer-Weinberg et al. [568] found that exposure to polyunsaturated fatty acids elevated oxidative stress biomarkers levels. In this context, it is plausible to hypothesize that the diverse polyunsaturated fatty acids may elicit distinct effects. Nonetheless, DHA has been reported to enhance the fluidity of the synaptic plasma membrane and induce the expression of other memory-related proteins [465]. Consequently, n3-fatty acids, particularly DHA, have gained significant scientific interest, leading to the development of nutraceuticals in the form of dietary supplements that incorporate these fatty acids.

In the preceding sections, the various roles of DHA in the physicochemical properties of membranes have been described. However, DHA also has crucial biological functions within membranes. DHA-enriched membranes have been suggested to influence membrane proteins by inducing curvature stress [569–571], affecting membrane thickness [473,570,572], and modulating fatty acid packing free volume [565]. These alterations in membrane properties can lead to modifications in the activity of most cellular proteins, affecting signal propagation. For instance, unsaturated fatty acids have been reported to interact with various proteins, including rhodopsin, ion channels (L-type Ca^{2+} and Na^{+}), protein kinase C (PKC), apoptosis-associated proteins, PPAR-γ, nuclear receptor Nur77, G-protein coupled receptor 40, mitogen-activated protein kinase, toll-like receptors, and nuclear factor kappa-light-chain-enhancer of activated B cells (NF-κB) [573–585]. However, the major relationships between DHA and cellular protein activities remain ambiguous due to the vast diversity of proteins, the complexity of protein interactions, and the limited number of studies. Despite being highly unsaturated (with six double bonds), DHA exhibits antioxidant properties in the liver [586], brain [587,588], and skeletal muscles [465]. This property is of particular significance for fertility, as Roqueta-Rivera et al. [589] observed that DHA supplementation effectively restored impaired spermatogenesis in male mice.

Both EPA and DHA have demonstrated the ability to counteract pro-inflammatory cytokines by down-regulating the NF-κB signaling pathway [590–592], a transcriptional pathway that regulates both innate and adaptive immune responses. In contrast, AA levels have been found to correlate positively with lipid peroxidation [593] and activation of the NF-κB signaling pathway [594], thereby promoting pro-inflammatory stimuli. AA can also up-regulate SMase activity [595], leading to increased levels of Cers, molecules that trigger apoptotic signals, which are derived from SM hydrolysis. Thus, AA is a biologically essential fatty acid, contributing to a wide array of functions either directly or through its bioactive metabolites. Hashidate-Yoshida et al. [596] demonstrated that AA facilitates the transportation of triglycerides to the lumen of the ER in hepatocytes and enterocytes.

The ratio between fatty acids within cellular membranes serves as a reflection of universal cellular signaling and inflammatory responses. Notably, EPA and DHA exhibit distinct signaling profiles compared to AA. Consequently, the ratio of EPA and DHA to AA can serve as an indirect indicator for assessing the inflammatory response and lipid peroxidation. It is worth emphasizing that these fatty acids serve as precursors for numerous bioactive mediators, contributing to a wide array of physiological functions. However, it is also important to recognize that many of the reported findings are likely attributed to the direct alterations of membrane physicochemical properties and membrane-associated proteins [481], along with the unidentified bioactive metabolites they generate. Polyunsaturated fatty acids can undergo chemical reactions with various molecules and cellular components, resulting in the formation of novel compounds with biological activity. Heshmati [597] has described interactions between n3-fatty acids and specific transcription factors in genes. Furthermore, an intriguing observation is the interaction of nitric oxide (NO) with polyunsaturated fatty acids, leading to the formation of nitroalkene derivatives. These plasma-identifiable derivatives have been demonstrated to promote vascular relaxation, inhibit neutrophil cell degranulation and superoxide production, and hinder platelet activation [598–600]. Nitroalkene derivatives possess inherent PPAR ligand activity and are known to degrade into NO in the bloodstream. These observations underscore the capacity of polyunsaturated fatty acids to engage in reactions with other non-lipidous cellular constituents, resulting in the formation of novel compounds with specific biological activities.

4.3. Bioactive Lipid Mediators Derived from Fatty Acids

Numerous classes of lipids, including LysoP, SLs, PA, DAG, inositol phosphate, *N*-acylethanolamine, fatty acids, and oxylipins, are renowned for their bioactive intracellular and extracellular signaling properties, acting as messengers/mediators. On the other hand, certain functions of polyunsaturated fatty acids necessitate their conversion into lipid mediators. These mediators serve as signaling molecules that modulate various biological processes, including the inflammatory response, gene transcription, and signal transduction pathways. For instance, the tissue hormone-like lipids referred to as "eicosanoids", which were initially identified in the prostate [601], possess the ability to regulate the function of various transcription factors, thus inducing alterations in gene expression. To comprehend the mechanisms underlying the generation of lipid mediators derived from fatty acids, this section elaborates on the cleavage mechanism of membrane fatty acids, the oxygenation mechanisms of deacylated fatty acids, and the biological functions of lipid mediators derived from fatty acids.

4.3.1. Enzyme-Mediated Cleavage of Fatty Acids from Membranes

Polyunsaturated fatty acids are abundant in biomembranes but can be enzymatically cleaved from *sn*-positions and *N*-acyl linkages of membrane lipids by lipase-type enzymes. These enzymes encompass PLA2, phospholipase B (PLB, an enzyme with both PLA1 and PLA2 activities), diacylglycerol lipase [602,603], CDase [604], glucosylceramide deacylase [605,606], and sphingomyelin deacylase [605,607]. Other phospholipase enzymes, such as PLA1, PLC, and PLD, play a lesser role in the generation of polyunsaturated fatty acid-derived mediators, as they cleave the highly saturated chains at the *sn*-1 position [608], the phosphate group at the *sn*-3 position [609], and the head group from the phosphorus group [610], respectively.

Among these enzymes, PLA2 has received considerable attention due to the biological importance of its substrates. Over recent decades, six isoforms of PLA2, which hydrolyze the ester bond at the *sn*-2 position, have been identified [611]. Each of these isoforms exhibits selectivity for specific fatty acids on phospholipids. For example, cytosolic PLA2α (cPLA2α) acts on phospholipids rich in AA [612], calcium-independent PLA2β (iPLA2β) acts on phospholipids rich in DHA [613,614], and secretory PLA2 (sPLA2) acts on phospholipids containing various fatty acids, including AA, EPA, and DHA [615,616]. On the other hand, PLB possesses both hydrolase activity, cleaving ester bonds on the *sn*-1 and *sn*-2 positions of phospholipids, and acyltransferase activity, acylating fatty acid to form LysoP, and, as a result, may contribute less to the production of oxylipins compared to PLA2. Following the removal of fatty acids from complex membrane lipids, various events, including reacylation and/or oxidation, may occur.

4.3.2. Fatty Acid Oxygenation

CHOL and liberated polyunsaturated fatty acids can undergo oxidation through enzymatic processes and non-enzymatic agents, such as reactive oxygen species (ROS). This oxidative transformation leads to the production of oxysterols and oxylipins, respectively. Notably, these compounds may also originate from dietary sources. It is of particular significance that polyunsaturated fatty acids frequently engage in metabolic competition with each other, a phenomenon specifically mediated by enzymes such as PLA2 and oxidative enzymes. The extent of competition among different fatty acids depends on their respective concentrations within the cell and their relative affinities for oxidative enzymes and reactive molecules [617,618]. These oxidized lipid metabolites serve as pivotal mediators in cell signaling. For instance, oxysterols have the capacity to interact with nuclear receptors and, as a consequence, modulate gene expression [619,620]. This section primarily focuses on the enzymatic pathways involved in generating these bioactive lipid mediators.

Numerous bioactive oxylipins have been identified as products of enzymatic pathways, including those facilitated by cyclooxygenase (COX) and subsequent synthases, lipoxygenase (LOX), and cytochrome P450 (CYP) mixed-function oxidase enzymes [621]. These oxylipins are further categorized based on the chain length of their respective substrates (see Figure 7), resulting in octadecanoids (derived from C18 fatty acids), eicosanoids (derived from C20 fatty acids), docosanoids (derived from C22 fatty acids), and elovanoids (derived from C32 or C34 fatty acids).

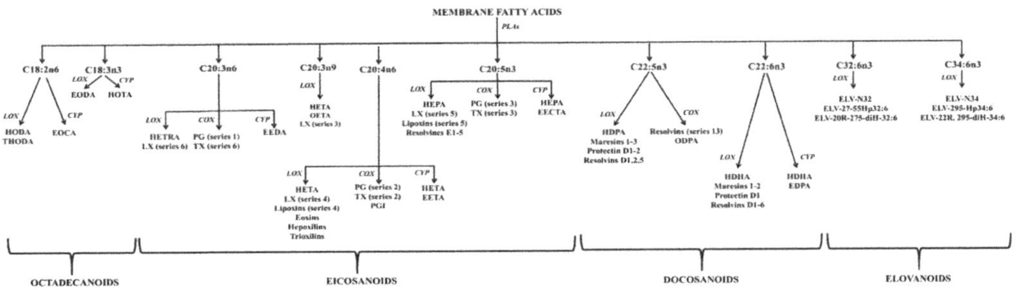

Figure 7. Diagram illustrating the various lipid mediators (including octadecanoids, eicosanoids, docosanoids, and elovanoids) synthesized from fatty acids such as LA, ALA, DGLA, Mead acid, AA, EPA, DPA-n3, DHA, C32:6n3, and C34:6n3. Abbreviations: COX, cyclooxygenase; EDPA, epoxy-docosapentaenoic acid; EECTA, epoxy-ecosatetraenoic acid; EEDA, epoxy-eicosadienoic acid; EETA, epoxy-eicosatrienoic acid; ELV, elovanoids; EOCA, epoxy-octadecenoic acid; EODA, epoxy-octadecadienoic acid; CYP, cytochrome P450; HDHA, hydroxy-docosahexaenoic acid; HDPA, hydroxy-docosapentaenoic acid; HEPA, hydroxy-eicosapentaenoic acid; HETA, eicosatetraenoic acid; HETRA, hydroxy-eicosatrienoic acid; HODA, hydroxy-octadecadienoic acid; HOTA, hydroxy-octadecatrienoic acid; LOX, lipoxygenase; LX, leukotrienes; ODPA, oxodocosapentaenoic acid; OETA, oxoeicosatetraenoic acid; PG, prostaglandin; PGI, prostacyclins; PLAs, phospholipases; THODA, trihydroxy-octadecenoic acid; TX, thromboxane.

In mammals, COXs, also known as housekeeping enzymes, comprise three isoforms as follows: COX-1, COX-2, and COX-3 isoforms [622], with COX-3 being considered a variant of COX-1 [623]. These enzymes are heme-containing and possess the dual capacity to function as both oxygenases and peroxidases. Notably, these enzymes are constitutively expressed and are subject to modulation by inflammatory signals. Their main role involves catalyzing the oxygenation of various unsaturated fatty acids, culminating in the generation of bioactive end-products collectively referred to as prostanoids. These prostanoids encompass the prostaglandin series (PGD, PGE (dinoprostone), PGF (carboprost), and PGI (prostacy-clins)), thromboxanes, hydroxy fatty acids, resolvins (series 13), and oxo-fatty acids [624,625].

LOXs, which comprise six genes within the human genome, represent a class of non-heme iron-containing dioxygenases. These enzymes possess the capability to oxygenate a broad range of unsaturated fatty acids. It is noteworthy that LOX enzymes typically exist in an inactive form at their base state, necessitating activation facilitated by hydroperoxides. Subsequently, they act on a diverse array of substrates and engage in various modes of action, including dioxygenase activity, functioning as catalysts in processes characterized by the involvement of free radicals [626,627]. These catalytic actions lead to the formation of bioactive end-products recognized as hydroperoxyl fatty acids and their metabolites, including leukotrienes, lipoxins, resolvins, protectins, maresins, and elovanoids [628,629].

On the other hand, CYPs are enzymes encoded by an extensive set of up to 57 genes within the human genome, representing a class of monooxygenases widely distributed in mammals. These enzymes exhibit elevated activity levels in numerous tissues, including but not limited to the liver, brain, kidneys, and lungs [629,630]. CYPs are renowned for their involvement in various modes of action, including hydroxylation, heteroatom oxidation, allylic oxidation reactions, group migration, and various other enzymatic reactions [631–633]. They display the capability to act on a diverse range of unsaturated fatty acids and sterols [629], thereby generating a wide array of lipid mediators. In particular, these lipid mediators consist of hydroxyl and epoxy fatty acids, which play critical roles in the induction of various signaling pathways.

4.3.3. Functions of Bioactive Lipid Mediators

Bioactive lipid mediators go beyond being inert components of cellular membranes. Instead, they serve as dynamic signaling agents and are capable of modulating a wide range of signaling pathways, gene regulation, and immune responses. The unique characteristics and functions of these mediators have led to extensive research efforts aimed at harnessing their therapeutic potential for developing innovative treatment/preventive approaches. Therefore, comprehending the functions of bioactive lipid mediators holds great importance in the domains of biological and medical research.

Octadecanoids

Essential fatty acids and their extended metabolites have well-documented diverse biological effects and implications in various disease mechanisms. The effects of lipid mediators can vary depending on the type of cells and metabolic factors, leading to both beneficial and detrimental metabolic outcomes (see Figure 8).

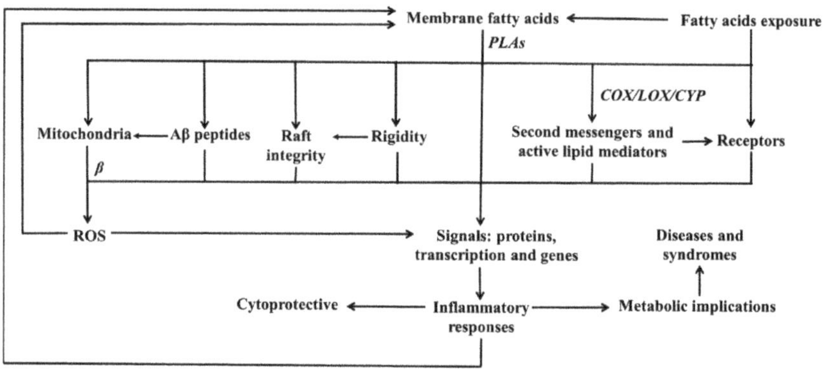

Figure 8. Illustration of the influence of bilayer-cleaved fatty acids on cellular signaling pathways and inflammation responses. Abbreviations: β, β-oxidation; COX, cyclooxygenase; CYP, cytochrome P450; LOX, lipoxygenase; PLAs, phospholipases; ROS, reactive oxygen species.

LA and ALA play an essential role in the generation of lipid mediators. As essential fatty acids, dietary levels of LA and ALA contribute to their proportions within cellular membranes, potentially playing a crucial role in modulating the extent/degree/severity of inflammation development. Epoxy-octadecadienoic acid and hydroxy-octadecatrienoic acid are lipid mediators derived from ALA through the enzymatic actions of LOX and CYP, respectively [624]. However, further research is needed to fully comprehend the bioactive functions of octadecanoids derived from ALA. Notably, Kumar et al. [634] have suggested that these mediators primarily exert anti-inflammatory effects. On the other hand, oxidized LA metabolites, including hydroxy, trihydroxy, and epoxy fatty acids, are produced under the effects of LOXs and CYPs [546,624]. These metabolites have been implicated in various biological pathways, including brain dysfunction [635], the inhibition of platelet adhesion in endothelial cells [636], the induction of inflammation signals [637–639], the maintenance of skin barrier integrity [536], the inhibition of pain thresholds [640,641], and the promotion of metabolic syndromes and cancer [638,642]. Consequently, these LA-derived mediators may greatly contribute to the inflammatory processes and the progression of diseases.

Both LA and ALA serve as essential fatty acids and precursors for extended polyunsaturated fatty acids. Thus, their dietary concentrations can alter the levels of long and very long chain polyunsaturated fatty acids within cellular biomembranes. However, the extent of this influence may vary depending on the specific substrate and metabolic pathways. Notably, a high dietary intake of LA tends to not significantly elevate the proportion of AA or the associated inflammatory cascades in humans [643]. In contrast, a high dietary intake of ALA has been shown to increase EPA and DHA concentrations [644]. Nevertheless, it is essential to recognize that the de novo pathways for elongating essential fatty acids exhibit variations among species. For instance, the conversion rate of ALA to its extended polyunsaturated fatty acids is lower in humans [645] than in marine species. These findings underscore the potential variability in the biological functions of ALA, with specific implications in distinct species to fulfill particular physiological functions.

Eicosanoids

Eicosanoids are bioactive lipid mediators primarily derived from unesterified fatty acids and are characterized by their autocrine/paracrine hormone activities. They mediate local signals and reactions, including processes related to homeostasis, inflammation, and anti-inflammation. Eicosanoids comprise various structures, such as PGs, thromboxanes, leukotrienes, lipoxins, and resolvins. Despite the fact that most mammalian cells are capable of synthesizing eicosanoids, the specific pathways and responses can vary by cell type [646]. DGLA, rapidly extended from γ-linolenic acid, serves as a substrate for enzymes such as COX, which yields series 1 prostaglandins and thromboxanes, 15-LOX, which yields 5-hydroxyeicosatrienoic acid, and CYP, which yields epoxy-eicosadienoic acid. Eicosanoids derived from DGLA are generally considered to be anti-inflammatory [647,648].

On the other hand, beyond the role of AA as a polyene fatty acid, it is unquestionably crucial in biomembranes as it is the primary target for most membrane-modifying effects. The activation of the PLA2 enzyme, which is responsible for cleaving AA from membrane phospholipids, often leads to membrane injury. According to Samuelsson [649], this enzyme rapidly (within seconds to minutes) responds to acute stimuli, releasing AA from membrane lipids. Liberated AA can be utilized as a precursor for the production of eicosanoids under the effects of COX (generating series 2 prostaglandins, prostacyclins, and thromboxanes), LOX (generating leukotrienes, lipoxins, eoxins, hepoxilins, and trioxilins), and CYP (generating hydroxyeicosatetraenoic acid and epoxyeicosatrienoic acid) [546,624]. AA-derived eicosanoids, often referred to as arachidonate or eicosanoid cascades, are involved in multiple systems, including vascular, inflammatory, renal, and neuronal signaling, as well as angiogenesis [650]. For instance, AA-derived eicosanoids have been shown to increase the permeability of the blood–brain barrier in humans [651], revealing the potential for drug modulation of this barrier.

Eicosanoids derived from Mead acid have displayed anti-inflammatory properties. For instance, oxygenated products of Mead acid via 5-lipoxygenase are produced during inflammation, providing potent activities [652]. However, the exact roles of this fatty acid are not yet clearly defined [652–655], necessitating further research. On the other hand, LOXs oxidize EPA to produce resolvins [656], which are renowned for their anti-inflammatory properties. EPA can also undergo oxygenation via COXs (yielding hydroxy-eicosapentaenoic acid and epoxy-eicosatetraenoic acid) and CYPs (yielding series 3 prostaglandins and leukotrienes) [624,657]. Overall, EPA-derived eicosanoids exhibit anti-inflammatory stimuli, such as the inhibition of platelet aggregation [658].

Eicosanoids play a remarkable role in the regulation of inflammatory responses by modulating pro-inflammatory cytokines, chemokines, and other signaling molecules. They have the potential to influence the recruitment, activation, and function of immune cells.

However, it is essential to recognize that eicosanoids can exhibit both pro-inflammatory and anti-inflammatory effects, with the ultimate effect determined by various factors, including mediator concentrations, timing of production, and the sensitivity of targeted cells/tissues [460,659]. Commonly, eicosanoids derived from different fatty acids, such as AA, Mead acid, and EPA, exhibit strikingly distinct biological effects, despite their closely resembling molecular structures.

Eicosanoids derived from n3-fatty acids are well-recognized for their anti-inflammatory properties, while those originating from n6-fatty acids are generally considered to be pro-inflammatory [660,661]. However, it is important to note that not all n6-fatty acids exert pro-inflammatory effects. Some prostanoids (PGs and thromboxanes), n6-fatty acid-derived lipoxins, as well as mediators derived from γ-linolenic and DGLA, along with adrenic acid, have been found to express anti-inflammatory properties and cytoprotective actions [662–669]. Imbalances in the production of eicosanoids have been implicated in numerous pathological processes, including inflammation, autoimmunity, allergy, cancer, atherosclerosis, and metabolic and degenerative diseases [650], by disrupting the normal lipid signaling pathways. In light of this, strategies that involve the suppression of COX, LOX, and CYP enzymes, which are responsible for the synthesis of active lipid mediators, may hold therapeutic potential for the management of disease-related inflammation and oxidative stress.

Docosanoids

DHA, likely the reason for the biological necessity of ALA, is a very long polyunsaturated fatty acid that accumulates abundantly in crucial tissues such as the brain, retina, and testis. Though EPA is known to produce pre-resolving mediators (resolvins), it is DHA that serves as the major precursor for these compounds [656,670]. Specialized pro-resolving mediators (SPMs), known as docosanoids, are primarily derived from the LOX oxidation of DHA and DPA-n3 [670–673]. However, COX activity on DPA-n3 can also generate SPMs [672], and CYP activity on DHA yields hydroxy-docosapentaenoic acid and epoxy-docosapentaenoic acid [624]. The pre-resolving family comprises various structures, including resolvins, docosatrienes, maresins, and protectins, all of which exhibit anti-inflammatory and pro-resolving properties, countering the effects of pro-inflammatory cascades [659,666,667,671–682]. These docosanoids, which are derived from DPA-n3 and DHA, play pivotal roles in the regulation of leukocyte trafficking, suppression of cytokine expression, inhibition of brain ischemia-reperfusion injury, maintenance of cellular homeostasis, mitigation of potential DNA oxidation, normalization of brain-derived neurotrophic factor levels, and promotion of the clearance of apoptotic cells and cellular debris by phagocytes. Thus, these mediators represent a promising therapeutic approach for resolving cellular inflammation and associated diseases.

Furthermore, EPA and DHA are known to limit pro-inflammatory cytokines and reduce inflammation, potentially by increasing peroxisome proliferator-activated receptor alpha (PPAR-α) mRNA and protein activities [683]. Remarkably, alternative lipid mediators with resembling impacts to resolvins have been identified. According to Dalli et al. [684], DPA-n3, an intermediate fatty acid during DHA synthesis, is transformed into novel immunoresolvents similar to resolvins in mice and human leukocytes during inflammation. However, it is important to acknowledge that the resolution of inflammation mediated by docosanoids is characterized by its complexity in restoring cellular homeostasis [656].

Elovanoids

In response to unmitigated oxidative stress, elovanoids exhibit a remarkable ability to enhance the intracellular synthesis of pro-survival signals, owing to their distinctive molecular structures. This class of bioactive lipids, initially discovered by Bazan's research group in the retinal pigment epithelium in 2017 [685], is derived from mono-hydroxyl-very long polyunsaturated fatty acids formed through the enzymatic activity of ELVOL4 and LOX. It is important to emphasize that very long polyunsaturated fatty acids are prominent constituents of critical tissues such as the brain, testis, and spermatozoa [686]. This observation suggests the potential formation of elovanoids in these tissues, where they might serve as mediators of specific signals. However, while the retina has been a focal point of research on elovanoids, studies examining neural signaling are comparatively limited.

Elovanoids play an indispensable role in the functions of the retina and neural signaling [290,685,687–691]. The protective effects of elovanoids in these tissues are most likely attributed to their role in mitigating the effects of oxidative stress. In events where oxidative stress remains unresolved, elovanoids serve as critical survival signals [685]. These authors have reported that dihydroxylated derivatives of C32:6n-3 and C34:6n-3 effectively protect retinal pigment epithelial cells from apoptosis induced by hydrogen peroxide. These derivatives have been shown to up-regulate the expression of pro-survival proteins, including Bcl-2 and Bcl-xL, while concurrently down-regulating the expression of pro-apoptotic proteins, such as Bax, Bim, and Bid. These findings underscore the ability of elovanoids to mitigate the cytotoxic effects of ROS on photoreceptor cells and contribute to their survival.

5. Conclusions and Future Perspectives

This review intends to provide an in-depth overview of the lipids of eukaryotic cell membrane lipids, with a particular emphasis on fatty acids. It introduces the extensive array of lipids present in biomembranes and delves into their composition within healthy organisms, thereby illustrating the intricate nature of lipid metabolism and its fundamental role within cells. This perspective underscores the remarkable adaptability and flexibility inherent in the fatty acid profiles of biomembranes, enabling organisms to rapidly respond to various stimuli, including alterations in environmental temperature, dietary factors, inflammatory processes, or diseases. Thus, the absence of a universally defined "physiologically normal fatty acid composition" underscores the natural variability in fatty acid composition. This natural phenomenon is, likewise, a continuous process of adaptation. This review further provides an in-depth exploration of fatty acid biosynthesis and post-synthetic modifications, such as elongation and desaturation. In addition, it highlights the preferences of fatty acids for incorporation into diverse complex membrane lipids and their roles in biological systems, encompassing both physicochemical properties and the regulation of biological signaling. This understanding holds significant implications across various disciplines, including lipid-based drug delivery, cell membrane engineering, and the advancement of lipid-based biomaterials. Nevertheless, further research remains essential to unveil the intricate mechanisms and regulatory pathways governing eukaryotic lipid metabolism and fatty acid composition. This includes investigations into the mechanisms underpinning cellular membrane adaptability, with the potential to shed light on the molecular foundations of cellular processes, diverse diseases, and the development of therapeutic strategies for lipid-related disorders.

Evidently, the pivotal role of fatty acids in biomembranes is ascending and is poised to exert a substantial influence across various disciplines, notably within the realms of nutrition and medicine. This review serves to illuminate the multifaceted roles and contributions of distinct membrane lipids, along with their associated fatty acids, with a specific focus on matters pertaining to health and the intricate aspects of inflammatory responses. Enhancing our comprehensive comprehension of the diverse repertoire of membrane lipids stands to be invaluable for assessing the overall health of organisms. The trajectory of the field nutrition is set to emphasize progressively specific fatty acids that are indispensable for organism health. In this context, the n3 and n6-fatty acids are assuming paramount significance due to their critical roles as precursors for bioactive lipids that play a pivotal role in the modulation of inflammatory processes. They also contribute indispensably to the development and sustenance of vital organ functions, exemplified by the brain, heart, lungs, liver, and kidneys. Elevated levels of these fatty acids have been consistently correlated with to the regulation of chronic maladies, encompassing diabetes, cardiovascular disorders, and certain forms of cancer. However, it is crucial to recognize that the optimization of fatty acid biosynthesis, the preservation of their stability, and a comprehensive understanding of their various roles in biological systems continue to remain areas ripe for exploration. Thus, the unwavering dedication to research and development in this domain holds the promise of unveiling the unlocking of novel approaches to incorporate these essential nutrients into the diets of organisms, thereby fostering enduring health and well-being.

Author Contributions: O.A. conceptualized the idea, wrote the manuscript and prepared the figures; A.S. reviewed and edited the manuscript. Both authors actively participated in discussion, manuscript revision, and final version contributions. All authors have read and agreed to the published version of the manuscript.

Funding: This work was supported by the following projects: EFOP-3.6.3-VEKOP-16-2017-00005 and GINOP-2.3.2-15-2016-00046. This work was further funded by the Hungarian Academy of Sciences (HUN-REN-MATE, Mycotoxins in the Food Chain research group) and by the Hungarian National Laboratory project RRF-2.3.1-21-2022-00007.

Acknowledgments: The authors would like to acknowledge András Dinnyés for his insightful suggestions, which improved the present review.

Conflicts of Interest: The authors declare no conflict of interest.

References

1. Harayama, T.; Riezman, H. Understanding the diversity of membrane lipid composition. *Nat. Rev. Mol. Cell Biol.* **2018**, *19*, 281–296. [PubMed]
2. Watson, H. Biological membranes. *Essays Biochem.* **2015**, *59*, 43–69. [PubMed]
3. Janmey, P.A.; Kinnunen, P.K.J. Biophysical properties of lipids and dynamic membranes. *Trends Cell Biol.* **2006**, *16*, 538–546. [PubMed]
4. Dynarowicz-Łątka, P.; Dhanabalan, A.; Oliveira, O.N. Modern physicochemical research on Langmuir monolayers. *Adv. Colloid Interface Sci.* **2001**, *91*, 221–293. [PubMed]
5. Singer, S.J.; Nicolson, G.L. The fluid mosaic model of the structure of cell membranes. *Science* **1972**, *175*, 720–731.
6. Simons, K.; Ikonen, E. Functional rafts in cell membranes. *Nature* **1997**, *387*, 569–572.
7. Ballweg, S.; Sezgin, E.; Doktorova, M.; Covino, R.; Reinhard, J.; Wunnicke, D.; Hänelt, I.; Levental, I.; Hummer, G.; Ernst, R. Regulation of lipid saturation without sensing membrane fluidity. *Nat. Commun.* **2020**, *11*, 756.
8. Nicolson, G.L. The fluid—Mosaic model of membrane structure: Still relevant to understanding the structure, function and dynamics of biological membranes after more than 40years. *Biochim. Biophys. Acta Biomembr.* **2014**, *1838*, 1451–1466.
9. Edidin, M. Lipids on the frontier: A century of cell-membrane bilayers. *Nat. Rev. Mol. Cell Biol.* **2003**, *4*, 414–418.
10. Wang, Y.; Gao, J.; Guo, X.; Tong, T.; Shi, X.; Li, L.; Qi, M.; Wang, Y.; Cai, M.; Jiang, J.; et al. Regulation of EGFR nanocluster formation by ionic protein-lipid interaction. *Cell Res.* **2014**, *24*, 959–976.
11. Luckey, M. *Membrane Structural Biology*, 1st ed.; Cambridge University Press: Cambridge, UK, 2008; ISBN 9780521856553.
12. Hanczyc, M.M.; Fujikawa, S.M.; Szostak, J.W. Experimental models of primitive cellular compartments: Encapsulation, growth, and division. *Science* **2003**, *302*, 618–622.
13. Chevreul, M.E. Sur plusieurs corps gras et particulièrement sur leurs combinaisons avec les alcalis. [On several fatty substances and particularly on their combinations with alkalis]. *Ann. Chim.* **1813**, *88*, 225–261.
14. Lombard, J. Once upon a time the cell membranes: 175 years of cell boundary research. *Biol. Direct* **2014**, *9*, 32.
15. Gorter, E.; Grendel, F. On bimolecular layers of lipoids on the chromocytes of the blood. *J. Exp. Med.* **1925**, *41*, 439–443. [CrossRef]
16. Glatz, J.F.C.; Luiken, J.J.F.P.; Bonen, A. Membrane fatty acid transporters as regulators of lipid metabolism: Implications for metabolic disease. *Physiol. Rev.* **2010**, *90*, 367–417. [PubMed]
17. Anheuser, S.; Breiden, B.; Sandhoff, K. Membrane lipids and their degradation compounds control GM2 catabolism at intralysosomal luminal vesicles. *J. Lipid Res.* **2019**, *60*, 1099–1111. [PubMed]
18. Settembre, C.; Ballabio, A. Lysosome: Regulator of lipid degradation pathways. *Trends Cell Biol.* **2014**, *24*, 743–750. [PubMed]
19. Kolter, T.; Sandhoff, K. Lysosomal degradation of membrane lipids. *FEBS Lett.* **2010**, *584*, 1700–1712. [PubMed]
20. Van Meer, G. Cellular lipidomics. *EMBO J.* **2005**, *24*, 3159–3165.
21. Sackmann, E. Biological membranes architecture and function. In *Structure and Dynamics of Membranes: I. From Cells to Vesicles/II. Generic and Specific Interactions*; Lipowsky, R., Sackman, T., Eds.; Elsevier: Amsterdam, The Netherlands, 1995; pp. 1–63.
22. Liebisch, G.; Fahy, E.; Aoki, J.; Dennis, E.A.; Durand, T.; Ejsing, C.S.; Fedorova, M.; Feussner, I.; Griffiths, W.J.; Köfeler, H.; et al. Update on LIPID MAPS classification, nomenclature, and shorthand notation for MS-derived lipid structures. *J. Lipid Res.* **2020**, *61*, 1539–1555.
23. Fahy, E.; Cotter, D.; Sud, M.; Subramaniam, S. Lipid classification, structures and tools. *Biochim. Biophys. Acta Mol. Cell Biol. Lipids* **2011**, *1811*, 637–647. [CrossRef] [PubMed]

24. Bogdanov, M.; Dowhan, W. Functional roles of lipids in biological membranes. In *Biochemistry of Lipids, Lipoproteins and Membranes*; Ridgway, N.D., McLeod, R.S., Eds.; Elsevier: Amsterdam, The Netherlands, 2021; pp. 1–51.
25. Drin, G. Topological regulation of lipid balance in cells. *Annu. Rev. Biochem.* **2014**, *83*, 51–77. [CrossRef] [PubMed]
26. Vauquelin, N.L. Analyse de la matière cérébrale de l'homme et de quelques animaux [Translation: Analysis of the Brain Matter of Man and Some Animals]. Ph.D. Thesis, Faculty of Medicine of Paris, Paris, France, 12 August 1811.
27. Van Meer, G.; Voelker, D.R.; Feigenson, G.W. Membrane lipids: Where they are and how they behave. *Nat. Rev. Mol. Cell Biol.* **2008**, *9*, 112–124. [PubMed]
28. Buckland, A.G.; Wilton, D.C. Anionic phospholipids, interfacial binding and the regulation of cell functions. *Biochim. Biophys. Acta Mol. Cell Biol. Lipids* **2000**, *1483*, 199–216.
29. Moolenaar, W.H.; Kruijer, W.; Tilly, B.C.; Verlaan, I.; Bierman, A.J.; de Laat, S.W. Growth factor-like action of phosphatidic acid. *Nature* **1986**, *323*, 171–173. [CrossRef] [PubMed]
30. Kennedy, E.P.; Lehninger, A.L. Oxidation of fatty acids and tricarboxylic acid cycle intermediates by isolated rat liver mitochondria. *J. Biol. Chem.* **1949**, *179*, 957–972. [CrossRef]
31. Tanguy, E.; Wang, Q.; Moine, H.; Vitale, N. Phosphatidic acid: From pleiotropic functions to neuronal pathology. *Front. Cell. Neurosci.* **2019**, *13*, 2.
32. Lee, J.; Ridgway, N.D. Substrate channeling in the glycerol-3-phosphate pathway regulates the synthesis, storage and secretion of glycerolipids. *Biochim. Biophys. Acta Mol. Cell Biol. Lipids* **2020**, *1865*, 158438.
33. Tanguy, E.; Costé de Bagneaux, P.; Kassas, N.; Ammar, M.-R.; Wang, Q.; Haeberlé, A.-M.; Raherindratsara, J.; Fouillen, L.; Renard, P.-Y.; Montero-Hadjadje, M.; et al. Mono- and poly-unsaturated phosphatidic acid regulate distinct steps of regulated exocytosis in neuroendocrine cells. *Cell Rep.* **2020**, *32*, 108026. [CrossRef]
34. Zegarlińska, J.; Piaścik, M.; Sikorski, A.F.; Czogalla, A. Phosphatidic acid—A simple phospholipid with multiple faces. *Acta Biochim. Pol.* **2018**, *65*, 163–171. [CrossRef]
35. Ammar, M.-R.; Kassas, N.; Bader, M.-F.; Vitale, N. Phosphatidic acid in neuronal development: A node for membrane and cytoskeleton rearrangements. *Biochimie* **2014**, *107*, 51–57. [CrossRef] [PubMed]
36. Stace, C.; Ktistakis, N. Phosphatidic acid- and phosphatidylserine-binding proteins. *Biochim. Biophys. Acta Mol. Cell Biol. Lipids* **2006**, *1761*, 913–926. [CrossRef]
37. Benson, A.A.; Maruo, B. Plant phospholipids I. Identification of the phosphatidyl glycerols. *Biochim. Biophys. Acta* **1958**, *27*, 189–195. [CrossRef] [PubMed]
38. Ridgway, N.D. Phospholipid synthesis in mammalian cells. In *Biochemistry of Lipids, Lipoproteins and Membranes*; Ridgway, N.D., McLeod, R.S., Eds.; Elsevier: Amsterdam, The Netherlands, 2021; pp. 227–258.
39. Agassandian, M.; Mallampalli, R.K. Surfactant phospholipid metabolism. *Biochim. Biophys. Acta Mol. Cell Biol. Lipids* **2013**, *1831*, 612–625. [CrossRef]
40. Hallman, M.; Enhorning, G.; Possmayer, F. Composition and surface activity of normal and phosphatidylglycerol-deficient lung surfactant. *Pediatr. Res.* **1985**, *19*, 286–292. [CrossRef]
41. Furse, S. Is phosphatidylglycerol essential for terrestrial life? *J. Chem. Biol.* **2017**, *10*, 1–9. [CrossRef]
42. Kanoh, H.; Kondoh, H.; Ono, T. Diacylglycerol kinase from pig brain. Purification and phospholipid dependencies. *J. Biol. Chem.* **1983**, *258*, 1767–1774. [CrossRef]
43. Laurinavičius, S.; Käkelä, R.; Bamford, D.H.; Somerharju, P. The origin of phospholipids of the enveloped bacteriophage phi6. *Virology* **2004**, *326*, 182–190. [CrossRef]
44. Bamford, D.H.; Romantschuk, M.; Somerharju, P.J. Membrane fusion in prokaryotes: Bacteriophage phi 6 membrane fuses with the Pseudomonas syringae outer membrane. *EMBO J.* **1987**, *6*, 1467–1473. [CrossRef] [PubMed]
45. Sands, J.A.; Lowlicht, R.A. Temporal origin of viral phospholipids of the enveloped bacteriophage ϕ6. *Can. J. Microbiol.* **1976**, *22*, 154–158. [CrossRef]
46. Numata, M.; Kandasamy, P.; Nagashima, Y.; Posey, J.; Hartshorn, K.; Woodland, D.; Voelker, D.R. Phosphatidylglycerol suppresses influenza A virus infection. *Am. J. Respir. Cell Mol. Biol.* **2012**, *46*, 479–487. [CrossRef] [PubMed]
47. Kandasamy, P.; Zarini, S.; Chan, E.D.; Leslie, C.C.; Murphy, R.C.; Voelker, D.R. Pulmonary surfactant phosphatidylglycerol inhibits Mycoplasma pneumoniae-stimulated eicosanoid production from Human and mouse macrophages. *J. Biol. Chem.* **2011**, *286*, 7841–7853. [CrossRef] [PubMed]
48. Kuronuma, K.; Mitsuzawa, H.; Takeda, K.; Nishitani, C.; Chan, E.D.; Kuroki, Y.; Nakamura, M.; Voelker, D.R. Anionic pulmonary surfactant phospholipids inhibit inflammatory responses from alveolar macrophages and U937 cells by binding the lipopolysaccharide-interacting proteins CD14 and MD-2. *J. Biol. Chem.* **2009**, *284*, 25488–25500. [CrossRef] [PubMed]
49. Bollag, W.B.; Gonzales, J.N. Phosphatidylglycerol and surfactant: A potential treatment for COVID-19? *Med. Hypotheses* **2020**, *144*, 110277. [CrossRef]
50. Zeisel, S.H. A brief history of choline. *Ann. Nutr. Metab.* **2012**, *61*, 254–258. [CrossRef]
51. Lassègue, B.; Alexander, R.W.; Clark, M.; Akers, M.; Griendling, K.K. Phosphatidylcholine is a major source of phosphatidic acid and diacylglycerol in angiotensin II-stimulated vascular smooth-muscle cells. *Biochem. J.* **1993**, *292*, 509–517. [CrossRef]
52. Bozelli, J.C.; Azher, S.; Epand, R.M. Plasmalogens and chronic inflammatory diseases. *Front. Physiol.* **2021**, *12*, 730829. [CrossRef]
53. Koivuniemi, A. The biophysical properties of plasmalogens originating from their unique molecular architecture. *FEBS Lett.* **2017**, *591*, 2700–2713. [CrossRef]

54. Goss, V.; Hunt, A.N.; Postle, A.D. Regulation of lung surfactant phospholipid synthesis and metabolism. *Biochim. Biophys. Acta Mol. Cell Biol. Lipids* **2013**, *1831*, 448–458. [CrossRef]
55. Escribá, P.V.; González-Ros, J.M.; Goñi, F.M.; Kinnunen, P.K.J.; Vigh, L.; Sánchez-Magraner, L.; Fernández, A.M.; Busquets, X.; Horváth, I.; Barceló-Coblijn, G. Membranes: A meeting point for lipids, proteins and therapies. *J. Cell. Mol. Med.* **2008**, *12*, 829–875. [CrossRef]
56. Cui, Z.; Houweling, M. Phosphatidylcholine and cell death. *Biochim. Biophys. Acta Mol. Cell Biol. Lipids* **2002**, *1585*, 87–96. [CrossRef] [PubMed]
57. Robertson, B. Lung surfactant for replacement therapy. *Clin. Physiol.* **1983**, *3*, 97–110. [CrossRef] [PubMed]
58. Li, G.; Kim, J.; Huang, Z.; St. Clair, J.R.; Brown, D.A.; London, E. Efficient replacement of plasma membrane outer leaflet phospholipids and sphingolipids in cells with exogenous lipids. *Proc. Natl. Acad. Sci. USA* **2016**, *113*, 14025–14030. [CrossRef] [PubMed]
59. Sarri, E.; Sicart, A.; Lázaro-Diéguez, F.; Egea, G. Phospholipid synthesis participates in the regulation of diacylglycerol required for membrane trafficking at the Golgi complex. *J. Biol. Chem.* **2011**, *286*, 28632–28643. [CrossRef] [PubMed]
60. Vance, D.E.; Choy, P.C. How is phosphatidylcholine biosynthesis regulated? *Trends Biochem. Sci.* **1979**, *4*, 145–148. [CrossRef]
61. Kennedy, E.P. The synthesis of cytidine diphosphate choline, cytidine diphosphate ethanolamine, and related compounds. *J. Biol. Chem.* **1956**, *222*, 185–191. [CrossRef]
62. Cornell, R.B.; Ridgway, N.D. CTP: Phosphocholine cytidylyltransferase: Function, regulation, and structure of an amphitropic enzyme required for membrane biogenesis. *Prog. Lipid Res.* **2015**, *59*, 147–171. [CrossRef]
63. Vance, D.E. Glycerolipid biosynthesis in eukaryotes. In *Biochemistry of Lipids, Lipoproteins and Membranes*; Vance, D.E., Vance, J.E., Eds.; Elsevier: Amsterdam, The Netherlands, 1996; pp. 153–181.
64. Dawson, G. Measuring brain lipids. *Biochim. Biophys. Acta Mol. Cell Biol. Lipids* **2015**, *1851*, 1026–1039. [CrossRef]
65. Van der Veen, J.N.; Lingrell, S.; Vance, D.E. The membrane lipid phosphatidylcholine is an unexpected source of triacylglycerol in the liver. *J. Biol. Chem.* **2012**, *287*, 23418–23426. [CrossRef]
66. Kanno, K.; Wu, M.K.; Scapa, E.F.; Roderick, S.L.; Cohen, D.E. Structure and function of phosphatidylcholine transfer protein (PC-TP)/StarD2. *Biochim. Biophys. Acta Mol. Cell Biol. Lipids* **2007**, *1771*, 654–662. [CrossRef]
67. Lagarde, M.; Hachem, M.; Picq, M.; Guichardant, M.; Bernoud-Hubac, N. AceDoPC, a structured phospholipid to target the brain with docosahexaenoic acid. *OCL* **2016**, *23*, D102. [CrossRef]
68. Thudichum, J.L.W. A treatise on the chemical constitution of the brain: Based throughout upon original researches. *Glas. Med. J.* **1884**, *22*, 363–364.
69. Patel, D.; Witt, S.N. Ethanolamine and phosphatidylethanolamine: Partners in health and disease. *Oxid. Med. Cell. Longev.* **2017**, *2017*, 4829180. [CrossRef] [PubMed]
70. Vance, J.E.; Tasseva, G. Formation and function of phosphatidylserine and phosphatidylethanolamine in mammalian cells. *Biochim. Biophys. Acta Mol. Cell Biol. Lipids* **2013**, *1831*, 543–554. [CrossRef]
71. Borkenhagen, L.F.; Kennedy, E.P.; Fielding, L. Enzymatic formation and decarboxylation of phosphatidylserine. *J. Biol. Chem.* **1961**, *236*, PC28–PC30. [CrossRef]
72. Sundler, R.; Åkesson, B.; Nilsson, Å. Quantitative role of base exchange in phosphatidylethanolamine synthesis in isolated rat hepatocytes. *FEBS Lett.* **1974**, *43*, 303–307. [CrossRef]
73. Van Veldhoven, P.P.; Gijsbers, S.; Mannaerts, G.P.; Vermeesch, J.R.; Brys, V. Human sphingosine-1-phosphate lyase: cDNA cloning, functional expression studies and mapping to chromosome 10q2211DNA sequence was deposited in the EMBL database (AJ011304). *Biochim. Biophys. Acta Mol. Cell Biol. Lipids* **2000**, *1487*, 128–134. [CrossRef]
74. Riekhof, W.R.; Wu, J.; Jones, J.L.; Voelker, D.R. Identification and characterization of the major lysophosphatidylethanolamine acyltransferase in saccharomyces cerevisiae. *J. Biol. Chem.* **2007**, *282*, 28344–28352. [CrossRef]
75. Bouchet, A.M.; Frías, M.A.; Lairion, F.; Martini, F.; Almaleck, H.; Gordillo, G.; Disalvo, E.A. Structural and dynamical surface properties of phosphatidylethanolamine containing membranes. *Biochim. Biophys. Acta Biomembr.* **2009**, *1788*, 918–925. [CrossRef]
76. McIntosh, T.J. Differences in hydrocarbon chain tilt between hydrated phosphatidylethanolamine and phosphatidylcholine bilayers. A molecular packing model. *Biophys. J.* **1980**, *29*, 237–245. [CrossRef]
77. Dawaliby, R.; Trubbia, C.; Delporte, C.; Noyon, C.; Ruysschaert, J.-M.; Van Antwerpen, P.; Govaerts, C. Phosphatidylethanolamine is a key regulator of membrane fluidity in eukaryotic cells. *J. Biol. Chem.* **2016**, *291*, 3658–3667. [CrossRef] [PubMed]
78. Van der Veen, J.N.; Kennelly, J.P.; Wan, S.; Vance, J.E.; Vance, D.E.; Jacobs, R.L. The critical role of phosphatidylcholine and phosphatidylethanolamine metabolism in health and disease. *Biochim. Biophys. Acta Biomembr.* **2017**, *1859*, 1558–1572. [CrossRef] [PubMed]
79. Kreutzberger, A.J.B.; Kiessling, V.; Liang, B.; Yang, S.-T.; Castle, J.D.; Tamm, L.K. Asymmetric phosphatidylethanolamine distribution controls fusion pore lifetime and probability. *Biophys. J.* **2017**, *113*, 1912–1915. [CrossRef] [PubMed]
80. Pohl, E.E.; Jovanovic, O. The role of phosphatidylethanolamine adducts in modification of the activity of membrane proteins under oxidative stress. *Molecules* **2019**, *24*, 4545. [CrossRef]
81. Strandberg, E.; Tiltak, D.; Ehni, S.; Wadhwani, P.; Ulrich, A.S. Lipid shape is a key factor for membrane interactions of amphipathic helical peptides. *Biochim. Biophys. Acta Biomembr.* **2012**, *1818*, 1764–1776. [CrossRef]
82. Tsuboi, K.; Uyama, T.; Okamoto, Y.; Ueda, N. Endocannabinoids and related N-acylethanolamines: Biological activities and metabolism. *Inflamm. Regen.* **2018**, *38*, 28. [CrossRef]

83. Jin, X.-H.; Okamoto, Y.; Morishita, J.; Tsuboi, K.; Tonai, T.; Ueda, N. Discovery and characterization of a Ca^{2+}-independent phosphatidylethanolamine N-acyltransferase generating the anandamide precursor and its congeners. *J. Biol. Chem.* **2007**, *282*, 3614–3623. [CrossRef]
84. Folch, J.; Schneider, H.A. an amino acid constituent of the Ox brain cephalin. *J. Biol. Chem.* **1941**, *137*, 51–62. [CrossRef]
85. Vance, J.E. Historical perspective: Phosphatidylserine and phosphatidylethanolamine from the 1800s to the present. *J. Lipid Res.* **2018**, *59*, 923–944. [CrossRef]
86. Svennerholm, L. Distribution and fatty acid composition of phosphoglycerides in normal human brain. *J. Lipid Res.* **1968**, *9*, 570–579. [CrossRef]
87. Gardner, R.G.; Hampton, R.Y. A highly conserved signal controls degradation of 3-hydroxy-3-methylglutaryl-coenzyme A (HMG-CoA) reductase in eukaryotes. *J. Biol. Chem.* **1999**, *274*, 31671–31678. [CrossRef] [PubMed]
88. Kanfer, J.; Kennedy, E.P. Metabolism and function of bacterial lipids. *J. Biol. Chem.* **1964**, *239*, 1720–1726. [CrossRef] [PubMed]
89. Hübscher, G.; Dils, R.R.; Pover, W.F.R. Studies on the biosynthesis of phosphatidyl serine. *Biochim. Biophys. Acta* **1959**, *36*, 518–528. [CrossRef]
90. Van Meer, G. Dynamic transbilayer lipid asymmetry. *Cold Spring Harb. Perspect. Biol.* **2011**, *3*, a004671. [CrossRef] [PubMed]
91. Fadok, V.A.; Voelker, D.R.; Campbell, P.A.; Cohen, J.J.; Bratton, D.L.; Henson, P.M. Exposure of phosphatidylserine on the surface of apoptotic lymphocytes triggers specific recognition and removal by macrophages. *J. Immunol.* **1992**, *148*, 2207–2216. [CrossRef] [PubMed]
92. Vallabhapurapu, S.D.; Blanco, V.M.; Sulaiman, M.K.; Vallabhapurapu, S.L.; Chu, Z.; Franco, R.S.; Qi, X. Variation in human cancer cell external phosphatidylserine is regulated by flippase activity and intracellular calcium. *Oncotarget* **2015**, *6*, 34375–34388. [CrossRef]
93. Skotland, T.; Sandvig, K. The role of PS 18:0/18:1 in membrane function. *Nat. Commun.* **2019**, *10*, 2752. [CrossRef]
94. Ma, X.; Li, X.; Wang, W.; Zhang, M.; Yang, B.; Miao, Z. Phosphatidylserine, inflammation, and central nervous system diseases. *Front. Aging Neurosci.* **2022**, *14*, 975176. [CrossRef]
95. Naeini, M.B.; Bianconi, V.; Pirro, M.; Sahebkar, A. The role of phosphatidylserine recognition receptors in multiple biological functions. *Cell. Mol. Biol. Lett.* **2020**, *25*, 23. [CrossRef]
96. Merolli, A.; Santin, M. Role of phosphatidyl-serine in bone repair and its technological exploitation. *Molecules* **2009**, *14*, 5367–5381. [CrossRef]
97. Lentz, B.R. Exposure of platelet membrane phosphatidylserine regulates blood coagulation. *Prog. Lipid Res.* **2003**, *42*, 423–438. [CrossRef]
98. Anderson, R.J. The chemistry of the lipoids of Tubercle bacilli. XIV. The occurence of inosite in the phosphatide from human Tubercle bacilli1. *J. Am. Chem. Soc.* **1930**, *52*, 1607–1608. [CrossRef]
99. Pizer, F.L.; Ballou, C.E. Studies on myo-Inositol phosphates of natural origin. *J. Am. Chem. Soc.* **1959**, *81*, 915–921. [CrossRef]
100. Dickson, E.J.; Hille, B. Understanding phosphoinositides: Rare, dynamic, and essential membrane phospholipids. *Biochem. J.* **2019**, *476*, 1–23. [CrossRef] [PubMed]
101. D'Souza, K.; Epand, R.M. Enrichment of phosphatidylinositols with specific acyl chains. *Biochim. Biophys. Acta Biomembr.* **2014**, *1838*, 1501–1508. [CrossRef] [PubMed]
102. Payrastre, B.; Missy, K.; Giuriato, S.; Bodin, S.; Plantavid, M.; Gratacap, M.-P. Phosphoinositides: Key players in cell signalling, in time and space. *Cell. Signal.* **2001**, *13*, 377–387. [CrossRef]
103. Blunsom, N.J.; Cockcroft, S. Phosphatidylinositol synthesis at the endoplasmic reticulum. *Biochim. Biophys. Acta Mol. Cell Biol. Lipids* **2020**, *1865*, 158471. [CrossRef]
104. Daniels, C.J.; Palmer, F.B.S.C. Biosynthesis of phosphatidylinositol in *Crithidia fasciculata*. *Biochim. Biophys. Acta Lipids Lipid Metab.* **1980**, *618*, 263–272. [CrossRef]
105. Jorge, C.D.; Borges, N.; Santos, H. A novel pathway for the synthesis of inositol phospholipids uses cytidine diphosphate (CDP)-inositol as donor of the polar head group. *Environ. Microbiol.* **2015**, *17*, 2492–2504. [CrossRef]
106. Balla, T. Phosphoinositides: Tiny lipids with giant impact on cell regulation. *Physiol. Rev.* **2013**, *93*, 1019–1137. [CrossRef]
107. Anderson, K.E.; Juvin, V.; Clark, J.; Stephens, L.R.; Hawkins, P.T. Investigating the effect of arachidonate supplementation on the phosphoinositide content of MCF10a breast epithelial cells. *Adv. Biol. Regul.* **2016**, *62*, 18–24. [CrossRef] [PubMed]
108. Anderson, K.E.; Kielkowska, A.; Durrant, T.N.; Juvin, V.; Clark, J.; Stephens, L.R.; Hawkins, P.T. Lysophosphatidylinositol-Acyltransferase-1 (LPIAT1) Is Required to Maintain Physiological Levels of PtdIns and PtdInsP2 in the Mouse. *PLoS ONE* **2013**, *8*, e58425. [CrossRef] [PubMed]
109. Lee, J.Y.; Yang, J.S.; Park, S.M.; Byeon, S.K.; Moon, M.H. On-line high speed lipid extraction for nanoflow liquid chromatography-tandem mass spectrometry. *J. Chromatogr. A* **2016**, *1464*, 12–20. [CrossRef] [PubMed]
110. Yui, K.; Imataka, G.; Nakamura, H.; Ohara, N.; Naito, Y. Eicosanoids derived from arachidonic acid and their family prostaglandins and cyclooxygenase in psychiatric disorders. *Curr. Neuropharmacol.* **2015**, *13*, 776–785. [CrossRef]
111. Piñeiro, R.; Falasca, M. Lysophosphatidylinositol signalling: New wine from an old bottle. *Biochim. Biophys. Acta Mol. Cell Biol. Lipids* **2012**, *1821*, 694–705. [CrossRef]
112. Ferguson, M.A.J.; Williams, A.F. Cell-surface anchoring of proteins via glycosyl-phosphatidylinositol structures. *Annu. Rev. Biochem.* **1988**, *57*, 285–320. [CrossRef]
113. Posor, Y.; Jang, W.; Haucke, V. Phosphoinositides as membrane organizers. *Nat. Rev. Mol. Cell Biol.* **2022**, *23*, 797–816. [CrossRef]

114. Pangborn, M.C. Isolation and purification of a serologically active phospholipid from beef heart. *J. Biol. Chem.* **1942**, *143*, 247–256. [CrossRef]
115. Schlame, M.; Haldar, D. Cardiolipin is synthesized on the matrix side of the inner membrane in rat liver mitochondria. *J. Biol. Chem.* **1993**, *268*, 74–79. [CrossRef]
116. Acoba, M.G.; Senoo, N.; Claypool, S.M. Phospholipid ebb and flow makes mitochondria go. *J. Cell Biol.* **2020**, *219*, e202003131. [CrossRef]
117. Luévano-Martínez, L.A.; Duncan, A.L. Origin and diversification of the cardiolipin biosynthetic pathway in the Eukarya domain. *Biochem. Soc. Trans.* **2020**, *48*, 1035–1046. [CrossRef] [PubMed]
118. Hatch, G.M. Cell biology of cardiac mitochondrial phospholipids. *Biochem. Cell Biol.* **2004**, *82*, 99–112. [CrossRef] [PubMed]
119. Jiang, Z.; Shen, T.; Huynh, H.; Fang, X.; Han, Z.; Ouyang, K. Cardiolipin regulates mitochondrial ultrastructure and function in mammalian cells. *Genes* **2022**, *13*, 1889. [CrossRef] [PubMed]
120. Poulaki, A.; Giannouli, S. Mitochondrial lipids: From membrane organization to apoptotic facilitation. *Int. J. Mol. Sci.* **2022**, *23*, 3738. [CrossRef] [PubMed]
121. Osman, C.; Voelker, D.R.; Langer, T. Making heads or tails of phospholipids in mitochondria. *J. Cell Biol.* **2011**, *192*, 7–16. [CrossRef] [PubMed]
122. Kobayashi, T.; Beuchat, M.-H.; Lindsay, M.; Frias, S.; Palmiter, R.D.; Sakuraba, H.; Parton, R.G.; Gruenberg, J. Late endosomal membranes rich in lysobisphosphatidic acid regulate cholesterol transport. *Nat. Cell Biol.* **1999**, *1*, 113–118. [CrossRef]
123. Kobayashi, T.; Stang, E.; Fang, K.S.; de Moerloose, P.; Parton, R.G.; Gruenberg, J. A lipid associated with the antiphospholipid syndrome regulates endosome structure and function. *Nature* **1998**, *392*, 193–197. [CrossRef]
124. Wherrett, J.R.; Huterer, S. Enrichment of Bis-(monoacylglyceryl) phosphate in lysosomes from rat liver. *J. Biol. Chem.* **1972**, *247*, 4114–4120. [CrossRef]
125. Body, D.R.; Gray, G.M. The isolation and characterisation of phosphatidylglycerol and a structural isomer from pig lung. *Chem. Phys. Lipids* **1967**, *1*, 254–263. [CrossRef]
126. Brotherus, J.; Renkonen, O. Isolation and characterisation of bis-phosphatidic acid and its partially deacylated derivatives from cultured BHK-cells. *Chem. Phys. Lipids* **1974**, *13*, 11–20. [CrossRef]
127. Hullin-Matsuda, F.; Kawasaki, K.; Delton-Vandenbroucke, I.; Xu, Y.; Nishijima, M.; Lagarde, M.; Schlame, M.; Kobayashi, T. De novo biosynthesis of the late endosome lipid, bis(monoacylglycero)phosphate. *J. Lipid Res.* **2007**, *48*, 1997–2008. [CrossRef]
128. Waite, M.; Roddick, V.; Thornburg, T.; King, L.; Cochran, F. Conversion of phosphatidylglycerol to lyso(bis)phosphatidic acid by alveolar macrophages. *FASEB J.* **1987**, *1*, 318–325. [CrossRef]
129. Poorthuis, B.J.; Hostetler, K.Y. Conversion of diphosphatidylglycerol to bis(monoacylglyceryl)phosphate by lysosomes. *J. Lipid Res.* **1978**, *19*, 309–315. [CrossRef]
130. Showalter, M.R.; Berg, A.L.; Nagourney, A.; Heil, H.; Carraway, K.L.; Fiehn, O. The emerging and diverse roles of Bis(monoacylglycero) Phosphate lipids in cellular physiology and disease. *Int. J. Mol. Sci.* **2020**, *21*, 8067. [CrossRef]
131. Hostetler, K.Y. Chapter 6 Polyglycerophospholipids: Phosphatidylglycerol, diphosphatidylglycerol and bis (monoacylglycero) phosphate. In *New Comprehensive Biochemistry*; Hawthorne, N.J., Ansell, G.B., Eds.; Elsevier: Amsterdam, The Netherlands, 1982; Volume 4, pp. 215–261.
132. Akgoc, Z.; Iosim, S.; Seyfried, T.N. Bis(monoacylglycero)phosphate as a macrophage enriched phospholipid. *Lipids* **2015**, *50*, 907–912. [CrossRef] [PubMed]
133. Matsuzawa, Y.; Hostetler, K.Y. Studies on drug-induced lipidosis: Subcellular localization of phospholipid and cholesterol in the liver of rats treated with chloroquine or 4,4′-bis (diethylaminoethoxy)alpha, beta-diethyldiphenylethane. *J. Lipid Res.* **1980**, *21*, 202–214. [CrossRef] [PubMed]
134. Rouser, G.; Kritchevsky, G.; Knudson, A.G.; Simon, G. Accumulation of a glycerolphospholipid in classical niemann-pick disease. *Lipids* **1968**, *3*, 287–290. [CrossRef] [PubMed]
135. Gallala, H.D.; Sandhoff, K. Biological function of the cellular lipid BMP—BMP as a key activator for cholesterol sorting and membrane digestion. *Neurochem. Res.* **2011**, *36*, 1594–1600. [CrossRef]
136. Snyder, F.; Lee, T.; Wykle, R.L. Ether-linked glycerolipids and their bioactive species: Enzymes and metabolic regulation. In *The Enzymes of Biological Membranes*; Martonosi, A.N., Ed.; Springer: Boston, MA, USA, 1985; pp. 1–58.
137. Benveniste, J.; Henson, P.M.; Cochrane, C.G. Leukocyte-dependent histamine release from rabbit platelets. The role of IgE, basophils, and a platelet-activating factor. *J. Exp. Med.* **1972**, *136*, 1356–1377. [CrossRef]
138. Prescott, S.M.; Zimmerman, G.A.; Stafforini, D.M.; McIntyre, T.M. Platelet-activating factor and related lipid mediators. *Annu. Rev. Biochem.* **2000**, *69*, 419–445. [CrossRef]
139. Hishikawa, D.; Hashidate, T.; Shimizu, T.; Shindou, H. Diversity and function of membrane glycerophospholipids generated by the remodeling pathway in mammalian cells. *J. Lipid Res.* **2014**, *55*, 799–807. [CrossRef] [PubMed]
140. Tsoupras, A.; Lordan, R.; Zabetakis, I. Inflammation, not cholesterol, is a cause of chronic disease. *Nutrients* **2018**, *10*, 604. [CrossRef] [PubMed]
141. Roudebush, W.E. Function of platelet-activating factor in spermatozoa motility and fertility potential. *Glob. J. Reprod. Med.* **2022**, *8*, 5556749. [CrossRef]
142. Imaizumi, T.-A.; Stafforini, D.M.; Yamada, Y.; Mcintyre, T.M.; Prescott, S.M.; Zimmerman, G.A. Platelet-activating factor: A mediator for clinicians. *J. Intern. Med.* **1995**, *238*, 5–20. [CrossRef]

143. Schlondorff, D.; Neuwirth, R. Platelet-activating factor and the kidney. *Am. J. Physiol. Physiol.* **1986**, *251*, F1–F11. [CrossRef]
144. Travers, J.B.; Rohan, J.G.; Sahu, R.P. New insights into the pathologic roles of the platelet-activating factor system. *Front. Endocrinol.* **2021**, *12*, 624132. [CrossRef]
145. Zimmerman, G.A.; McIntyre, T.M.; Prescott, S.M.; Stafforini, D.M. The platelet-activating factor signaling system and its regulators in syndromes of inflammation and thrombosis. *Crit. Care Med.* **2002**, *30*, S294–S301. [CrossRef]
146. Makide, K.; Kitamura, H.; Sato, Y.; Okutani, M.; Aoki, J. Emerging lysophospholipid mediators, lysophosphatidylserine, lysophosphatidylthreonine, lysophosphatidylethanolamine and lysophosphatidylglycerol. *Prostaglandins Other Lipid Mediat.* **2009**, *89*, 135–139. [CrossRef]
147. Meyer zu Heringdorf, D.; Jakobs, K.H. Lysophospholipid receptors: Signalling, pharmacology and regulation by lysophospholipid metabolism. *Biochim. Biophys. Acta Biomembr.* **2007**, *1768*, 923–940. [CrossRef]
148. Hannun, Y.A.; Bell, R.M. Lysosphingolipids inhibit protein kinase C: Implications for the sphingolipidoses. *Science* **1987**, *235*, 670–674. [CrossRef]
149. Anliker, B.; Chun, J. Lysophospholipid G protein-coupled receptors. *J. Biol. Chem.* **2004**, *279*, 20555–20558. [CrossRef] [PubMed]
150. Law, S.-H.; Chan, M.-L.; Marathe, G.K.; Parveen, F.; Chen, C.-H.; Ke, L.-Y. An updated review of lysophosphatidylcholine metabolism in human diseases. *Int. J. Mol. Sci.* **2019**, *20*, 1149. [CrossRef] [PubMed]
151. Engel, K.M.; Schiller, J.; Galuska, C.E.; Fuchs, B. Phospholipases and Reactive Oxygen Species Derived Lipid Biomarkers in Healthy and Diseased Humans and Animals—A Focus on Lysophosphatidylcholine. *Front. Physiol.* **2021**, *12*, 732319. [CrossRef] [PubMed]
152. Hellenthal, K.E.M.; Brabenec, L.; Gross, E.R.; Wagner, N.-M. TRP channels as sensors of aldehyde and oxidative stress. *Biomolecules* **2021**, *11*, 1401. [CrossRef]
153. Aoki, J. Two pathways for lysophosphatidic acid production. *Biochim. Biophys. Acta Mol. Cell Biol. Lipids* **2008**, *1781*, 513–518. [CrossRef]
154. Lin, Y.-H.; Lin, Y.-C.; Chen, C.-C. Lysophosphatidic acid receptor antagonists and cancer: The current trends, clinical implications, and trials. *Cells* **2021**, *10*, 1629. [CrossRef]
155. Lin, M.-E.; Herr, D.R.; Chun, J. Lysophosphatidic acid (LPA) receptors: Signaling properties and disease relevance. *Prostaglandins Other Lipid Mediat.* **2010**, *91*, 130–138. [CrossRef]
156. Yeagle, P.L. Cholesterol and the cell membrane. *Biochim. Biophys. Acta Rev. Biomembr.* **1985**, *822*, 267–287. [CrossRef]
157. Haas, E.; Kim, Y.; Stanley, D. Why can insects not biosynthesize cholesterol? *Arch. Insect Biochem. Physiol.* **2023**, *112*, e21983. [CrossRef]
158. Bittman, R. Has nature designed the cholesterol side chain for optimal interaction with phospholipids? In *Cholesterol: Subcellular Biochemistry*; Bittman, R., Ed.; Springer: Boston, MA, USA, 1997; pp. 145–171.
159. Kroon, P.A.; Kainosho, M.; Chan, S.I. State of molecular motion of cholesterol in lecithin bilayers. *Nature* **1975**, *256*, 582–584. [CrossRef]
160. Lönnfors, M.; Långvik, O.; Björkbom, A.; Slotte, J.P. Cholesteryl phosphocholine—A study on its interactions with ceramides and other membrane lipids. *Langmuir* **2013**, *29*, 2319–2329. [CrossRef] [PubMed]
161. Schroeder, F.; Nemecz, G.; Gibson Wood, W.; Joiner, C.; Morrot, G.; Ayraut-Jarrier, M.; Devaux, P.F. Transmembrane distribution of sterol in the human erythrocyte. *Biochim. Biophys. Acta Biomembr.* **1991**, *1066*, 183–192. [CrossRef] [PubMed]
162. Mondal, M.; Mesmin, B.; Mukherjee, S.; Maxfield, F.R. Sterols are mainly in the cytoplasmic leaflet of the plasma membrane and the endocytic recycling compartment in CHO cells. *Mol. Biol. Cell* **2009**, *20*, 581–588. [CrossRef] [PubMed]
163. Kuijpers, P.M.J.C. History in medicine: The story of cholesterol, lipids and cardiology. *e-J. Cardiol. Pract.* **2021**, *19*, 1–5.
164. Cockcroft, S. Mammalian lipids: Structure, synthesis and function. *Essays Biochem.* **2021**, *65*, 813–845. [CrossRef]
165. Luo, J.; Jiang, L.-Y.; Yang, H.; Song, B.-L. Intracellular cholesterol transport by sterol transfer proteins at membrane contact sites. *Trends Biochem. Sci.* **2019**, *44*, 273–292. [CrossRef]
166. Frallicciardi, J.; Melcr, J.; Siginou, P.; Marrink, S.J.; Poolman, B. Membrane thickness, lipid phase and sterol type are determining factors in the permeability of membranes to small solutes. *Nat. Commun.* **2022**, *13*, 1605. [CrossRef]
167. De Oliveira Andrade, L. Understanding the role of cholesterol in cellular biomechanics and regulation of vesicular trafficking: The power of imaging. *Biomed. Spectrosc. Imaging* **2016**, *5*, S101–S117. [CrossRef]
168. Demel, R.A.; Van Deenen, L.L.M.; Pethica, B.A. Monolayer interactions of phospholipids and cholesterol. *Biochim. Biophys. Acta Biomembr.* **1967**, *135*, 11–19. [CrossRef]
169. Niu, S.-L.; Litman, B.J. Determination of membrane cholesterol partition coefficient using a lipid vesicle–cyclodextrin binary system: Effect of phospholipid acyl chain unsaturation and headgroup composition. *Biophys. J.* **2002**, *83*, 3408–3415. [CrossRef]
170. Shaikh, S.R.; Cherezov, V.; Caffrey, M.; Soni, S.P.; LoCascio, D.; Stillwell, W.; Wassall, S.R. Molecular organization of cholesterol in unsaturated phosphatidylethanolamines: X-ray diffraction and solid state 2 H NMR reveal differences with phosphatidylcholines. *J. Am. Chem. Soc.* **2006**, *128*, 5375–5383. [CrossRef] [PubMed]
171. Brzustowicz, M.R.; Cherezov, V.; Zerouga, M.; Caffrey, M.; Stillwell, W.; Wassall, S.R. Controlling membrane cholesterol content. A role for polyunsaturated (docosahexaenoate) phospholipids. *Biochemistry* **2002**, *41*, 12509–12519. [CrossRef] [PubMed]
172. Siminovitch, D.J.; Jeffrey, K.R. Orientational order in the choline headgroup of sphingomyelin: A 14N-NMR study. *Biochim. Biophys. Acta Biomembr.* **1981**, *645*, 270–278. [CrossRef] [PubMed]

173. Róg, T.; Pasenkiewicz-Gierula, M. Cholesterol-sphingomyelin interactions: A molecular dynamics simulation study. *Biophys. J.* **2006**, *91*, 3756–3767. [CrossRef] [PubMed]
174. Lönnfors, M.; Doux, J.P.F.; Killian, J.A.; Nyholm, T.K.M.; Slotte, J.P. Sterols have higher affinity for sphingomyelin than for phosphatidylcholine bilayers even at equal acyl-chain order. *Biophys. J.* **2011**, *100*, 2633–2641. [CrossRef]
175. Chiantia, S.; London, E. Acyl chain length and saturation modulate interleaflet coupling in asymmetric bilayers: Effects on dynamics and structural order. *Biophys. J.* **2012**, *103*, 2311–2319. [CrossRef]
176. Slotte, J.P. The importance of hydrogen bonding in sphingomyelin's membrane interactions with co-lipids. *Biochim. Biophys. Acta Biomembr.* **2016**, *1858*, 304–310. [CrossRef]
177. Sohrabi, Y.; Reinecke, H.; Godfrey, R. Altered cholesterol and lipid synthesis Mediates hyperinflammation in COVID-19. *Trends Endocrinol. Metab.* **2021**, *32*, 132–134. [CrossRef]
178. King, R.J.; Singh, P.K.; Mehla, K. The cholesterol pathway: Impact on immunity and cancer. *Trends Immunol.* **2022**, *43*, 78–92. [CrossRef]
179. Kopecka, J.; Trouillas, P.; Gašparović, A.Č.; Gazzano, E.; Assaraf, Y.G.; Riganti, C. Phospholipids and cholesterol: Inducers of cancer multidrug resistance and therapeutic targets. *Drug Resist. Updat.* **2020**, *49*, 100670. [CrossRef]
180. Hu, J.; Zhang, Z.; Shen, W.-J.; Azhar, S. Cellular cholesterol delivery, intracellular processing and utilization for biosynthesis of steroid hormones. *Nutr. Metab.* **2010**, *7*, 47. [CrossRef] [PubMed]
181. Panveliwalla, D.; Lewis, B.; Wootton, I.D.P.; Tabaqchali, S. Determination of individual bile acids in biological fluids by thin-layer chromatography and fluorimetry. *J. Clin. Pathol.* **1970**, *23*, 309–314. [CrossRef] [PubMed]
182. Kuo, T.; McQueen, A.; Chen, T.-C.; Wang, J.-C. Regulation of glucose homeostasis by glucocorticoids. In *Glucocorticoid Signaling from Molecules to Mice to Man*; Wang, J.-C., Harris, C., Eds.; Springer: New York, NY, USA, 2015; pp. 99–126.
183. Wollam, J.; Antebi, A. Sterol regulation of metabolism, homeostasis, and development. *Annu. Rev. Biochem.* **2011**, *80*, 885–916. [CrossRef] [PubMed]
184. Christiansen, J.J.; Djurhuus, C.B.; Gravholt, C.H.; Iversen, P.; Christiansen, J.S.; Schmitz, O.; Weeke, J.; Jørgensen, J.O.L.; Møller, N. Effects of cortisol on carbohydrate, lipid, and protein metabolism: Studies of acute cortisol withdrawal in adrenocortical failure. *J. Clin. Endocrinol. Metab.* **2007**, *92*, 3553–3559. [CrossRef]
185. Schwartz, C.C.; VandenBroek, J.M.; Cooper, P.S. Lipoprotein cholesteryl ester production, transfer, and output in vivo in humans. *J. Lipid Res.* **2004**, *45*, 1594–1607. [CrossRef]
186. Breslow, D.K. Sphingolipid homeostasis in the endoplasmic reticulum and beyond. *Cold Spring Harb. Perspect. Biol.* **2013**, *5*, a013326. [CrossRef]
187. Carter, H.E.; Glick, F.J.; Norris, W.P.; Phillips, G.E. Biochemistry of the sphingolipids. *J. Biol. Chem.* **1947**, *170*, 285–294. [CrossRef]
188. Tidhar, R.; Futerman, A.H. The complexity of sphingolipid biosynthesis in the endoplasmic reticulum. *Biochim. Biophys. Acta Mol. Cell Res.* **2013**, *1833*, 2511–2518. [CrossRef]
189. Eliot, A.C.; Kirsch, J.F. Pyridoxal phosphate enzymes: Mechanistic, structural, and evolutionary considerations. *Annu. Rev. Biochem.* **2004**, *73*, 383–415. [CrossRef]
190. Eichler, F.S.; Hornemann, T.; McCampbell, A.; Kuljis, D.; Penno, A.; Vardeh, D.; Tamrazian, E.; Garofalo, K.; Lee, H.-J.; Kini, L.; et al. Overexpression of the Wild-Type SPT1 Subunit Lowers Desoxysphingolipid Levels and Rescues the Phenotype of HSAN1. *J. Neurosci.* **2009**, *29*, 14646–14651. [CrossRef]
191. Rosen, H.; Gonzalez-Cabrera, P.J.; Sanna, M.G.; Brown, S. Sphingosine 1-phosphate receptor signaling. *Annu. Rev. Biochem.* **2009**, *78*, 743–768. [CrossRef] [PubMed]
192. Cuvillier, O.; Rosenthal, D.S.; Smulson, M.E.; Spiegel, S. Sphingosine 1-Phosphate Inhibits Activation of Caspases that Cleave Poly(ADP-ribose) Polymerase and Lamins during Fas- and Ceramide-mediated Apoptosis in Jurkat T Lymphocytes. *J. Biol. Chem.* **1998**, *273*, 2910–2916. [CrossRef] [PubMed]
193. Hait, N.C.; Oskeritzian, C.A.; Paugh, S.W.; Milstien, S.; Spiegel, S. Sphingosine kinases, sphingosine 1-phosphate, apoptosis and diseases. *Biochim. Biophys. Acta Biomembr.* **2006**, *1758*, 2016–2026. [CrossRef] [PubMed]
194. Merrill, A.H. De novo sphingolipid biosynthesis: A necessary, but dangerous, pathway. *J. Biol. Chem.* **2002**, *277*, 25843–25846. [CrossRef]
195. Stoffel, W.; Bauer, E.; Stahl, J. The metabolism of sphingosine bases in Tetrahymena pyriformis. Sphingosine kinase and sphingosine-1-phosphate lyase. *Biol. Chem.* **1974**, *355*, 61–74. [CrossRef]
196. Hannun, Y.A.; Obeid, L.M. Sphingolipids and their metabolism in physiology and disease. *Nat. Rev. Mol. Cell Biol.* **2018**, *19*, 175–191. [CrossRef]
197. Zupancic, E.; Carreira, A.C.; de Almeida, R.F.M.; Silva, L.C. Biophysical implications of sphingosine accumulation in membrane properties at neutral and acidic pH. *J. Phys. Chem. B* **2014**, *118*, 4858–4866. [CrossRef]
198. Merrill, A.H. Sphingolipid and glycosphingolipid metabolic pathways in the era of sphingolipidomics. *Chem. Rev.* **2011**, *111*, 6387–6422. [CrossRef]
199. Katoh, N. Modulation by sphingosine of phosphorylation of substrate proteins by protein kinase C in nuclei from cow mammary gland. *J. Vet. Med. Sci.* **2004**, *66*, 1237–1242. [CrossRef]
200. Natarajan, V.; Jayaram, H.N.; Scribner, W.M.; Garcia, J.G. Activation of endothelial cell phospholipase D by sphingosine and sphingosine-1-phosphate. *Am. J. Respir. Cell Mol. Biol.* **1994**, *11*, 221–229. [CrossRef]

201. Tomonaga, N.; Tsuduki, T.; Manabe, Y.; Sugawara, T. Sphingoid bases of dietary ceramide 2-aminoethylphosphonate, a marine sphingolipid, absorb into lymph in rats. *J. Lipid Res.* **2019**, *60*, 333–340. [CrossRef] [PubMed]
202. Riley, R.T.; Wang, E.; Schroeder, J.J.; Smith, E.R.; Plattner, R.D.; Abbas, H.; Yoo, H.-S.; Merrill, A.H. Evidence for disruption of sphingolipid metabolism as a contributing factor in the toxicity and carcinogenicity of fumonisins. *Nat. Toxins* **1996**, *4*, 3–15. [CrossRef] [PubMed]
203. Iwabuchi, K.; Nakayama, H.; Oizumi, A.; Suga, Y.; Ogawa, H.; Takamori, K. Role of ceramide from glycosphingolipids and Its metabolites in immunological and inflammatory responses in humans. *Mediat. Inflamm.* **2015**, *2015*, 120748. [CrossRef] [PubMed]
204. Fanani, M.L.; Maggio, B. The many faces (and phases) of ceramide and sphingomyelin I—Single lipids. *Biophys. Rev.* **2017**, *9*, 589–600. [CrossRef]
205. Fürst, W.; Sandhoff, K. Activator proteins and topology of lysosomal sphingolipid catabolism. *Biochim. Biophys. Acta Lipids Lipid Metab.* **1992**, *1126*, 1–16. [CrossRef]
206. Kitatani, K.; Idkowiak-Baldys, J.; Hannun, Y.A. The sphingolipid salvage pathway in ceramide metabolism and signaling. *Cell. Signal.* **2008**, *20*, 1010–1018. [CrossRef]
207. Pewzner-Jung, Y.; Ben-Dor, S.; Futerman, A.H. When Do Lasses (Longevity Assurance Genes) Become CerS (Ceramide Synthases)? *J. Biol. Chem.* **2006**, *281*, 25001–25005. [CrossRef]
208. Ho, Q.W.C.; Zheng, X.; Ali, Y. Ceramide acyl chain length and its relevance to intracellular lipid regulation. *Int. J. Mol. Sci.* **2022**, *23*, 9697. [CrossRef]
209. Lachkar, F.; Ferré, P.; Foufelle, F.; Papaioannou, A. Dihydroceramides: Their emerging physiological roles and functions in cancer and metabolic diseases. *Am. J. Physiol. Metab.* **2021**, *320*, E122–E130. [CrossRef]
210. Shimeno, H.; Soeda, S.; Sakamoto, M.; Kouchi, T.; Kowakame, T.; Kihara, T. Partial purification and characterization of sphingosine N-acyltransferase (ceramide synthase) from bovine liver mitochondrion-rich fraction. *Lipids* **1998**, *33*, 601–605. [CrossRef]
211. Zuellig, R.A.; Hornemann, T.; Othman, A.; Hehl, A.B.; Bode, H.; Güntert, T.; Ogunshola, O.O.; Saponara, E.; Grabliauskaite, K.; Jang, J.-H.; et al. Deoxysphingolipids, Novel Biomarkers for Type 2 Diabetes, Are Cytotoxic for Insulin-Producing Cells. *Diabetes* **2014**, *63*, 1326–1339. [CrossRef] [PubMed]
212. Zitomer, N.C.; Mitchell, T.; Voss, K.A.; Bondy, G.S.; Pruett, S.T.; Garnier-Amblard, E.C.; Liebeskind, L.S.; Park, H.; Wang, E.; Sullards, M.C.; et al. Ceramide synthase inhibition by Fumonisin B1 causes accumulation of 1-deoxysphinganine. *J. Biol. Chem.* **2009**, *284*, 4786–4795. [CrossRef] [PubMed]
213. Eto, M.; Bennouna, J.; Hunter, O.C.; Hershberger, P.A.; Kanto, T.; Johnson, C.S.; Lotze, M.T.; Amoscato, A.A. C16 ceramide accumulates following androgen ablation in LNCaP prostate cancer cells. *Prostate* **2003**, *57*, 66–79. [CrossRef] [PubMed]
214. Koybasi, S.; Senkal, C.E.; Sundararaj, K.; Spassieva, S.; Bielawski, J.; Osta, W.; Day, T.A.; Jiang, J.C.; Jazwinski, S.M.; Hannun, Y.A.; et al. Defects in cell growth regulation by C18:0-ceramide and longevity assurance gene 1 in human head and neck squamous cell carcinomas. *J. Biol. Chem.* **2004**, *279*, 44311–44319. [CrossRef] [PubMed]
215. Grösch, S.; Schiffmann, S.; Geisslinger, G. Chain length-specific properties of ceramides. *Prog. Lipid Res.* **2012**, *51*, 50–62.
216. Kalinichenko, L.S.; Gulbins, E.; Kornhuber, J.; Müller, C.P. Sphingolipid control of cognitive functions in health and disease. *Prog. Lipid Res.* **2022**, *86*, 101162.
217. Chiantia, S.; Ries, J.; Chwastek, G.; Carrer, D.; Li, Z.; Bittman, R.; Schwille, P. Role of ceramide in membrane protein organization investigated by combined AFM and FCS. *Biochim. Biophys. Acta Biomembr.* **2008**, *1778*, 1356–1364.
218. Goñi, F.M.; Alonso, A. Effects of ceramide and other simple sphingolipids on membrane lateral structure. *Biochim. Biophys. Acta Biomembr.* **2009**, *1788*, 169–177. [CrossRef]
219. Hanada, K.; Kumagai, K.; Yasuda, S.; Miura, Y.; Kawano, M.; Fukasawa, M.; Nishijima, M. Molecular machinery for non-vesicular trafficking of ceramide. *Nature* **2003**, *426*, 803–809. [CrossRef]
220. Fukasawa, M.; Nishijima, M.; Hanada, K. Genetic Evidence for ATP-dependent Endoplasmic Reticulum-to-Golgi Apparatus Trafficking of Ceramide for Sphingomyelin Synthesis in Chinese Hamster Ovary Cells. *J. Cell Biol.* **1999**, *144*, 673–685. [CrossRef]
221. Blom, T.; Somerharju, P.; Ikonen, E. Synthesis and biosynthetic trafficking of membrane lipids. *Cold Spring Harb. Perspect. Biol.* **2011**, *3*, a004713. [CrossRef] [PubMed]
222. Murakami, M.; Nakatani, Y.; Kudo, I. Type II secretory phospholipase A2 associated with cell surfaces via C-terminal heparin-binding lysine residues augments stimulus-initiated delayed prostaglandin generation. *J. Biol. Chem.* **1996**, *271*, 30041–30051. [PubMed]
223. Ullman, M.D.; Radin, N.S. The enzymatic formation of sphingomyelin from ceramide and lecithin in mouse liver. *J. Biol. Chem.* **1974**, *249*, 1506–1512.
224. Huitema, K.; van den Dikkenberg, J.; Brouwers, J.F.H.M.; Holthuis, J.C.M. Identification of a family of animal sphingomyelin synthases. *EMBO J.* **2004**, *23*, 33–44. [CrossRef] [PubMed]
225. Futerman, A.H. Sphingolipids. In *Biochemistry of Lipids, Lipoproteins and Membranes*; Elsevier: Amsterdam, The Netherlands, 2021; pp. 281–316.
226. Bielawski, J.; Pierce, J.S.; Snider, J.; Rembiesa, B.; Szulc, Z.M.; Bielawska, A. Sphingolipid analysis by high performance liquid chromatography-tandem mass spectrometry (HPLC-MS/MS). In *Sphingolipids as Signaling and Regulatory Molecules. Advances in Experimental Medicine and Biology*; Chalfant, C., Poeta, M.D., Eds.; Springer: New York, NY, USA, 2010; pp. 46–59.
227. Lucki, N.C.; Sewer, M.B. Nuclear sphingolipid metabolism. *Annu. Rev. Physiol.* **2012**, *74*, 131–151.

228. Birbes, H.; Luberto, C.; Hsu, Y.-T.; El Bawab, S.; Hannun, Y.A.; Obeid, L.M. A mitochondrial pool of sphingomyelin is involved in TNFα-induced Bax translocation to mitochondria. *Biochem. J.* **2005**, *386*, 445–451. [PubMed]
229. Albi, E.; Mersel, M.; Leray, C.; Tomassoni, M.L.; Viola-Magni, M.P. Rat liver chromatin phospholipids. *Lipids* **1994**, *29*, 715–719. [CrossRef]
230. Van Helvoort, A.; Giudici, M.L.; Thielemans, M.; Meer, G. van Transport of sphingomyelin to the cell surface is inhibited by brefeldin A and in mitosis, where C6-NBD-sphingomyelin is translocated across the plasma membrane by a multidrug transporter activity. *J. Cell Sci.* **1997**, *110*, 75–83.
231. Kornhuber, J.; Rhein, C.; Müller, C.P.; Mühle, C. Secretory sphingomyelinase in health and disease. *Biol. Chem.* **2015**, *396*, 707–736.
232. Yang, F.; Chen, G. The nutritional functions of dietary sphingomyelin and its applications in food. *Front. Nutr.* **2022**, *9*, 1002574.
233. Chakraborty, M.; Jiang, X.-C. Sphingomyelin and Its Role in Cellular Signaling. In *Lipid-Mediated Protein Signaling: Advances in Experimental Medicine and Biology*; Capelluto, D., Ed.; Springer: Dordrecht, The Netherlands, 2013; pp. 1–14.
234. Slotte, J.P. Biological functions of sphingomyelins. *Prog. Lipid Res.* **2013**, *52*, 424–437. [PubMed]
235. Hannun, Y.A.; Obeid, L.M. Principles of bioactive lipid signalling: Lessons from sphingolipids. *Nat. Rev. Mol. Cell Biol.* **2008**, *9*, 139–150. [PubMed]
236. Hannun, Y.A.; Bell, R.M. Regulation of protein kinase C by sphingosine and lysosphingolipids. *Clin. Chim. Acta* **1989**, *185*, 333–345.
237. Subbaiah, P.V.; Gesquiere, L.R.; Wang, K. Regulation of the selective uptake of cholesteryl esters from high density lipoproteins by sphingomyelin. *J. Lipid Res.* **2005**, *46*, 2699–2705. [PubMed]
238. Dingjan, T.; Futerman, A.H. The role of the 'sphingoid motif' in shaping the molecular interactions of sphingolipids in biomembranes. *Biochim. Biophys. Acta Biomembr.* **2021**, *1863*, 183701. [CrossRef]
239. García-Arribas, A.B.; Alonso, A.; Goñi, F.M. Cholesterol interactions with ceramide and sphingomyelin. *Chem. Phys. Lipids* **2016**, *199*, 26–34. [PubMed]
240. Lingwood, D.; Simons, K. Lipid rafts as a membrane-organizing principle. *Science* **2010**, *327*, 46–50. [PubMed]
241. Prinetti, A.; Chigorno, V.; Prioni, S.; Loberto, N.; Marano, N.; Tettamanti, G.; Sonnino, S. Changes in the lipid turnover, composition, and organization, as sphingolipid-enriched membrane domains, in rat cerebellar granule cells developing in vitro. *J. Biol. Chem.* **2001**, *276*, 21136–21145.
242. Futerman, A.H.; Pagano, R.E. Determination of the intracellular sites and topology of glucosylceramide synthesis in rat liver. *Biochem. J.* **1991**, *280*, 295–302. [CrossRef]
243. Carruthers, A.; Carey, E.M. UDP-galactose: Ceramide galactosyl transferase of isolated oligodendroglia. *J. Neurochem.* **1983**, *41*, 22–29. [CrossRef]
244. Yu, R.K.; Yanagisawa, M.; Ariga, T. Glycosphingolipid structures. In *Comprehensive Glycoscience: Chemistry to Systems Biology*; Kamerling, H., Ed.; Elsevier: Amsterdam, The Netherlands, 2007; pp. 73–122. ISBN 13: 978-0-444-52746-2.
245. Merrill, A.H.; Sandhoff, K. Chapter 14 Sphingolipids: Metabolism and cell signaling. In *Biochemistry of Lipids, Lipoproteins and Membranes*; Vance, D.E., Vance, J.E., Eds.; Elsevier: Amsterdam, The Netherlands, 2002; pp. 373–407.
246. Leray, C. *Introduction to Lipidomics: From Human to Bacteria*; CRC Press, Taylor and Francis Group: Boca Raton, FL, USA, 2013.
247. Farwanah, H.; Kolter, T. Lipidomics of glycosphingolipids. *Metabolites* **2012**, *2*, 134–164. [CrossRef] [PubMed]
248. Tettamanti, G. Ganglioside/glycosphingolipid turnover: New concepts. *Glycoconj. J.* **2003**, *20*, 301–317. [CrossRef] [PubMed]
249. Lingwood, C.A. Glycosphingolipid functions. *Cold Spring Harb. Perspect. Biol.* **2011**, *3*, a004788. [CrossRef] [PubMed]
250. He, Q.; Chen, Y.; Wang, Z.; He, H.; Yu, P. Cellular uptake, metabolism and sensing of long-chain fatty acids. *Front. Biosci.* **2023**, *28*, 10. [CrossRef] [PubMed]
251. Zhang, C.; Rodriguez, E.; Bi, C.; Zheng, X.; Suresh, D.; Suh, K.; Li, Z.; Elsebaei, F.; Hage, D.S. High performance affinity chromatography and related separation methods for the analysis of biological and pharmaceutical agents. *Analyst* **2018**, *143*, 374–391. [CrossRef]
252. Rodriguez-Cuenca, S.; Pellegrinelli, V.; Campbell, M.; Oresic, M.; Vidal-Puig, A. Sphingolipids and glycerophospholipids—The "ying and yang" of lipotoxicity in metabolic diseases. *Prog. Lipid Res.* **2017**, *66*, 14–29. [CrossRef]
253. Hakomori, S. Structure and function of glycosphingolipids and sphingolipids: Recollections and future trends. *Biochim. Biophys. Acta Gen. Subj.* **2008**, *1780*, 325–346. [CrossRef]
254. Shaikh, S.R.; Edidin, M. Polyunsaturated fatty acids, membrane organization, T cells, and antigen presentation. *Am. J. Clin. Nutr.* **2006**, *84*, 1277–1289. [CrossRef]
255. Sud, M.; Fahy, E.; Cotter, D.; Brown, A.; Dennis, E.A.; Glass, C.K.; Merrill, A.H.; Murphy, R.C.; Raetz, C.R.H.; Russell, D.W.; et al. LMSD: LIPID MAPS structure database. *Nucleic Acids Res.* **2007**, *35*, D527–D532. [CrossRef]
256. Jalil, A.; Bourgeois, T.; Ménégaut, L.; Lagrost, L.; Thomas, C.; Masson, D. Revisiting the role of LXRs in PUFA metabolism and phospholipid homeostasis. *Int. J. Mol. Sci.* **2019**, *20*, 3787. [CrossRef]
257. Repa, J.J.; Liang, G.; Ou, J.; Bashmakov, Y.; Lobaccaro, J.-M.A.; Shimomura, I.; Shan, B.; Brown, M.S.; Goldstein, J.L.; Mangelsdorf, D.J. Regulation of mouse sterol regulatory element-binding protein-1c gene (SREBP-1c) by oxysterol receptors, LXRα and LXRβ. *Genes Dev.* **2000**, *14*, 2819–2830. [CrossRef]
258. Seo, J.B.; Moon, H.M.; Kim, W.S.; Lee, Y.S.; Jeong, H.W.; Yoo, E.J.; Ham, J.; Kang, H.; Park, M.-G.; Steffensen, K.R.; et al. Activated liver X receptors stimulate adipocyte differentiation through induction of peroxisome proliferator-activated receptor γ expression. *Mol. Cell. Biol.* **2004**, *24*, 3430–3444. [CrossRef] [PubMed]

259. Cha, J.-Y.; Repa, J.J. The liver X receptor (LXR) and hepatic lipogenesis. *J. Biol. Chem.* **2007**, *282*, 743–751. [CrossRef] [PubMed]
260. Schörken, U.; Kempers, P. Lipid biotechnology: Industrially relevant production processes. *Eur. J. Lipid Sci. Technol.* **2009**, *111*, 627–645.
261. Shanklin, J.; Cahoon, E.B. Desaturation and related modifications of fatty acids. *Annu. Rev. Plant Physiol. Plant Mol. Biol.* **1998**, *49*, 611–641. [CrossRef]
262. Jayakumar, A.; Tai, M.H.; Huang, W.Y.; Al-Feel, W.; Hsu, M.; Abu-Elheiga, L.; Chirala, S.S.; Wakil, S.J. Human fatty acid synthase: Properties and molecular cloning. *Proc. Natl. Acad. Sci. USA* **1995**, *92*, 8695–8699. [CrossRef]
263. Semenkovich, C.F.; Coleman, T.; Fiedorek, F.T. Human fatty acid synthase mRNA: Tissue distribution, genetic mapping, and kinetics of decay after glucose deprivation. *J. Lipid Res.* **1995**, *36*, 1507–1521.
264. Nowinski, S.M.; Van Vranken, J.G.; Dove, K.K.; Rutter, J. Impact of mitochondrial fatty acid synthesis on mitochondrial biogenesis. *Curr. Biol.* **2018**, *28*, R1212–R1219. [CrossRef]
265. Strawford, A.; Antelo, F.; Christiansen, M.; Hellerstein, M.K. Adipose tissue triglyceride turnover, de novo lipogenesis, and cell proliferation in humans measured with ^2H_2O. *Am. J. Physiol. Metab.* **2004**, *286*, E577–E588.
266. Turner, S.M.; Murphy, E.J.; Neese, R.A.; Antelo, F.; Thomas, T.; Agarwal, A.; Go, C.; Hellerstein, M.K. Measurement of TG synthesis and turnover in vivo by 2 H 2 O incorporation into the glycerol moiety and application of MIDA. *Am. J. Physiol. Metab.* **2003**, *285*, E790–E803. [CrossRef]
267. Bauman, D.E.; Mellenberger, R.W.; Derrig, R.G. Fatty acid synthesis in sheep mammary tissue. *J. Dairy Sci.* **1973**, *56*, 1312–1318. [PubMed]
268. Chandel, N.S. Glycolysis. *Cold Spring Harb. Perspect. Biol.* **2021**, *13*, a040535. [CrossRef] [PubMed]
269. McCommis, K.S.; Finck, B.N. Mitochondrial pyruvate transport: A historical perspective and future research directions. *Biochem. J.* **2015**, *466*, 443–454. [PubMed]
270. Martínez-Reyes, I.; Chandel, N.S. Mitochondrial TCA cycle metabolites control physiology and disease. *Nat. Commun.* **2020**, *11*, 102.
271. Wallace, M.; Green, C.R.; Roberts, L.S.; Lee, Y.M.; McCarville, J.L.; Sanchez-Gurmaches, J.; Meurs, N.; Gengatharan, J.M.; Hover, J.D.; Phillips, S.A.; et al. Enzyme promiscuity drives branched-chain fatty acid synthesis in adipose tissues. *Nat. Chem. Biol.* **2018**, *14*, 1021–1031.
272. Bressler, R.; Wakil, S.J. Studies on the mechanism of fatty acid synthesis. *J. Biol. Chem.* **1962**, *237*, 1441–1448.
273. Kuhajda, F.P. Fatty-acid synthase and human cancer: New perspectives on its role in tumor biology. *Nutrition* **2000**, *16*, 202–208.
274. Wakil, S.J. A malonic acid derivative as an intermediate in fatty acid synthesis. *J. Am. Chem. Soc.* **1958**, *80*, 6465. [CrossRef]
275. Wakil, S.J.; Titchener, E.B.; Gibson, D.M. Evidence for the participation of biotin in the enzymic synthesis of fatty acids. *Biochim. Biophys. Acta* **1958**, *29*, 225–226. [CrossRef]
276. O'Neill, L.M.; Miyazaki, M.; Bond, L.M.; Lewis, S.A.; Ding, F.; Liu, Z.; Ntambi, J.M. Fatty acid desaturation and elongation in mammals. In *Biochemistry of Lipids, Lipoproteins and Membranes*; Ridgway, N.D., McLeod, R.S., Eds.; Elsevier: Amsterdam, The Netherlands, 2021; pp. 201–226.
277. Wakil, S.J. Fatty acid synthase, a proficient multifunctional enzyme. *Biochemistry* **1989**, *28*, 4523–4530. [CrossRef]
278. Cook, H.W. Fatty acid desaturation and chain elongation in eucaryotes. In *Biochemistry of Lipids and Membranes*; Vance, D.E., Vance, J.E., Eds.; Elsevier Science: Amsterdam, The Netherlands, 1985; pp. 181–212.
279. Heil, C.S.; Wehrheim, S.S.; Paithankar, K.S.; Grininger, M. Fatty acid biosynthesis: Chain-length regulation and control. *ChemBioChem* **2019**, *20*, 2298–2321. [PubMed]
280. Kim, K.-H. Regulation of mammalian acetyl-coenzyme A carboxylase. *Annu. Rev. Nutr.* **1997**, *17*, 77–99. [CrossRef] [PubMed]
281. Wang, Y.; Yu, W.; Li, S.; Guo, D.; He, J.; Wang, Y. Acetyl-CoA carboxylases and diseases. *Front. Oncol.* **2022**, *12*, 836058. [PubMed]
282. Paiva, P.; Medina, F.E.; Viegas, M.; Ferreira, P.; Neves, R.P.P.; Sousa, J.P.M.; Ramos, M.J.; Fernandes, P.A. Animal fatty acid synthase: A chemical nanofactory. *Chem. Rev.* **2021**, *121*, 9502–9553.
283. Leibundgut, M.; Maier, T.; Jenni, S.; Ban, N. The multienzyme architecture of eukaryotic fatty acid synthases. *Curr. Opin. Struct. Biol.* **2008**, *18*, 714–725. [CrossRef]
284. Smith, S. The animal fatty acid synthase: One gene, one polypeptide, seven enzymes. *FASEB J.* **1994**, *8*, 1248–1259. [CrossRef]
285. Magnuson, K.; Jackowski, S.; Rock, C.O.; Cronan, J.E. Regulation of fatty acid biosynthesis in Escherichia coli. *Microbiol. Rev.* **1993**, *57*, 522–542. [CrossRef]
286. Bazan, H.E.P.; Careaga, M.M.; Sprecher, H.; Bazan, N.G. Chain elongation and desaturation of eicosapentaenoate to docosahexaenoate and phospholipid labeling in the rat retina in vivo. *Biochim. Biophys. Acta Lipids Lipid Metab.* **1982**, *712*, 123–128. [CrossRef]
287. Cinti, D.L.; Cook, L.; Nagi, M.N.; Suneja, S.K. The fatty acid chain elongation system of mammalian endoplasmic reticulum. *Prog. Lipid Res.* **1992**, *31*, 1–51.
288. Jump, D.B. Mammalian fatty acid elongases. In *Lipidomics: Methods in Molecular Biology*; Armstrong, D., Ed.; Humana Press: Totowa, NJ, USA, 2009; pp. 375–389.
289. Oboh, A. Investigating the Long-Chain Polyunsaturated Fatty Acid Biosynthesis of the African Catfish *Clarias gariepinus* (Burchell, 1822). Ph.D. Thesis, University of Stirling, Stirling, UK, 2018.
290. Yeboah, G.K.; Lobanova, E.S.; Brush, R.S.; Agbaga, M.-P. Very long chain fatty acid-containing lipids: A decade of novel insights from the study of ELOVL4. *J. Lipid Res.* **2021**, *62*, 100030.

291. Robinson, B.S.; Johnson, D.W.; Poulos, A. Unique molecular species of phosphatidylcholine containing very-long-chain (C24-C38) polyenoic fatty acids in rat brain. *Biochem. J.* **1990**, *265*, 763–767. [CrossRef]
292. Řezanka, T. Very-long-chain fatty acids from the animal and plant kingdoms. *Prog. Lipid Res.* **1989**, *28*, 147–187. [CrossRef]
293. Poulos, A.; Sharp, P.; Johnson, D.; Easton, C. The occurrence of polyenoic very long chain fatty acids with greater than 32 carbon atoms in molecular species of phosphatidylcholine in normal and peroxisome-deficient (Zellweger's syndrome) brain. *Biochem. J.* **1988**, *253*, 645–650. [CrossRef]
294. Aveldaño, M.I.; Sprecher, H. Very long chain (C24 to C36) polyenoic fatty acids of the n-3 and n-6 series in dipolyunsaturated phosphatidylcholines from bovine retina. *J. Biol. Chem.* **1987**, *262*, 1180–1186. [CrossRef] [PubMed]
295. Furland, N.E.; Maldonado, E.N.; Aveldaño, M.I. Very long chain PUFA in murine testicular triglycerides and cholesterol esters. *Lipids* **2003**, *38*, 73–80. [CrossRef] [PubMed]
296. Aveldaño, M.I.; Robinson, B.S.; Johnson, D.W.; Poulos, A. Long and very long chain polyunsaturated fatty acids of the n-6 series in rat seminiferous tubules. Active desaturation of 24:4n-6 to 24:5n-6 and concomitant formation of odd and even chain tetraenoic and pentaenoic fatty acids up to C32. *J. Biol. Chem.* **1993**, *268*, 11663–11669. [CrossRef] [PubMed]
297. Furland, N.E.; Zanetti, S.R.; Oresti, G.M.; Maldonado, E.N.; Aveldaño, M.I. Ceramides and sphingomyelins with high proportions of very long-chain polyunsaturated fatty acids in mammalian germ cells. *J. Biol. Chem.* **2007**, *282*, 18141–18150. [CrossRef]
298. Torrissen, M.; Ytteborg, E.; Svensen, H.; Stoknes, I.; Nilsson, A.; Østbye, T.-K.; Berge, G.M.; Bou, M.; Ruyter, B. Investigation of the functions of n-3 very-long-chain PUFAs in skin using in vivo Atlantic salmon and in vitro human and fish skin models. *Br. J. Nutr.* **2023**, 1–17. [CrossRef]
299. Butovich, I.A. Cholesteryl esters as a depot for very long chain fatty acids in human meibum. *J. Lipid Res.* **2009**, *50*, 501–513. [CrossRef]
300. Butovich, I.A.; Uchiyama, E.; McCulley, J.P. Lipids of human meibum: Mass-spectrometric analysis and structural elucidation. *J. Lipid Res.* **2007**, *48*, 2220–2235. [CrossRef]
301. Rissmann, R.; Groenink, H.W.W.; Weerheim, A.M.; Hoath, S.B.; Ponec, M.; Bouwstra, J.A. New insights into ultrastructure, lipid composition and organization of Vernix Caseosa. *J. Investig. Dermatol.* **2006**, *126*, 1823–1833. [CrossRef] [PubMed]
302. Hiltunen, J.K.; Schonauer, M.S.; Autio, K.J.; Mittelmeier, T.M.; Kastaniotis, A.J.; Dieckmann, C.L. Mitochondrial fatty acid synthesis type II: More than just fatty acids. *J. Biol. Chem.* **2009**, *284*, 9011–9015. [CrossRef] [PubMed]
303. Chuman, L.; Brody, S. Acyl carrier protein is present in the mitochondria of plants and eucaryotic micro-organisms. *Eur. J. Biochem.* **1989**, *184*, 643–649. [CrossRef] [PubMed]
304. Monteuuis, G.; Suomi, F.; Kerätär, J.M.; Masud, A.J.; Kastaniotis, A.J. A conserved mammalian mitochondrial isoform of acetyl-CoA carboxylase ACC1 provides the malonyl-CoA essential for mitochondrial biogenesis in tandem with ACSF3. *Biochem. J.* **2017**, *474*, 3783–3797. [CrossRef]
305. Wongkittichote, P.; Ah Mew, N.; Chapman, K.A. Propionyl-CoA carboxylase—A review. *Mol. Genet. Metab.* **2017**, *122*, 145–152. [CrossRef]
306. Seubert, W.; Podack, E.R. Mechanisms and physiological roles of fatty acid chain elongation in microsomes and mitochondria. *Mol. Cell. Biochem.* **1973**, *1*, 29–40. [CrossRef]
307. Lu, Y.-J.; Zhang, Y.-M.; Grimes, K.D.; Qi, J.; Lee, R.E.; Rock, C.O. Acyl-phosphates initiate membrane phospholipid synthesis in gram-positive pathogens. *Mol. Cell* **2006**, *23*, 765–772. [CrossRef]
308. Rezaei Zonooz, S.; Hasani, M.; Morvaridzadeh, M.; Beatriz Pizarro, A.; Heydari, H.; Yosaee, S.; Rezamand, G.; Heshmati, J. Effect of alpha-lipoic acid on oxidative stress parameters: A systematic review and meta-analysis. *J. Funct. Foods* **2021**, *87*, 104774. [CrossRef]
309. Petersen Shay, K.; Moreau, R.F.; Smith, E.J.; Hagen, T.M. Is α-lipoic acid a scavenger of reactive oxygen speciesin vivo? Evidence for its initiation of stress signaling pathways that promote endogenous antioxidant capacity. *IUBMB Life* **2008**, *60*, 362–367. [CrossRef]
310. Stoll, S.; Hartmann, H.; Cohen, S.A.; Müller, W.E. The potent free radical scavenger α-lipoic acid improves memory in aged mice: Putative relationship to NMDA receptor deficits. *Pharmacol. Biochem. Behav.* **1993**, *46*, 799–805. [CrossRef]
311. Capece, U.; Moffa, S.; Improta, I.; Di Giuseppe, G.; Nista, E.C.; Cefalo, C.M.A.; Cinti, F.; Pontecorvi, A.; Gasbarrini, A.; Giaccari, A.; et al. Alpha-lipoic acid and glucose metabolism: A comprehensive update on biochemical and therapeutic features. *Nutrients* **2022**, *15*, 18. [CrossRef]
312. Hiltunen, J.K.; Autio, K.J.; Schonauer, M.S.; Kursu, V.A.S.; Dieckmann, C.L.; Kastaniotis, A.J. Mitochondrial fatty acid synthesis and respiration. *Biochim. Biophys. Acta Bioenerg.* **2010**, *1797*, 1195–1202. [CrossRef] [PubMed]
313. Angerer, H.; Schönborn, S.; Gorka, J.; Bahr, U.; Karas, M.; Wittig, I.; Heidler, J.; Hoffmann, J.; Morgner, N.; Zickermann, V. Acyl modification and binding of mitochondrial ACP to multiprotein complexes. *Biochim. Biophys. Acta Mol. Cell Res.* **2017**, *1864*, 1913–1920. [CrossRef] [PubMed]
314. Das, A.K.; Uhler, M.D.; Hajra, A.K. Molecular cloning and expression of mammalian peroxisomaltrans-2-enoyl-coenzyme A reductase cDNAs. *J. Biol. Chem.* **2000**, *275*, 24333–24340. [CrossRef] [PubMed]
315. Buist, P.H. Fatty acid desaturases: Selecting the dehydrogenation channel. *Nat. Prod. Rep.* **2004**, *21*, 249. [CrossRef]
316. Smith, S.; Tsai, S.-C. The type I fatty acid and polyketide synthases: A tale of two megasynthases. *Nat. Prod. Rep.* **2007**, *24*, 1041. [CrossRef]

317. Kaulmann, U.; Hertweck, C. Biosynthesis of polyunsaturated fatty acids by polyketide synthases. *Angew. Chem. Int. Ed.* **2002**, *41*, 1866. [CrossRef]
318. Napier, J. Plumbing the depths of PUFA biosynthesis: A novel polyketide synthase-like pathway from marine organisms. *Trends Plant Sci.* **2002**, *7*, 51–54. [CrossRef]
319. Metz, J.G.; Roessler, P.; Facciotti, D.; Levering, C.; Dittrich, F.; Lassner, M.; Valentine, R.; Lardizabal, K.; Domergue, F.; Yamada, A.; et al. Production of polyunsaturated fatty acids by polyketide synthases in both prokaryotes and eukaryotes. *Science* **2001**, *293*, 290–293. [CrossRef]
320. Chen, Z.; Chen, H.; Li, X.; Yuan, Q.; Su, J.; Yang, L.; Ning, L.; Lei, H. Fumonisin B1 damages the barrier functions of porcine intestinal epithelial cells in vitro. *J. Biochem. Mol. Toxicol.* **2019**, *33*, e22397. [CrossRef]
321. Bentley, R.; Bennett, J.W. Constructing polyketides: From collie to combinatorial biosynthesis. *Annu. Rev. Microbiol.* **1999**, *53*, 411–446. [CrossRef] [PubMed]
322. Cerone, M.; Smith, T.K. Desaturases: Structural and mechanistic insights into the biosynthesis of unsaturated fatty acids. *IUBMB Life* **2022**, *74*, 1036–1051. [CrossRef] [PubMed]
323. Gostinčar, C.; Turk, M.; Gunde-Cimerman, N. The evolution of fatty acid desaturases and cytochrome b5 in eukaryotes. *J. Membr. Biol.* **2010**, *233*, 63–72. [CrossRef] [PubMed]
324. Sprecher, H.; Luthria, D.L.; Mohammed, B.S.; Baykousheva, S.P. Reevaluation of the pathways for the biosynthesis of polyunsaturated fatty acids. *J. Lipid Res.* **1995**, *36*, 2471–2477. [CrossRef]
325. Paton, C.M.; Ntambi, J.M. Biochemical and physiological function of stearoyl-CoA desaturase. *Am. J. Physiol. Metab.* **2009**, *297*, E28–E37. [CrossRef]
326. Nakamura, M.T.; Nara, T.Y. Structure, function, and dietary regulation of delta6, delta5, and delta9 desaturases. *Annu. Rev. Nutr.* **2004**, *24*, 345–376. [CrossRef]
327. Meesapyodsuk, D.; Qiu, X. Structure Determinants for the Substrate Specificity of Acyl-CoA Δ9 Desaturases from a Marine Copepod. *ACS Chem. Biol.* **2014**, *9*, 922–934. [CrossRef]
328. Haritos, V.S.; Horne, I.; Damcevski, K.; Glover, K.; Gibb, N. Unexpected functional diversity in the fatty acid desaturases of the flour beetle Tribolium castaneum and identification of key residues determining activity. *Insect Biochem. Mol. Biol.* **2014**, *51*, 62–70. [CrossRef]
329. Bonamore, A.; Macone, A.; Colotti, G.; Matarese, R.M.; Boffi, A. The desaturase from Bacillus subtilis, a promising tool for the selective olefination of phospholipids. *J. Biotechnol.* **2006**, *121*, 49–53. [CrossRef]
330. Heilmann, I.; Mekhedov, S.; King, B.; Browse, J.; Shanklin, J. Identification of the Arabidopsis Palmitoyl-Monogalactosyldiacy lglycerol Δ7-Desaturase Gene FAD5, and Effects of Plastidial Retargeting of Arabidopsis Desaturases on the fad5 Mutant Phenotype. *Plant Physiol.* **2004**, *136*, 4237–4245. [CrossRef]
331. Knipple, D.C.; Rosenfield, C.-L.; Miller, S.J.; Liu, W.; Tang, J.; Ma, P.W.K.; Roelofs, W.L. Cloning and functional expression of a cDNA encoding a pheromone gland-specific acyl-CoA Δ11-desaturase of the cabbage looper moth, Trichoplusia ni. *Proc. Natl. Acad. Sci. USA* **1998**, *95*, 15287–15292. [CrossRef] [PubMed]
332. Ohnishi, M.; Thompson, G.A. Biosynthesis of the unique trans-Δ3-hexadecenoic acid component of chloroplast phosphatidylglycerol: Evidence concerning its site and mechanism of formation. *Arch. Biochem. Biophys.* **1991**, *288*, 591–599. [CrossRef] [PubMed]
333. Weiss-Hersh, K.; Garcia, A.L.; Marosvölgyi, T.; Szklenár, M.; Decsi, T.; Rühl, R. Saturated and monounsaturated fatty acids in membranes are determined by the gene expression of their metabolizing enzymes SCD1 and ELOVL6 regulated by the intake of dietary fat. *Eur. J. Nutr.* **2020**, *59*, 2759–2769. [CrossRef] [PubMed]
334. Miyazaki, M.; Jacobson, M.J.; Man, W.C.; Cohen, P.; Asilmaz, E.; Friedman, J.M.; Ntambi, J.M. Identification and characterization of murine SCD4, a novel heart-specific stearoyl-CoA desaturase isoform regulated by leptin and dietary factors. *J. Biol. Chem.* **2003**, *278*, 33904–33911. [CrossRef]
335. Human Gene Nomenclature Containing (HGNC) Gene Group: Fatty Acid Desaturases (FADS). Available online: https://www.genenames.org/data/genegroup/#!/group/553 (accessed on 15 September 2023).
336. Nagao, K.; Murakami, A.; Umeda, M. Structure and Function of Δ9-Fatty Acid Desaturase. *Chem. Pharm. Bull.* **2019**, *67*, 327–332. [CrossRef]
337. Grajchen, E.; Loix, M.; Baeten, P.; Côrte-Real, B.F.; Hamad, I.; Vanherle, S.; Haidar, M.; Dehairs, J.; Broos, J.Y.; Ntambi, J.M.; et al. Fatty acid desaturation by stearoyl-CoA desaturase-1 controls regulatory T cell differentiation and autoimmunity. *Cell. Mol. Immunol.* **2023**, *20*, 666–679. [CrossRef]
338. Tang, B.; Qiu, J.; Hu, S.; Li, L.; Wang, J. Role of stearyl-coenzyme A desaturase 1 in mediating the effects of palmitic acid on endoplasmic reticulum stress, inflammation, and apoptosis in goose primary hepatocytes. *Anim. Biosci.* **2021**, *34*, 1210–1220. [CrossRef]
339. Koeberle, A.; Löser, K.; Thürmer, M. Stearoyl-CoA desaturase-1 and adaptive stress signaling. *Biochim. Biophys. Acta Mol. Cell Biol. Lipids* **2016**, *1861*, 1719–1726. [CrossRef]
340. Kucharski, M.; Kaczor, U. Stearoyl-CoA desaturase—The lipid metabolism regulator. *Adv. Hyg. Exp. Med.* **2014**, *68*, 334–342. [CrossRef]
341. Liu, X.; Strable, M.S.; Ntambi, J.M. Stearoyl CoA Desaturase 1: Role in Cellular Inflammation and Stress. *Adv. Nutr.* **2011**, *2*, 15–22. [CrossRef]

342. Dobrzyn, P.; Dobrzyn, A.; Miyazaki, M.; Cohen, P.; Asilmaz, E.; Hardie, D.G.; Friedman, J.M.; Ntambi, J.M. Stearoyl-CoA desaturase 1 deficiency increases fatty acid oxidation by activating AMP-activated protein kinase in liver. *Proc. Natl. Acad. Sci. USA* **2004**, *101*, 6409–6414. [CrossRef] [PubMed]
343. Park, W.J.; Kothapalli, K.S.D.; Reardon, H.T.; Lawrence, P.; Qian, S.-B.; Brenna, J.T. A novel FADS1 isoform potentiates FADS2-mediated production of eicosanoid precursor fatty acids. *J. Lipid Res.* **2012**, *53*, 1502–1512. [CrossRef] [PubMed]
344. Park, W.J.; Reardon, H.T.; Tyburczy, C.; Kothapalli, K.S.D.; Brenna, J.T. Alternative splicing generates a novel FADS2 alternative transcript in baboons. *Mol. Biol. Rep.* **2010**, *37*, 2403–2406. [CrossRef] [PubMed]
345. Park, W.J.; Kothapalli, K.S.D.; Reardon, H.T.; Kim, L.Y.; Brenna, J.T. Novel fatty acid desaturase 3 (FADS3) transcripts generated by alternative splicing. *Gene* **2009**, *446*, 28–34. [CrossRef] [PubMed]
346. Srikanth, K.; Kwan, A.; Lee, E.; Kim, S.; Lim, Y.; Chung, H. Associations of single nucleotide polymorphisms in the bovine FADS6 gene with fatty acid composition in Hanwoo (Korean Cattle). *Open J. Genet.* **2015**, *05*, 137–144. [CrossRef]
347. Chen, H.; Hao, G.; Wang, L.; Wang, H.; Gu, Z.; Liu, L.; Zhang, H.; Chen, W.; Chen, Y.Q. Identification of a critical determinant that enables efficient fatty acid synthesis in oleaginous fungi. *Sci. Rep.* **2015**, *5*, 11247. [CrossRef]
348. Stroud, C.K.; Nara, T.Y.; Roqueta-Rivera, M.; Radlowski, E.C.; Lawrence, P.; Zhang, Y.; Cho, B.H.; Segre, M.; Hess, R.A.; Brenna, J.T.; et al. Disruption of FADS2 gene in mice impairs male reproduction and causes dermal and intestinal ulceration. *J. Lipid Res.* **2009**, *50*, 1870–1880. [CrossRef]
349. Zhu, K.-C.; Song, L.; Guo, H.-Y.; Guo, L.; Zhang, N.; Liu, B.-S.; Jiang, S.-G.; Zhang, D.-C. Identification of fatty acid desaturase 6 in Golden Pompano Trachinotus Ovatus (Linnaeus 1758) and its regulation by the PPARαb transcription factor. *Int. J. Mol. Sci.* **2018**, *20*, 23. [CrossRef]
350. Marquardt, A.; Stöhr, H.; White, K.; Weber, B.H.F. cDNA cloning, genomic structure, and chromosomal localization of three members of the human fatty acid desaturase family. *Genomics* **2000**, *66*, 175–183. [CrossRef]
351. Rioux, V.; Pédrono, F.; Blanchard, H.; Duby, C.; Boulier-Monthéan, N.; Bernard, L.; Beauchamp, E.; Catheline, D.; Legrand, P. Trans-vaccenate is Δ13-desaturated by FADS3 in rodents. *J. Lipid Res.* **2013**, *54*, 3438–3452. [CrossRef]
352. Zhang, J.Y.; Qin, X.; Liang, A.; Kim, E.; Lawrence, P.; Park, W.J.; Kothapalli, K.S.D.; Brenna, J.T. Fads3 modulates docosahexaenoic acid in liver and brain. *Prostaglandins Leukot. Essent. Fat. Acids* **2017**, *123*, 25–32. [CrossRef] [PubMed]
353. Karsai, G.; Lone, M.; Kutalik, Z.; Brenna, J.T.; Li, H.; Pan, D.; von Eckardstein, A.; Hornemann, T. FADS3 is a Δ14Z sphingoid base desaturase that contributes to gender differences in the human plasma sphingolipidome. *J. Biol. Chem.* **2020**, *295*, 1889–1897. [CrossRef] [PubMed]
354. Castro, L.F.C.; Tocher, D.R.; Monroig, O. Long-chain polyunsaturated fatty acid biosynthesis in chordates: Insights into the evolution of Fads and Elovl gene repertoire. *Prog. Lipid Res.* **2016**, *62*, 25–40. [CrossRef]
355. Lee, J.; Lee, H.; Kang, S.; Park, W. Fatty acid desaturases, polyunsaturated fatty acid regulation, and biotechnological advances. *Nutrients* **2016**, *8*, 23. [CrossRef] [PubMed]
356. Guillou, H.; Zadravec, D.; Martin, P.G.P.; Jacobsson, A. The key roles of elongases and desaturases in mammalian fatty acid metabolism: Insights from transgenic mice. *Prog. Lipid Res.* **2010**, *49*, 186–199. [CrossRef] [PubMed]
357. Spector, A.A.; Kim, H.-Y. Discovery of essential fatty acids. *J. Lipid Res.* **2015**, *56*, 11–21. [CrossRef]
358. Kaur, N.; Chugh, V.; Gupta, A.K. Essential fatty acids as functional components of foods—A review. *J. Food Sci. Technol.* **2014**, *51*, 2289–2303. [CrossRef]
359. Burr, G.O.; Burr, M.M. On the nature and role of the fatty acids essential in nutrition. *J. Biol. Chem.* **1930**, *86*, 587–621. [CrossRef]
360. Cho, H.P.; Nakamura, M.; Clarke, S.D. Cloning, expression, and fatty acid regulation of the human Δ-5 desaturase. *J. Biol. Chem.* **1999**, *274*, 37335–37339. [CrossRef]
361. Cho, H.P.; Nakamura, M.T.; Clarke, S.D. Cloning, expression, and nutritional regulation of the mammalian Δ-6 desaturase. *J. Biol. Chem.* **1999**, *274*, 471–477. [CrossRef]
362. Scott, B.L.; Bazan, N.G. Membrane docosahexaenoate is supplied to the developing brain and retina by the liver. *Proc. Natl. Acad. Sci. USA* **1989**, *86*, 2903–2907. [CrossRef] [PubMed]
363. De Antueno, R.J.; Knickle, L.C.; Smith, H.; Elliot, M.L.; Allen, S.J.; Nwaka, S.; Winther, M.D. Activity of human Δ5 and Δ6 desaturases on multiple n-3 and n-6 polyunsaturated fatty acids. *FEBS Lett.* **2001**, *509*, 77–80. [CrossRef] [PubMed]
364. Vagner, M.; Santigosa, E. Characterization and modulation of gene expression and enzymatic activity of delta-6 desaturase in teleosts: A review. *Aquaculture* **2011**, *315*, 131–143. [CrossRef]
365. Arshad, Z.; Rezapour-Firouzi, S.; Ebrahimifar, M.; Mosavi-Jarrahi, A.; Mohammadian, M. Association of delta-6-desaturase expression with aggressiveness of cancer, diabetes mellitus, and multiple sclerosis: A narrative review. *Asian Pac. J. Cancer Prev.* **2019**, *20*, 1005–1018. [CrossRef]
366. Tosi, F.; Sartori, F.; Guarini, P.; Olivieri, O.; Martinelli, N. Delta-5 and delta-6 desaturases: Crucial enzymes in polyunsaturated fatty acid-related pathways with pleiotropic influences in health and disease. In *Oxidative Stress and Inflammation in Non-Communicable Diseases—Molecular Mechanisms and Perspectives in Therapeutics: Advances in Experimental Medicine and Biology*; Camps, J., Ed.; Springer: Cham, Switzerland, 2014; pp. 61–81.
367. Mead, J.F.; Slaton, W.H. Metabolism of essential fatty acids. *J. Biol. Chem.* **1956**, *219*, 705–709. [CrossRef]
368. Siguel, E.N.; Chee, K.M.; Gong, J.X.; Schaefer, E.J. Criteria for essential fatty acid deficiency in plasma as assessed by capillary column gas-liquid chromatography. *Clin. Chem.* **1987**, *33*, 1869–1873. [CrossRef]

369. Holman, R.T. The ratio of trienoic: Tetraenoic acids in tissue lipids as a measure of essential fatty acid requirement. *J. Nutr.* **1960**, *70*, 405–410. [CrossRef]
370. Park, H.G.; Engel, M.G.; Vogt-Lowell, K.; Lawrence, P.; Kothapalli, K.S.; Brenna, J.T. The role of fatty acid desaturase (FADS) genes in oleic acid metabolism: FADS1 Δ7 desaturates 11-20:1 to 7,11-20:2. *Prostaglandins Leukot. Essent. Fat. Acids* **2018**, *128*, 21–25. [CrossRef]
371. Park, H.G.; Park, W.J.; Kothapalli, K.S.D.; Brenna, J.T. The fatty acid desaturase 2 (FADS2) gene product catalyzes Δ4 desaturation to yield n-3 docosahexaenoic acid and n-6 docosapentaenoic acid in human cells. *FASEB J.* **2015**, *29*, 3911–3919. [CrossRef]
372. Li, Y.; Monroig, O.; Zhang, L.; Wang, S.; Zheng, X.; Dick, J.R.; You, C.; Tocher, D.R. Vertebrate fatty acyl desaturase with Δ4 activity. *Proc. Natl. Acad. Sci. USA* **2010**, *107*, 16840–16845. [CrossRef]
373. Martinez, M.; Ichaso, N.; Setien, F.; Durany, N.; Qiu, X.; Roesler, W. The Δ4-desaturation pathway for DHA biosynthesis is operative in the human species: Differences between normal controls and children with the Zellweger syndrome. *Lipids Health Dis.* **2010**, *9*, 98. [CrossRef] [PubMed]
374. Qiu, X.; Hong, H.; MacKenzie, S.L. Identification of a Δ4 fatty acid desaturase from *Thraustochytrium* sp. involved in the biosynthesis of docosahexanoic acid by Heterologous Expression in *Saccharomyces cerevisiae* and *Brassica juncea*. *J. Biol. Chem.* **2001**, *276*, 31561–31566. [CrossRef] [PubMed]
375. Schenck, P.A.; Rakoff, H.; Emken, E.A. δ8 desaturationin vivo of deuterated eicosatrienoic acid by mouse liver. *Lipids* **1996**, *31*, 593–600. [CrossRef]
376. Cook, H.; Byers, D.; Palmer, F.; Spence, M.; Rakoff, H.; Duval, S.; Emken, E. Alternate pathways in the desaturation and chain elongation of linolenic acid, 18:3(n-3), in cultured glioma cells. *J. Lipid Res.* **1991**, *32*, 1265–1273. [CrossRef]
377. Park, W.J.; Kothapalli, K.S.D.; Lawrence, P.; Tyburczy, C.; Brenna, J.T. An alternate pathway to long-chain polyunsaturates: The FADS2 gene product Δ8-desaturates 20:2n-6 and 20:3n-3. *J. Lipid Res.* **2009**, *50*, 1195–1202. [CrossRef] [PubMed]
378. Kabeya, N.; Fonseca, M.M.; Ferrier, D.E.K.; Navarro, J.C.; Bay, L.K.; Francis, D.S.; Tocher, D.R.; Castro, L.F.C.; Monroig, Ó. Genes for de novo biosynthesis of omega-3 polyunsaturated fatty acids are widespread in animals. *Sci. Adv.* **2018**, *4*, eaar6849. [CrossRef]
379. Xue, Z.; He, H.; Hollerbach, D.; Macool, D.J.; Yadav, N.S.; Zhang, H.; Szostek, B.; Zhu, Q. Identification and characterization of new Δ-17 fatty acid desaturases. *Appl. Microbiol. Biotechnol.* **2013**, *97*, 1973–1985. [CrossRef]
380. Jump, D.B.; Clarke, S.D.; Thelen, A.; Liimatta, M.; Ren, B.; Badin, M. Dietary polyunsaturated fatty acid regulation of gene transcription. *Prog. Lipid Res.* **1996**, *35*, 227–241. [CrossRef]
381. Blake, W.L.; Clarke, S.D. Suppression of rat hepatic fatty acid synthase and S14 gene transcription by dietary polyunsaturated fat. *J. Nutr.* **1990**, *120*, 1727–1729. [CrossRef]
382. Mullen, T.D.; Hannun, Y.A.; Obeid, L.M. Ceramide synthases at the centre of sphingolipid metabolism and biology. *Biochem. J.* **2012**, *441*, 789–802. [CrossRef]
383. Brush, R.S.; Tran, J.-T.A.; Henry, K.R.; McClellan, M.E.; Elliott, M.H.; Mandal, M.N.A. Retinal sphingolipids and their very-long-chain fatty acid–containing species. *Investig. Opthalmology Vis. Sci.* **2010**, *51*, 4422. [CrossRef]
384. Sandhoff, R. Very long chain sphingolipids: Tissue expression, function and synthesis. *FEBS Lett.* **2010**, *584*, 1907–1913. [CrossRef] [PubMed]
385. Nachtschatt, M.; Okada, S.; Speight, R. Integral membrane fatty acid desaturases: A review of biochemical, structural, and biotechnological advances. *Eur. J. Lipid Sci. Technol.* **2020**, *122*, 2000181. [CrossRef]
386. Zhu, G.; Koszelak-Rosenblum, M.; Connelly, S.M.; Dumont, M.E.; Malkowski, M.G. The crystal structure of an integral membrane fatty acid α-hydroxylase. *J. Biol. Chem.* **2015**, *290*, 29820–29833. [CrossRef] [PubMed]
387. Vacchina, P.; Tripodi, K.E.J.; Escalante, A.M.; Uttaro, A.D. Characterization of bifunctional sphingolipid Δ4-desaturases/C4-hydroxylases of trypanosomatids by liquid chromatography–electrospray tandem mass spectrometry. *Mol. Biochem. Parasitol.* **2012**, *184*, 29–38. [CrossRef] [PubMed]
388. Chen, M.; Markham, J.E.; Cahoon, E.B. Sphingolipid Δ8 unsaturation is important for glucosylceramide biosynthesis and low-temperature performance in Arabidopsis. *Plant J.* **2012**, *69*, 769–781. [PubMed]
389. López-Lara, I.M.; Soto, M.J. Fatty acid synthesis and regulation. In *Biogenesis of Fatty Acids, Lipids and Membranes*; Springer International Publishing: Cham, Switzerland, 2019; pp. 391–407.
390. Yamashita, A.; Hayashi, Y.; Nemoto-Sasaki, Y.; Ito, M.; Oka, S.; Tanikawa, T.; Waku, K.; Sugiura, T. Acyltransferases and transacylases that determine the fatty acid composition of glycerolipids and the metabolism of bioactive lipid mediators in mammalian cells and model organisms. *Prog. Lipid Res.* **2014**, *53*, 18–81. [PubMed]
391. Kent, C. Eukaryotic phospholipid biosynthesis. *Annu. Rev. Biochem.* **1995**, *64*, 315–343. [CrossRef] [PubMed]
392. Quehenberger, O.; Armando, A.M.; Brown, A.H.; Milne, S.B.; Myers, D.S.; Merrill, A.H.; Bandyopadhyay, S.; Jones, K.N.; Kelly, S.; Shaner, R.L.; et al. Lipidomics reveals a remarkable diversity of lipids in human plasma. *J. Lipid Res.* **2010**, *51*, 3299–3305. [CrossRef]
393. Shindou, H.; Hishikawa, D.; Harayama, T.; Yuki, K.; Shimizu, T. Recent progress on acyl CoA: Lysophospholipid acyltransferase research. *J. Lipid Res.* **2009**, *50*, S46–S51. [CrossRef]
394. Coleman, R.A.; Lewin, T.M.; Van Horn, C.G.; Gonzalez-Baró, M.R. Do long-chain acyl-CoA synthetases regulate fatty acid Entry into Synthetic Versus degradative pathways? *J. Nutr.* **2002**, *132*, 2123–2126. [PubMed]

395. MacDonald, J.I.S.; Sprecher, H. Phospholipid fatty acid remodeling in mammalian cells. *Biochim. Biophys. Acta Lipids Lipid Metab.* **1991**, *1084*, 105–121.
396. Lands, W.E.M. Stories about acyl chains. *Biochim. Biophys. Acta Mol. Cell Biol. Lipids* **2000**, *1483*, 1–14.
397. Lands, W.E.M.; Merkl, I. Metabolism of glycerolipids. *J. Biol. Chem.* **1963**, *238*, 898–904. [CrossRef]
398. Markham, J.E.; Molino, D.; Gissot, L.; Bellec, Y.; Hématy, K.; Marion, J.; Belcram, K.; Palauqui, J.-C.; Satiat-JeuneMaître, B.; Faure, J.-D. Sphingolipids containing very-long-chain fatty acids define a secretory pathway for specific polar plasma membrane protein targeting in arabidopsis. *Plant Cell* **2011**, *23*, 2362–2378. [CrossRef]
399. Samovski, D.; Jacome-Sosa, M.; Abumrad, N.A. Fatty Acid transport and signaling: Mechanisms and physiological implications. *Annu. Rev. Physiol.* **2023**, *85*, 317–337. [CrossRef]
400. Farkas, T. Adaptation of fatty acid compositions to temperature—A study on planktonic crustaceans. *Comp. Biochem. Physiol. Part B Comp. Biochem.* **1979**, *64*, 71–76. [CrossRef]
401. Dymond, M.K. Mammalian phospholipid homeostasis: Evidence that membrane curvature elastic stress drives homeoviscous adaptation in vivo. *J. R. Soc. Interface* **2016**, *13*, 20160228. [CrossRef]
402. Kulig, W.; Pasenkiewicz-Gierula, M.; Róg, T. Cis and trans unsaturated phosphatidylcholine bilayers: A molecular dynamics simulation study. *Chem. Phys. Lipids* **2016**, *195*, 12–20.
403. Gillan, F.T.; Johns, R.B.; Verheyen, T.V.; Volkman, J.K.; Bavor, H.J. Trans-monounsaturated acids in a marine bacterial isolate. *Appl. Environ. Microbiol.* **1981**, *41*, 849–856. [CrossRef]
404. Hanahan, D.J.; Brockerhoff, H.; Barron, E.J. The Site of Attack of Phospholipase (Lecithinase) A on Lecithin: A Re-evaluation. *J. Biol. Chem.* **1960**, *235*, 1917–1923. [CrossRef] [PubMed]
405. Tattrie, N.H. Positional distribution of saturated and unsaturated fatty acids on egg lecithin. *J. Lipid Res.* **1959**, *1*, 60–65. [CrossRef]
406. Pearson, R.H.; Pascher, I. The molecular structure of lecithin dihydrate. *Nature* **1979**, *281*, 499–501. [CrossRef] [PubMed]
407. Büldt, G.; Gally, H.U.; Seelig, J.; Zaccai, G. Neutron diffraction studies on phosphatidylcholine model membranes. *J. Mol. Biol.* **1979**, *134*, 673–691. [CrossRef] [PubMed]
408. Sanders, R.L.; Longmore, W.J. Phosphatidylglycerol in rat lung. II. Comparison of occurrence, composition, and metabolism in surfactant and residual lung fractions. *Biochemistry* **1975**, *14*, 835–840. [CrossRef]
409. Xie, D.; Seremwe, M.; Edwards, J.G.; Podolsky, R.; Bollag, W.B. Distinct Effects of Different Phosphatidylglycerol Species on Mouse Keratinocyte Proliferation. *PLoS ONE* **2014**, *9*, e107119.
410. O'Donnell, V.B. New appreciation for an old pathway: The Lands Cycle moves into new arenas in health and disease. *Biochem. Soc. Trans.* **2022**, *50*, 1–11.
411. Nakanishi, H.; Iida, Y.; Shimizu, T.; Taguchi, R. Separation and quantification of sn-1 and sn-2 fatty acid positional isomers in phosphatidylcholine by RPLC-ESIMS/MS. *J. Biochem.* **2010**, *147*, 245–256. [CrossRef]
412. Wood, R.; Harlow, R.D. Structural studies of neutral glycerides and phosphoglycerides of rat liver. *Arch. Biochem. Biophys.* **1969**, *131*, 495–501. [CrossRef]
413. Kuksis, A.; Breckenridge, W.C.; Marai, L.; Stachnyk, O. Molecular species of lecithins of rat heart, kidney, and plasma. *J. Lipid Res.* **1969**, *10*, 25–32. [CrossRef]
414. Yabuuchi, H.; O'Brien, J.S. Positional distribution of fatty acids in glycerophosphatides of bovine gray matter. *J. Lipid Res.* **1968**, *9*, 65–67. [CrossRef] [PubMed]
415. O'Brien, J.S.; Rouser, G. The fatty acid composition of brain sphingolipids: Sphingomyelin, ceramide, cerebroside, and cerebroside sulfate. *J. Lipid Res.* **1964**, *5*, 339–342. [CrossRef] [PubMed]
416. Kuksis, A.; Marai, L. Determination of the complete structure of natural lecithins. *Lipids* **1967**, *2*, 217–224. [CrossRef]
417. Holub, B.J.; Kuksis, A. Molecular species of phosphatidyl ethanolamine from egg yolk. *Lipids* **1969**, *4*, 466–472. [CrossRef]
418. Doğru Pekiner, B. Fatty acid composition of red blood cell membrane phosphatidylethanolamine and phosphatidylcholine in the rat, rabbit, human and dog: Sıçan, tavşan, insan ve köpek eritrosit membranı fosfatidiletanol. *Ankara Univ. Eczac. Fak. Derg.* **2002**, *31*, 169–182. [CrossRef]
419. Kim, H.-Y.; Huang, B.X.; Spector, A.A. Phosphatidylserine in the brain: Metabolism and function. *Prog. Lipid Res.* **2014**, *56*, 1–18. [PubMed]
420. Hamilton, J.; Greiner, R.; Salem, N.; Kim, H.-Y. n-3 Fatty acid deficiency decreases phosphatidylserine accumulation selectively in neuronal tissues. *Lipids* **2000**, *35*, 863–869. [CrossRef]
421. Clark, S.R.; Thomas, C.P.; Hammond, V.J.; Aldrovandi, M.; Wilkinson, G.W.; Hart, K.W.; Murphy, R.C.; Collins, P.W.; O'Donnell, V.B. Characterization of platelet aminophospholipid externalization reveals fatty acids as molecular determinants that regulate coagulation. *Proc. Natl. Acad. Sci. USA* **2013**, *110*, 5875–5880. [CrossRef]
422. Thompson, W.; MacDonald, G. Isolation and characterization of cytidine diphosphate diglyceride from beef liver. *J. Biol. Chem.* **1975**, *250*, 6779–6785. [CrossRef]
423. Holub, B.J.; Kuksis, A.; Thompson, W. Molecular species of mono-, di-, and triphosphoinositides of bovine brain. *J. Lipid Res.* **1970**, *11*, 558–564. [CrossRef]
424. Ulmann, L.; Mimouni, V.; Roux, S.; Porsolt, R.; Poisson, J.-P. Brain and hippocampus fatty acid composition in phospholipid classes of aged-relative cognitive deficit rats. *Prostaglandins Leukot. Essent. Fat. Acids* **2001**, *64*, 189–195.
425. Gijón, M.A.; Riekhof, W.R.; Zarini, S.; Murphy, R.C.; Voelker, D.R. Lysophospholipid Acyltransferases and Arachidonate Recycling in Human Neutrophils. *J. Biol. Chem.* **2008**, *283*, 30235–30245. [PubMed]

426. Jia, D.; Zhang, J.; Nie, J.; Andersen, J.-P.; Rendon, S.; Zheng, Y.; Liu, X.; Tian, Z.; Shi, Y. Cardiolipin remodeling by ALCAT1 links hypoxia to coronary artery disease by promoting mitochondrial dysfunction. *Mol. Ther.* **2021**, *29*, 3498–3511. [PubMed]
427. Pennington, E.R.; Funai, K.; Brown, D.A.; Shaikh, S.R. The role of cardiolipin concentration and acyl chain composition on mitochondrial inner membrane molecular organization and function. *Biochim. Biophys. Acta Mol. Cell Biol. Lipids* **2019**, *1864*, 1039–1052.
428. Li, J.; Romestaing, C.; Han, X.; Li, Y.; Hao, X.; Wu, Y.; Sun, C.; Liu, X.; Jefferson, L.S.; Xiong, J.; et al. Cardiolipin remodeling by ALCAT1 links oxidative stress and mitochondrial dysfunction to obesity. *Cell Metab.* **2010**, *12*, 154–165.
429. Stefanyk, L.E.; Coverdale, N.; Roy, B.D.; Peters, S.J.; LeBlanc, P.J. Skeletal muscle type comparison of subsarcolemmal mitochondrial membrane phospholipid fatty acid composition in rat. *J. Membr. Biol.* **2010**, *234*, 207–215.
430. Schlame, M.; Ren, M.; Xu, Y.; Greenberg, M.L.; Haller, I. Molecular symmetry in mitochondrial cardiolipins. *Chem. Phys. Lipids* **2005**, *138*, 38–49.
431. Berger, A.; Gershwin, M.E.; German, J.B. Effects of various dietary fats on cardiolipin acyl composition during ontogeny of mice. *Lipids* **1992**, *27*, 605–612.
432. Wolff, R.L.; Entressangles, B. Compositional changes of fatty acids in the 1(1″)- and 2(2″)-positions of cardiolipin from liver, heart, and kidney mitochondria of rats fed a low-fat diet. *Biochim. Biophys. Acta Lipids Lipid Metab.* **1991**, *1082*, 136–142.
433. Anderson, D.M.G.; Ablonczy, Z.; Koutalos, Y.; Hanneken, A.M.; Spraggins, J.M.; Calcutt, M.W.; Crouch, R.K.; Caprioli, R.M.; Schey, K.L. Bis(monoacylglycero)phosphate lipids in the retinal pigment epithelium implicate lysosomal/endosomal dysfunction in a model of Stargardt disease and human retinas. *Sci. Rep.* **2017**, *7*, 17352.
434. Besson, N.; Hullin-Matsuda, F.; Makino, A.; Murate, M.; Lagarde, M.; Pageaux, J.-F.; Kobayashi, T.; Delton-Vandenbroucke, I. Selective incorporation of docosahexaenoic acid into lysobisphosphatidic acid in cultured THP-1 macrophages. *Lipids* **2006**, *41*, 189–196. [PubMed]
435. Luquain, C.; Dolmazon, R.; Enderlin, J.M.; Laugier, C.; Lagarde, M.; Pageaux, J.F. Bis(monoacylglycerol) phosphate in rat uterine stromal cells: Structural characterization and specific esterification of docosahexaenoic acid. *Biochem. J.* **2000**, *351 Pt 3*, 795–804. [PubMed]
436. Huterer, S.; Wherrett, J. Metabolism of bis(monoacylglycero)phosphate in macrophages. *J. Lipid Res.* **1979**, *20*, 966–973. [PubMed]
437. Wherrett, J.R.; Huterer, S. Bis-(monoacylglyceryl)-phosphate of rat and human liver: Fatty acid composition and NMR spectroscopy. *Lipids* **1973**, *8*, 531–533.
438. Holbrook, P.G.; Pannell, L.K.; Murata, Y.; Daly, J.W. Bis(monoacylglycero) phosphate from PC12 cells, a phospholipid that can comigrate with phosphatidic acid: Molecular species analysis by fast atom bombardment mass spectrometry. *Biochim. Biophys. Acta Lipids Lipid Metab.* **1992**, *1125*, 330–334. [CrossRef]
439. Ryan, S.D.; Harris, C.S.; Carswell, C.L.; Baenziger, J.E.; Bennett, S.A.L. Heterogeneity in the sn-1 carbon chain of platelet-activating factor glycerophospholipids determines pro- or anti-apoptotic signaling in primary neurons. *J. Lipid Res.* **2008**, *49*, 2250–2258.
440. Nakagawa, Y.; Horrocks, L.A. Different metabolic rates for arachidonoyl molecular species of ethanolamine glycerophospholipids in rat brain. *J. Lipid Res.* **1988**, *27*, 629–636.
441. Pruett, S.T.; Bushnev, A.; Hagedorn, K.; Adiga, M.; Haynes, C.A.; Sullards, M.C.; Liotta, D.C.; Merrill, A.H. Thematic review series: Sphingolipids. Biodiversity of sphingoid bases ("sphingosines") and related amino alcohols. *J. Lipid Res.* **2008**, *49*, 1621–1639.
442. Merrill, A.H.; Nixon, D.W.; Williams, R.D. Activities of serine palmitoyltransferase (3-ketosphinganine synthase) in microsomes from different rat tissues. *J. Lipid Res.* **1985**, *26*, 617–622.
443. Ardail, D.; Popa, I.; Alcantara, K.; Pons, A.; Zanetta, J.; Louisot, P.; Thomas, L.; Portoukalian, J. Occurrence of ceramides and neutral glycolipids with unusual long-chain base composition in purified rat liver mitochondria. *FEBS Lett.* **2001**, *488*, 160–164. [CrossRef]
444. Eckhardt, M. Fatty Acid 2-Hydroxylase and 2-Hydroxylated Sphingolipids: Metabolism and Function in Health and Diseases. *Int. J. Mol. Sci.* **2023**, *24*, 4908. [CrossRef] [PubMed]
445. Marquês, J.T.; Marinho, H.S.; de Almeida, R.F.M. Sphingolipid hydroxylation in mammals, yeast and plants—An integrated view. *Prog. Lipid Res.* **2018**, *71*, 18–42. [CrossRef] [PubMed]
446. Sessa, L.; Nardiello, A.M.; Santoro, J.; Concilio, S.; Piotto, S. Hydroxylated fatty acids: The role of the sphingomyelin synthase and the origin of selectivity. *Membranes* **2021**, *11*, 787. [CrossRef] [PubMed]
447. Barenholz, Y. Sphingomyelin-lecithin balance in membranes: Composition, structure, and function relationships. In *Physiology of Membrane Fluidity*; Shinitzky, M., Ed.; CRC Press: Boca Raton, FL, USA, 1984; Volume 1, pp. 131–173.
448. Ramstedt, B.; Leppimäki, P.; Axberg, M.; Slotte, J.P. Analysis of natural and synthetic sphingomyelins using high-performance thin-layer chromatography. *Eur. J. Biochem.* **1999**, *266*, 997–1002. [CrossRef]
449. Karlsson, A.Å.; Michélsen, P.; Odham, G. Molecular species of sphingomyelin: Determination by high-performance liquid chromatography/mass spectrometry with electrospray and high-performance liquid chromatography/tandem mass spectrometry with atmospheric pressure chemical ionization. *J. Mass Spectrom.* **1998**, *33*, 1192–1198. [CrossRef]
450. Calhoun, W.I.; Shipley, G.G. Fatty acid composition and thermal behavior of natural sphingomyelins. *Biochim. Biophys. Acta Biomembr.* **1979**, *555*, 436–441. [CrossRef]
451. Poulos, A.; Sharp, P.; Johnson, D.; White, I.; Fellenberg, A. The occurrence of polyenoic fatty acids with greater than 22 carbon atoms in mammalian spermatozoa. *Biochem. J.* **1986**, *240*, 891–895. [CrossRef]

452. Robinson, B.S.; Johnson, D.W.; Poulos, A. Novel molecular species of sphingomyelin containing 2-hydroxylated polyenoic very-long-chain fatty acids in mammalian testes and spermatozoa. *J. Biol. Chem.* **1992**, *267*, 1746–1751. [CrossRef]
453. Skotland, T.; Sandvig, K. Need for more focus on lipid species in studies of biological and model membranes. *Prog. Lipid Res.* **2022**, *86*, 101160. [CrossRef]
454. Fujiwara, Y.; Hama, K.; Yokoyama, K. Mass spectrometry in combination with a chiral column and multichannel-MRM allows comprehensive analysis of glycosphingolipid molecular species from mouse brain. *Carbohydr. Res.* **2020**, *490*, 107959. [CrossRef]
455. Baumann, N.; Pham-Dinh, D. Biology of oligodendrocyte and myelin in the mammalian central nervous system. *Physiol. Rev.* **2001**, *81*, 871–927. [CrossRef] [PubMed]
456. Klenk, E.; Hendricks, U.W.; Gielen, W. β-D-Galaktosido-(1->3)-N-acetyl-D-galaktosamin, ein kristallisiertes Disaccharid aus menschlichen Gehirngangliosiden. *Biol. Chem.* **1962**, *330*, 140–144. [CrossRef] [PubMed]
457. Sastry, P.S. Lipids of nervous tissue: Composition and metabolism. *Prog. Lipid Res.* **1985**, *24*, 69–176. [PubMed]
458. Burr, G.O.; Burr, M.M. A new deficiency disease produced by the rigid exclusion of fat from the diet. *J. Biol. Chem.* **1929**, *82*, 345–367. [CrossRef]
459. von Euler, U.S. History and development of prostaglandins. *Gen. Pharmacol. Vasc. Syst.* **1983**, *14*, 3–6. [CrossRef]
460. Rodriguez-Estrada, M.T.; Tatay, A.C.; Cardenia, V.; Garcia-Llatas, G. Fats and sterols. In *Reference Module in Biomedical Sciences*; Elsevier: Amsterdam, The Netherlands, 2014.
461. Covino, R.; Ballweg, S.; Stordeur, C.; Michaelis, J.B.; Puth, K.; Wernig, F.; Bahrami, A.; Ernst, A.M.; Hummer, G.; Ernst, R. A Eukaryotic sensor for membrane lipid saturation. *Mol. Cell* **2016**, *63*, 49–59. [CrossRef]
462. Radhakrishnan, A.; Goldstein, J.L.; McDonald, J.G.; Brown, M.S. Switch-like control of SREBP-2 transport triggered by small changes in ER cholesterol: A delicate Balance. *Cell Metab.* **2008**, *8*, 512–521. [CrossRef]
463. Matveyenka, M.; Rizevsky, S.; Kurouski, D. Length and unsaturation of fatty acids of phosphatidic acid determines the aggregation rate of insulin and modifies the structure and toxicity of insulin aggregates. *ACS Chem. Neurosci.* **2022**, *13*, 2483–2489. [CrossRef]
464. Baccouch, R.; Shi, Y.; Vernay, E.; Mathelié-Guinlet, M.; Taib-Maamar, N.; Villette, S.; Feuillie, C.; Rascol, E.; Nuss, P.; Lecomte, S.; et al. The impact of lipid polyunsaturation on the physical and mechanical properties of lipid membranes. *Biochim. Biophys. Acta Biomembr.* **2023**, *1865*, 184084. [CrossRef]
465. Hashimoto, M.; Hossain, S.; Al Mamun, A.; Matsuzaki, K.; Arai, H. Docosahexaenoic acid: One molecule diverse functions. *Crit. Rev. Biotechnol.* **2017**, *37*, 579–597. [CrossRef]
466. Ibarguren, M.; López, D.J.; Escribá, P.V. The effect of natural and synthetic fatty acids on membrane structure, microdomain organization, cellular functions and human health. *Biochim. Biophys. Acta Biomembr.* **2014**, *1838*, 1518–1528. [CrossRef] [PubMed]
467. Rustan, A.C.; Drevon, C.A. Fatty acids: Structures and properties. In *eLS*; John Wiley & Sons: Hoboken, NJ, USA, 2005.
468. Small, D.M. Lateral chain packing in lipids and membranes. *J. Lipid Res.* **1984**, *25*, 1490–1500. [CrossRef] [PubMed]
469. Mondal, D.; Dutta, R.; Banerjee, P.; Mukherjee, D.; Maiti, T.K.; Sarkar, N. Modulation of membrane fluidity performed on model phospholipid membrane and live cell membrane: Revealing through spatiotemporal approaches of FLIM, FAIM, and TRFS. *Anal. Chem.* **2019**, *91*, 4337–4345. [CrossRef] [PubMed]
470. Goñi, F.M. The basic structure and dynamics of cell membranes: An update of the Singer–Nicolson model. *Biochim. Biophys. Acta Biomembr.* **2014**, *1838*, 1467–1476. [CrossRef] [PubMed]
471. De Santis, A.; Varela, Y.; Sot, J.; D'Errico, G.; Goñi, F.M.; Alonso, A. Omega-3 polyunsaturated fatty acids do not fluidify bilayers in the liquid-crystalline state. *Sci. Rep.* **2018**, *8*, 16240. [CrossRef]
472. Lor, C.; Hirst, L. Effects of low concentrations of docosahexaenoic acid on the structure and phase behavior of model lipid membranes. *Membranes* **2015**, *5*, 857–874. [CrossRef]
473. Killian, J.A. Hydrophobic mismatch between proteins and lipids in membranes. *Biochim. Biophys. Acta Rev. Biomembr.* **1998**, *1376*, 401–416. [CrossRef]
474. Salmon, A.; Dodd, S.W.; Williams, G.D.; Beach, J.M.; Brown, M.F. Configurational statistics of acyl chains in polyunsaturated lipid bilayers from deuterium NMR. *J. Am. Chem. Soc.* **1987**, *109*, 2600–2609. [CrossRef]
475. Yang, X.; Sheng, W.; Sun, G.Y.; Lee, J.C.-M. Effects of fatty acid unsaturation numbers on membrane fluidity and α-secretase-dependent amyloid precursor protein processing. *Neurochem. Int.* **2011**, *58*, 321–329. [CrossRef]
476. Straume, M.; Litman, B.J. Equilibrium and dynamic structure of large, unilamellar, unsaturated acyl chain phosphatidylcholine vesicles. Higher order analysis of 1,6-diphenyl-1,3,5-hexatriene and 1-[4-(trimethylammonio)phenyl]-6-phenyl-1,3,5-hexatriene anisotropy decay. *Biochemistry* **1987**, *26*, 5113–5120. [CrossRef]
477. Roach, C.; Feller, S.E.; Ward, J.A.; Shaikh, S.R.; Zerouga, M.; Stillwell, W. Comparison of cis and trans fatty acid containing phosphatidylcholines on membrane properties. *Biochemistry* **2004**, *43*, 6344–6351. [CrossRef]
478. Christie, W.W.; Holman, R.T. Synthesis and characterization of the complete series of methylene-interrupted cis,cis-octadecadienoic acids. *Chem. Phys. Lipids* **1967**, *1*, 407–423. [CrossRef]
479. Tyler, A.I.I.; Greenfield, J.L.; Seddon, J.M.; Brooks, N.J.; Purushothaman, S. Coupling phase behavior of fatty acid containing membranes to membrane bio-mechanics. *Front. Cell Dev. Biol.* **2019**, *7*, 187. [CrossRef]
480. Brenner, R. Effect of unsaturated acids on membrane structure and enzyme kinetics. *Prog. Lipid Res.* **1984**, *23*, 69–96. [CrossRef]
481. Schumann, J.; Leichtle, A.; Thiery, J.; Fuhrmann, H. Fatty acid and peptide profiles in plasma membrane and membrane rafts of PUFA supplemented RAW264.7 macrophages. *PLoS ONE* **2011**, *6*, e24066. [CrossRef]
482. Stillwell, W. The role of polyunsaturated lipids in membrane raft function. *Scand. J. Food Nutr.* **2006**, *50*, 107–113. [CrossRef]

483. Bennett, W.F.D.; Shea, J.-E.; Tieleman, D.P. Phospholipid chain interactions with cholesterol drive domain formation in lipid membranes. *Biophys. J.* **2018**, *114*, 2595–2605. [CrossRef]
484. Martinez-Seara, H.; Róg, T.; Karttunen, M.; Vattulainen, I.; Reigada, R. Cholesterol induces specific spatial and orientational order in cholesterol/phospholipid membranes. *PLoS ONE* **2010**, *5*, e11162. [CrossRef]
485. Ye, S.; Tan, L.; Ma, J.; Shi, Q.; Li, J. Polyunsaturated docosahexaenoic acid suppresses oxidative stress induced endothelial cell calcium influx by altering lipid composition in membrane caveolar rafts. *Prostaglandins Leukot. Essent. Fat. Acids* **2010**, *83*, 37–43. [CrossRef]
486. Wassall, S.R.; Stillwell, W. Polyunsaturated fatty acid–cholesterol interactions: Domain formation in membranes. *Biochim. Biophys. Acta Biomembr.* **2009**, *1788*, 24–32. [CrossRef]
487. Chen, W.; Jump, D.B.; Esselman, W.J.; Busik, J.V. Inhibition of cytokine Signaling in human retinal endothelial cells through modification of caveolae/lipid rafts by docosahexaenoic acid. *Investig. Opthalmology Vis. Sci.* **2007**, *48*, 18. [CrossRef] [PubMed]
488. Pitman, M.C.; Suits, F.; MacKerell, A.D.; Feller, S.E. Molecular-level organization of saturated and polyunsaturated fatty acids in a phosphatidylcholine bilayer containing cholesterol. *Biochemistry* **2004**, *43*, 15318–15328. [CrossRef] [PubMed]
489. Hashimoto, M.; Hossain, M.S.; Shimada, T.; Yamasaki, H.; Fujii, Y.; Shido, O. Effects of docosahexaenoic acid on annular lipid fluidity of the rat bile canalicular plasma membrane. *J. Lipid Res.* **2001**, *42*, 1160–1168. [CrossRef] [PubMed]
490. Stulnig, T.M.; Huber, J.; Leitinger, N.; Imre, E.-M.; Angelisová, P.; Nowotny, P.; Waldhäusl, W. Polyunsaturated eicosapentaenoic acid displaces proteins from membrane rafts by altering raft lipid composition. *J. Biol. Chem.* **2001**, *276*, 37335–37340. [CrossRef]
491. Hashimoto, M.; Hossain, M.S.; Yamasaki, H.; Yazawa, K.; Masumura, S. Effects of eicosapentaenoic acid and docosahexaenoic acid on plasma membrane fluidity of aortic endothelial cells. *Lipids* **1999**, *34*, 1297–1304. [CrossRef]
492. Van Blitterswijk, W.J.; Van der Meer, B.W.; Hilkmann, H. Quantitative contributions of cholesterol and the individual classes of phospholipids and their degree of fatty acyl (un)saturation to membrane fluidity measured by fluorescence polarization. *Biochemistry* **1987**, *26*, 1746–1756. [CrossRef]
493. Cheng, V.; Rallabandi, R.; Gorusupudi, A.; Lucas, S.; Rognon, G.; Bernstein, P.S.; Rainier, J.D.; Conboy, J.C. Influence of very-long-chain polyunsaturated fatty acids on membrane structure and dynamics. *Biophys. J.* **2022**, *121*, 2730–2741. [CrossRef]
494. Armstrong, V.T.; Brzustowicz, M.R.; Wassall, S.R.; Jenski, L.J.; Stillwell, W. Rapid flip-flop in polyunsaturated (docosahexaenoate) phospholipid membranes. *Arch. Biochem. Biophys.* **2003**, *414*, 74–82. [CrossRef]
495. Jacobs, M.L.; Faizi, H.A.; Peruzzi, J.A.; Vlahovska, P.M.; Kamat, N.P. EPA and DHA differentially modulate membrane elasticity in the presence of cholesterol. *Biophys. J.* **2021**, *120*, 2317–2329. [CrossRef]
496. Rawicz, W.; Olbrich, K.C.; McIntosh, T.; Needham, D.; Evans, E. Effect of chain length and unsaturation on elasticity of lipid bilayers. *Biophys. J.* **2000**, *79*, 328–339. [CrossRef]
497. Beck, R.; Bertolino, S.; Abbot, S.E.; Aaronson, P.I.; Smirnov, S.V. Modulation of arachidonic acid release and membrane fluidity by Albumin in Vascular Smooth muscle and endothelial cells. *Circ. Res.* **1998**, *83*, 923–931. [CrossRef] [PubMed]
498. Treen, M.; Uauy, R.D.; Jameson, D.M.; Thomas, V.L.; Hoffman, D.R. Effect of docosahexaenoic acid on membrane fluidity and function in intact cultured Y-79 retinoblastoma cells. *Arch. Biochem. Biophys.* **1992**, *294*, 564–570. [CrossRef] [PubMed]
499. Hendriks, T.; Klompmakers, A.A.; Daemen, F.J.M.; Bonting, S.L. Biochemical aspects of the visual process XXXII. Movement of sodium ions through bilayers composed of retinal and rod outer segment lipids. *Biochim. Biophys. Acta Biomembr.* **1976**, *433*, 271–281. [CrossRef]
500. Demel, R.A.; Geurts van Kessel, W.S.M.; van Deenen, L.L.M. The properties of polyunsaturated lecithins in monolayers and liposomes and the interactions of these lecithins with cholesterol. *Biochim. Biophys. Acta Biomembr.* **1972**, *266*, 26–40. [CrossRef]
501. Sherratt, S.C.R.; Juliano, R.A.; Copland, C.; Bhatt, D.L.; Libby, P.; Mason, R.P. EPA and DHA containing phospholipids have contrasting effects on membrane structure. *J. Lipid Res.* **2021**, *62*, 100106. [CrossRef]
502. Ehringer, W.; Belcher, D.; Wassall, S.R.; Stillwell, W. A comparison of the effects of linolenic (18:3Ω3) and docosahexaenoic (22:6Ω3) acids on phospholipid bilayers. *Chem. Phys. Lipids* **1990**, *54*, 79–88. [CrossRef]
503. Mitchell, D.C.; Litman, B.J. Effect of cholesterol on molecular order and dynamics in highly polyunsaturated phospholipid bilayers. *Biophys. J.* **1998**, *75*, 896–908. [CrossRef]
504. Li, X.; Zhou, S.; Lin, X. Molecular view on the impact of DHA molecules on the physical properties of a model cell membrane. *J. Chem. Inf. Model.* **2022**, *62*, 2421–2431. [CrossRef]
505. Pinot, M.; Vanni, S.; Pagnotta, S.; Lacas-Gervais, S.; Payet, L.-A.; Ferreira, T.; Gautier, R.; Goud, B.; Antonny, B.; Barelli, H. Polyunsaturated phospholipids facilitate membrane deformation and fission by endocytic proteins. *Science* **2014**, *345*, 693–697. [CrossRef]
506. Stillwell, W.; Wassall, S.R. Docosahexaenoic acid: Membrane properties of a unique fatty acid. *Chem. Phys. Lipids* **2003**, *126*, 1–27. [CrossRef]
507. Williams, E.E.; Jenski, L.J.; Stillwell, W. Docosahexaenoic acid (DHA) alters the structure and composition of membranous vesicles exfoliated from the surface of a murine leukemia cell line. *Biochim. Biophys. Acta Biomembr.* **1998**, *1371*, 351–362. [CrossRef] [PubMed]
508. Koenig, B.W.; Strey, H.H.; Gawrisch, K. Membrane lateral compressibility determined by NMR and x-ray diffraction: Effect of acyl chain polyunsaturation. *Biophys. J.* **1997**, *73*, 1954–1966. [CrossRef] [PubMed]
509. Mitchell, D.C.; Litman, B.J. Modulation of receptor signaling by phospholipid acyl chain composition. In *Fatty Acids: Nutrition and Health*; Mostofsky, D.I., Yehuda, S., Salem, N., Eds.; Humana Press: Totowa, NJ, USA, 2001; pp. 23–40. ISBN 978-1-59259-119-0.

510. Joardar, A.; Pattnaik, G.P.; Chakraborty, H. Mechanism of membrane fusion: Interplay of lipid and peptide. *J. Membr. Biol.* **2022**, *255*, 211–224. [CrossRef] [PubMed]
511. Ellens, H.; Siegel, D.P.; Alford, D.; Yeagle, P.L.; Boni, L.; Lis, L.J.; Quinn, P.J.; Bentz, J. Membrane fusion and inverted phases. *Biochemistry* **1989**, *28*, 3692–3703. [CrossRef] [PubMed]
512. Ahkong, Q.F.; Fisher, D.; Tampion, W.; Lucy, J.A. The fusion of erythrocytes by fatty acids, esters, retinol and α-tocopherol. *Biochem. J.* **1973**, *136*, 147–155. [CrossRef]
513. Meers, P.; Hong, K.; Papahadjopoulos, D. Free fatty acid enhancement of cation-induced fusion of liposomes: Synergism with synexin and other promoters of vesicle aggregation. *Biochemistry* **1988**, *27*, 6784–6794. [CrossRef]
514. Creutz, C.E. cis-unsaturated fatty acids induce the fusion of chromaffin granules aggregated by synexin. *J. Cell Biol.* **1981**, *91*, 247–256. [CrossRef]
515. Funari, S.S.; Barceló, F.; Escribá, P.V. Effects of oleic acid and its congeners, elaidic and stearic acids, on the structural properties of phosphatidylethanolamine membranes. *J. Lipid Res.* **2003**, *44*, 567–575. [CrossRef]
516. Prades, J.; Funari, S.S.; Escribá, P.V.; Barceló, F. Effects of unsaturated fatty acids and triacylglycerols on phosphatidylethanolamine membrane structure. *J. Lipid Res.* **2003**, *44*, 1720–1727. [CrossRef]
517. Jenske, R.; Lindström, F.; Gröbner, G.; Vetter, W. Impact of free hydroxylated and methyl-branched fatty acids on the organization of lipid membranes. *Chem. Phys. Lipids* **2008**, *154*, 26–32. [CrossRef]
518. Niebylski, C.D.; Salem, N. A calorimetric investigation of a series of mixed-chain polyunsaturated phosphatidylcholines: Effect of sn-2 chain length and degree of unsaturation. *Biophys. J.* **1994**, *67*, 2387–2393. [CrossRef] [PubMed]
519. Monteiro, J.; Leslie, M.; Moghadasian, M.H.; Arendt, B.M.; Allard, J.P.; Ma, D.W.L. The role of n-6 and n-3 polyunsaturated fatty acids in the manifestation of the metabolic syndrome in cardiovascular disease and non-alcoholic fatty liver disease. *Food Funct.* **2014**, *5*, 426. [CrossRef] [PubMed]
520. Hashimoto, M.; Hossain, S. Neuroprotective and ameliorative actions of polyunsaturated fatty acids against neuronal diseases: Beneficial effect of docosahexaenoic acid on cognitive decline in Alzheimer's disease. *J. Pharmacol. Sci.* **2011**, *116*, 150–162. [CrossRef] [PubMed]
521. Eltweri, A.M.; Thomas, A.L.; Metcalfe, M.; Calder, P.C.; Dennison, A.R.; Bowrey, D.J. Potential applications of fish oils rich in omega-3 polyunsaturated fatty acids in the management of gastrointestinal cancer. *Clin. Nutr.* **2017**, *36*, 65–78. [CrossRef]
522. Dyerberg, J. Linolenate-derived polyunsaturated fatty acids and prevention of atherosclerosis. *Nutr. Rev.* **1986**, *44*, 125–134. [CrossRef]
523. Sassa, T.; Suto, S.; Okayasu, Y.; Kihara, A. A shift in sphingolipid composition from C24 to C16 increases susceptibility to apoptosis in HeLa cells. *Biochim. Biophys. Acta Mol. Cell Biol. Lipids* **2012**, *1821*, 1031–1037. [CrossRef]
524. Cao, H.; Gerhold, K.; Mayers, J.R.; Wiest, M.M.; Watkins, S.M.; Hotamisligil, G.S. Identification of a lipokine, a lipid hormone linking adipose tissue to systemic metabolism. *Cell* **2008**, *134*, 933–944. [CrossRef]
525. Carrillo, C.; Cavia, M.D.M.; Alonso-Torre, S.R. Antitumor effect of oleic acid; mechanisms of action. A review. *Nutr. Hosp.* **2012**, *27*, 1860–1865.
526. Park, W.-J.; Park, J.-W.; Merrill, A.H.; Storch, J.; Pewzner-Jung, Y.; Futerman, A.H. Hepatic fatty acid uptake is regulated by the sphingolipid acyl chain length. *Biochim. Biophys. Acta Mol. Cell Biol. Lipids* **2014**, *1841*, 1754–1766. [CrossRef]
527. Huang, J.M.; Xian, H.; Bacaner, M. Long-chain fatty acids activate calcium channels in ventricular myocytes. *Proc. Natl. Acad. Sci. USA* **1992**, *89*, 6452–6456. [CrossRef]
528. Chang, J.P.-C.; Su, K.-P.; Mondelli, V.; Pariante, C.M. Omega-3 Polyunsaturated Fatty Acids in Youths with Attention Deficit Hyperactivity Disorder: A Systematic Review and Meta-Analysis of Clinical Trials and Biological Studies. *Neuropsychopharmacology* **2018**, *43*, 534–545. [CrossRef] [PubMed]
529. Watkins, B. Bioactive fatty acids: Role in bone biology and bone cell function. *Prog. Lipid Res.* **2001**, *40*, 125–148. [CrossRef] [PubMed]
530. Holman, R.; Johnson, S.; Hatch, T. A case of human linolenic acid deficiency involving neurological abnormalities. *Am. J. Clin. Nutr.* **1982**, *35*, 617–623. [CrossRef] [PubMed]
531. Brown, W.R.; Hansen, A.E.; Burr, G.O.; McQuarrie, I. Effects of prolonged use of extremely low-fat diet on an adult human subject. *J. Nutr.* **1938**, *16*, 511–524. [CrossRef]
532. Burr, G.O.; Burr, M.M.; Miller, E.S. On the fatty acids essential in nutrition. III. *J. Biol. Chem.* **1932**, *97*, 1–9. [CrossRef]
533. Sambra, V.; Echeverria, F.; Valenzuela, A.; Chouinard-Watkins, R.; Valenzuela, R. Docosahexaenoic and arachidonic acids as neuroprotective nutrients throughout the life cycle. *Nutrients* **2021**, *13*, 986. [CrossRef]
534. Hadley, K.; Ryan, A.; Forsyth, S.; Gautier, S.; Salem, N. The essentiality of arachidonic acid in infant development. *Nutrients* **2016**, *8*, 216. [CrossRef]
535. Martinez, M. Tissue levels of polyunsaturated fatty acids during early human development. *J. Pediatr.* **1992**, *120*, S129–S138. [CrossRef]
536. Zheng, Y.; Yin, H.; Boeglin, W.E.; Elias, P.M.; Crumrine, D.; Beier, D.R.; Brash, A.R. Lipoxygenases mediate the effect of essential fatty acid in skin barrier formation. *J. Biol. Chem.* **2011**, *286*, 24046–24056. [CrossRef]
537. Casillas-Vargas, G.; Ocasio-Malavé, C.; Medina, S.; Morales-Guzmán, C.; Del Valle, R.G.; Carballeira, N.M.; Sanabria-Ríos, D.J. Antibacterial fatty acids: An update of possible mechanisms of action and implications in the development of the next-generation of antibacterial agents. *Prog. Lipid Res.* **2021**, *82*, 101093. [CrossRef]

538. Das, U.N. Essential fatty acids: Biochemistry, physiology and pathology. *Biotechnol. J.* **2006**, *1*, 420–439. [CrossRef] [PubMed]
539. Giamarellos-Bourboulis, E.J.; Mouktaroudi, M.; Adamis, T.; Koussoulas, V.; Baziaka, F.; Perrea, D.; Karayannacos, P.E.; Giamarellou, H. n-6 polyunsaturated fatty acids enhance the activities of ceftazidime and amikacin in experimental sepsis caused by multidrug-resistant Pseudomonas aeruginosa. *Antimicrob. Agents Chemother.* **2004**, *48*, 4713–4717. [CrossRef] [PubMed]
540. Sun, C.Q.; O'Connor, C.J.; Roberton, A.M. Antibacterial actions of fatty acids and monoglycerides against Helicobacter pylori. *FEMS Immunol. Med. Microbiol.* **2003**, *36*, 9–17. [CrossRef] [PubMed]
541. Kapoor, R.; Huang, Y.-S. Gamma linolenic acid: An antiinflammatory omega-6 fatty acid. *Curr. Pharm. Biotechnol.* **2006**, *7*, 531–534. [CrossRef] [PubMed]
542. Kawamura, A.; Ooyama, K.; Kojima, K.; Kachi, H.; Abe, T.; Amano, K.; Aoyama, T. Dietary supplementation of Gamma-linolenic acid improves skin parameters in subjects with dry skin and mild atopic dermatitis. *J. Oleo Sci.* **2011**, *60*, 597–607. [CrossRef]
543. Kendall, A.C.; Kiezel-Tsugunova, M.; Brownbridge, L.C.; Harwood, J.L.; Nicolaou, A. Lipid functions in skin: Differential effects of n-3 polyunsaturated fatty acids on cutaneous ceramides, in a human skin organ culture model. *Biochim. Biophys. Acta Biomembr.* **2017**, *1859*, 1679–1689. [CrossRef]
544. Thies, F.; Nebe-von-Caron, G.; Powell, J.R.; Yaqoob, P.; Newsholme, E.A.; Calder, P.C. Dietary supplementation with γ-linolenic acid or fish oil decreases T lymphocyte proliferation in healthy older humans. *J. Nutr.* **2001**, *131*, 1918–1927. [CrossRef]
545. Yaqoob; Pala; Cortina-Borja; Newsholme; Calder. Encapsulated fish oil enriched in α-tocopherol alters plasma phospholipid and mononuclear cell fatty acid compositions but not mononuclear cell functions. *Eur. J. Clin. Investig.* **2000**, *30*, 260–274. [CrossRef]
546. Innes, J.K.; Calder, P.C. Omega-6 fatty acids and inflammation. *Prostaglandins Leukot. Essent. Fat. Acids* **2018**, *132*, 41–48. [CrossRef]
547. Hume, E.M.; Nunn, L.C.A.; Smedley-Maclean, I.; Smith, H.H. Studies of the essential unsaturated fatty acids in their relation to the fat-deficiency disease of rats. *Biochem. J.* **1938**, *32*, 2162–2177. [CrossRef]
548. Rieckehoff, I.G.; Holman, R.T.; Burr, G.O. Polyethenoid Fatty Acid Metabolism. Effect of Dietary Fat on Polyethenoid Fatty Acids of Rat Tissues. *Nutr. Rev.* **1980**, *38*, 247–250. [CrossRef] [PubMed]
549. Klenk, E.; Bongard, W. Die Konstitution der ungesättigten C20- und C22-Fettsäuren der Glycerinphosphatide des Gehirns [Constitution of the unsaturated C20 and C22 fatty acids of the glycophosphatides of the brain]. *Biol. Chem.* **1952**, *291*, 104–118. [CrossRef]
550. Biran, L.A.; Bartley, W. Distribution of fatty acids in lipids of rat brain, brain mitochondria and microsomes. *Biochem. J.* **1961**, *79*, 159–176. [CrossRef]
551. Benolken, R.M.; Anderson, R.E.; Wheeler, T.G. Membrane fatty acids associated with the electrical response in visual excitation. *Science* **1973**, *182*, 1253–1254. [CrossRef] [PubMed]
552. Bieri, J.G.; Prival, E.L. Lipid composition of testes from various species. *Comp. Biochem. Physiol.* **1965**, *15*, 275–282. [CrossRef]
553. Schnebelen-Berthier, C.; Acar, N.; Simon, E.; Thabuis, C.; Bourdillon, A.; Mathiaud, A.; Dauchet, L.; Delcourt, C.; Benlian, P.; Crochet, M.; et al. The ALGOVUE clinical trial: Effects of the daily consumption of eggs enriched with lutein and docosahexaenoic acid on plasma composition and macular pigment optical density. *Nutrients* **2021**, *13*, 3347. [CrossRef]
554. Hishikawa, D.; Valentine, W.J.; Iizuka-Hishikawa, Y.; Shindou, H.; Shimizu, T. Metabolism and functions of docosahexaenoic acid-containing membrane glycerophospholipids. *FEBS Lett.* **2017**, *591*, 2730–2744. [CrossRef]
555. Antonny, B.; Vanni, S.; Shindou, H.; Ferreira, T. From zero to six double bonds: Phospholipid unsaturation and organelle function. *Trends Cell Biol.* **2015**, *25*, 427–436. [CrossRef]
556. Soubias, O.; Gawrisch, K. The role of the lipid matrix for structure and function of the GPCR rhodopsin. *Biochim. Biophys. Acta Biomembr.* **2012**, *1818*, 234–240. [CrossRef]
557. Eckert, G.P.; Chang, S.; Eckmann, J.; Copanaki, E.; Hagl, S.; Hener, U.; Müller, W.E.; Kögel, D. Liposome-incorporated DHA increases neuronal survival by enhancing non-amyloidogenic APP processing. *Biochim. Biophys. Acta Biomembr.* **2011**, *1808*, 236–243. [CrossRef]
558. Roqueta-Rivera, M.; Abbott, T.L.; Sivaguru, M.; Hess, R.A.; Nakamura, M.T. Deficiency in the omega-3 fatty acid pathway results in failure of acrosome biogenesis in mice. *Biol. Reprod.* **2011**, *85*, 721–732. [CrossRef] [PubMed]
559. Sato, H.; Taketomi, Y.; Isogai, Y.; Miki, Y.; Yamamoto, K.; Masuda, S.; Hosono, T.; Arata, S.; Ishikawa, Y.; Ishii, T.; et al. Group III secreted phospholipase A2 regulates epididymal sperm maturation and fertility in mice. *J. Clin. Investig.* **2010**, *120*, 1400–1414. [CrossRef] [PubMed]
560. Bennett, M.P.; Mitchell, D.C. Regulation of membrane proteins by dietary lipids: Effects of cholesterol and docosahexaenoic acid acyl chain-containing phospholipids on rhodopsin stability and function. *Biophys. J.* **2008**, *95*, 1206–1216. [CrossRef]
561. Fedorova, I.; Hussein, N.; Di Martino, C.; Moriguchi, T.; Hoshiba, J.; Majchrzak, S.; Salem, N. An n-3 fatty acid deficient diet affects mouse spatial learning in the Barnes circular maze. *Prostaglandins Leukot. Essent. Fat. Acids* **2007**, *77*, 269–277. [CrossRef] [PubMed]
562. Kawakita, E.; Hashimoto, M.; Shido, O. Docosahexaenoic acid promotes neurogenesis in vitro and in vivo. *Neuroscience* **2006**, *139*, 991–997. [CrossRef] [PubMed]
563. Rejraji, H.; Sion, B.; Prensier, G.; Carreras, M.; Motta, C.; Frenoux, J.-M.; Vericel, E.; Grizard, G.; Vernet, P.; Drevet, J.R. Lipid remodeling of murine epididymosomes and spermatozoa during epididymal maturation1. *Biol. Reprod.* **2006**, *74*, 1104–1113. [CrossRef]

564. Kitajka, K.; Puskás, L.G.; Zvara, Á.; Hackler, L.; Barceló-Coblijn, G.; Yeo, Y.K.; Farkas, T. The role of n-3 polyunsaturated fatty acids in brain: Modulation of rat brain gene expression by dietary n-3 fatty acids. *Proc. Natl. Acad. Sci. USA* **2002**, *99*, 2619–2624. [CrossRef]
565. Litman, B.J.; Mitchell, D.C. A role for phospholipid polyunsaturation in modulating membrane protein function. *Lipids* **1996**, *31*, S193–S197. [CrossRef]
566. Miljanich, G.P.; Sklar, L.A.; White, D.L.; Dratz, E.A. Disaturated and dipolyunsaturated phospholipids in the bovine retinal rod outer segment disk membrane. *Biochim. Biophys. Acta Biomembr.* **1979**, *552*, 294–306. [CrossRef]
567. Ding, L.; Zhang, T.; Che, H.; Zhang, L.; Xue, C.; Chang, Y.; Wang, Y. DHA-enriched phosphatidylcholine and DHA-enriched phosphatidylserine improve age-related lipid metabolic disorder through different metabolism in the senescence-accelerated mouse. *Eur. J. Lipid Sci. Technol.* **2018**, *120*, 1700490. [CrossRef]
568. Shefer-Weinberg, D.; Sasson, S.; Schwartz, B.; Argov-Argaman, N.; Tirosh, O. Deleterious effect of n-3 polyunsaturated fatty acids in non-alcoholic steatohepatitis in the fat-1 mouse model. *Clin. Nutr. Exp.* **2017**, *12*, 37–49. [CrossRef]
569. Epand, R.M.; D'Souza, K.; Berno, B.; Schlame, M. Membrane curvature modulation of protein activity determined by NMR. *Biochim. Biophys. Acta Biomembr.* **2015**, *1848*, 220–228. [CrossRef] [PubMed]
570. Andersen, O.S.; Koeppe, R.E. Bilayer thickness and membrane protein function: An energetic perspective. *Annu. Rev. Biophys. Biomol. Struct.* **2007**, *36*, 107–130. [CrossRef] [PubMed]
571. Gruner, S.M. Intrinsic curvature hypothesis for biomembrane lipid composition: A role for nonbilayer lipids. *Proc. Natl. Acad. Sci. USA* **1985**, *82*, 3665–3669. [CrossRef] [PubMed]
572. Johannsson, A.; Smith, G.A.; Metcalfe, J.C. The effect of bilayer thickness on the activity of (Na^+ + K^+)-ATPase. *Biochim. Biophys. Acta Biomembr.* **1981**, *641*, 416–421. [CrossRef]
573. Jalili, M.; Hekmatdoost, A. Dietary ω-3 fatty acids and their influence on inflammation via Toll-like receptor pathways. *Nutrition* **2021**, *85*, 111070. [CrossRef]
574. Cordero-Morales, J.F.; Vásquez, V. How lipids contribute to ion channel function, a fat perspective on direct and indirect interactions. *Curr. Opin. Struct. Biol.* **2018**, *51*, 92–98. [CrossRef]
575. Elsherbiny, M.; Chen, H.; Emara, M.; Godbout, R. ω-3 and ω-6 fatty acids modulate conventional and atypical protein kinase C activities in a brain fatty acid binding protein dependent manner in glioblastoma multiforme. *Nutrients* **2018**, *10*, 454. [CrossRef]
576. Senapati, S.; Gragg, M.; Samuels, I.S.; Parmar, V.M.; Maeda, A.; Park, P.S.-H. Effect of dietary docosahexaenoic acid on rhodopsin content and packing in photoreceptor cell membranes. *Biochim. Biophys. Acta Biomembr.* **2018**, *1860*, 1403–1413. [CrossRef]
577. Esposito, G.; Scuderi, C.; Valenza, M.; Togna, G.I.; Latina, V.; De Filippis, D.; Cipriano, M.; Carratù, M.R.; Iuvone, T.; Steardo, L. Cannabidiol reduces Aβ-induced neuroinflammation and promotes hippocampal neurogenesis through PPARγ involvement. *PLoS ONE* **2011**, *6*, e28668. [CrossRef]
578. Vinayavekhin, N.; Saghatelian, A. Discovery of a protein–metabolite interaction between unsaturated fatty acids and the nuclear receptor Nur77 using a metabolomics approach. *J. Am. Chem. Soc.* **2011**, *133*, 17168–17171. [CrossRef] [PubMed]
579. Akhtar Khan, N. Polyunsaturated fatty acids in the modulation of T-cell signalling. *Prostaglandins Leukot. Essent. Fat. Acids* **2010**, *82*, 179–187. [CrossRef] [PubMed]
580. Ma, D.; Zhang, M.; Larsen, C.P.; Xu, F.; Hua, W.; Yamashima, T.; Mao, Y.; Zhou, L. DHA promotes the neuronal differentiation of rat neural stem cells transfected with GPR40 gene. *Brain Res.* **2010**, *1330*, 1–8. [CrossRef] [PubMed]
581. Grossfield, A.; Feller, S.E.; Pitman, M.C. Contribution of omega-3 fatty acids to the thermodynamics of membrane protein solvation. *J. Phys. Chem. B* **2006**, *110*, 8907–8909. [CrossRef]
582. Kang, J.X.; Leaf, A. Evidence that free polyunsaturated fatty acids modify Na+ channels by directly binding to the channel proteins. *Proc. Natl. Acad. Sci. USA* **1996**, *93*, 3542–3546. [CrossRef]
583. Mosior, M.; Golini, E.S.; Epand, R.M. Chemical specificity and physical properties of the lipid bilayer in the regulation of protein kinase C by anionic phospholipids: Evidence for the lack of a specific binding site for phosphatidylserine. *Proc. Natl. Acad. Sci. USA* **1996**, *93*, 1907–1912. [CrossRef]
584. Pepe, S.; Bogdanov, K.; Hallaq, H.; Spurgeon, H.; Leaf, A.; Lakatta, E. Omega 3 polyunsaturated fatty acid modulates dihydropyridine effects on L-type Ca^{2+} channels, cytosolic Ca^{2+}, and contraction in adult rat cardiac myocytes. *Proc. Natl. Acad. Sci. USA* **1994**, *91*, 8832–8836. [CrossRef]
585. Xiao, Y.-F.; Gomez, A.M.; Morgan, J.P.; Lederer, W.J.; Leaf, A. Suppression of voltage-gated L-type Ca^{2+} currents by polyunsaturated fatty acids in adult and neonatal rat ventricular myocytes. *Proc. Natl. Acad. Sci. USA* **1997**, *94*, 4182–4187. [CrossRef]
586. Nanji, A.A.; Griniuviene, B.; Sadrzadeh, S.M.; Levitsky, S.; McCully, J.D. Effect of type of dietary fat and ethanol on antioxidant enzyme mRNA induction in rat liver. *J. Lipid Res.* **1995**, *36*, 736–744. [CrossRef]
587. Casañas-Sánchez, V.; Pérez, J.A.; Fabelo, N.; Quinto-Alemany, D.; Díaz, M.L. Docosahexaenoic (DHA) modulates phospholipid-hydroperoxide glutathione peroxidase (Gpx4) gene expression to ensure self-protection from oxidative damage in hippocampal cells. *Front. Physiol.* **2015**, *6*, 203. [CrossRef]
588. Hossain, M.S.; Hashimoto, M.; Gamoh, S.; Masumura, S. Antioxidative effects of docosahexaenoic acid in the cerebrum versus cerebellum and brainstem of aged hypercholesterolemic rats. *J. Neurochem.* **2008**, *72*, 1133–1138. [CrossRef]
589. Roqueta-Rivera, M.; Stroud, C.K.; Haschek, W.M.; Akare, S.J.; Segre, M.; Brush, R.S.; Agbaga, M.-P.; Anderson, R.E.; Hess, R.A.; Nakamura, M.T. Docosahexaenoic acid supplementation fully restores fertility and spermatogenesis in male delta-6 desaturase-null mice. *J. Lipid Res.* **2010**, *51*, 360–367. [CrossRef]

590. Hernández-Rodas, M.C.; Valenzuela, R.; Echeverría, F.; Rincón-Cervera, M.Á.; Espinosa, A.; Illesca, P.; Muñoz, P.; Corbari, A.; Romero, N.; Gonzalez-Mañan, D.; et al. Supplementation with docosahexaenoic acid and extra virgin olive oil prevents liver steatosis induced by a high-fat diet in mice through PPAR-α and Nrf2 upregulation with concomitant SREBP-1c and NF-kB downregulation. *Mol. Nutr. Food Res.* **2017**, *61*, 1700479. [CrossRef]
591. Allam-Ndoul, B.; Guénard, F.; Barbier, O.; Vohl, M.-C. Effect of n-3 fatty acids on the expression of inflammatory genes in THP-1 macrophages. *Lipids Health Dis.* **2016**, *15*, 69. [CrossRef]
592. Mullen, A.; Loscher, C.E.; Roche, H.M. Anti-inflammatory effects of EPA and DHA are dependent upon time and dose-response elements associated with LPS stimulation in THP-1-derived macrophages. *J. Nutr. Biochem.* **2010**, *21*, 444–450. [CrossRef]
593. Bondy, S.C.; Guo, S.X. Effect of ethanol treatment on indices of cumulative oxidative stress. *Eur. J. Pharmacol. Environ. Toxicol. Pharmacol.* **1994**, *270*, 349–355. [CrossRef]
594. Camandola, S.; Leonarduzzi, G.; Musso, T.; Varesio, L.; Carini, R.; Scavazza, A.; Chiarpotto, E.; Baeuerle, P.A.; Poli, G. Nuclear factor kB is activated by arachidonic acid but not by eicosapentaenoic Acid. *Biochem. Biophys. Res. Commun.* **1996**, *229*, 643–647. [CrossRef]
595. Jayadev, S.; Linardic, C.M.; Hannun, Y.A. Identification of arachidonic acid as a mediator of sphingomyelin hydrolysis in response to tumor necrosis factor alpha. *J. Biol. Chem.* **1994**, *269*, 5757–5763. [CrossRef]
596. Hashidate-Yoshida, T.; Harayama, T.; Hishikawa, D.; Morimoto, R.; Hamano, F.; Tokuoka, S.M.; Eto, M.; Tamura-Nakano, M.; Yanobu-Takanashi, R.; Mukumoto, Y.; et al. Fatty acid remodeling by LPCAT3 enriches arachidonate in phospholipid membranes and regulates triglyceride transport. *eLife* **2015**, *4*, e06328. [CrossRef]
597. Heshmati, J. Effect of omega-3 fatty acid supplementation on gene expression of inflammation, oxidative stress and cardiometabolic parameters: Systematic review and meta-analysis. *J. Funct. Foods* **2021**, *85*, 104619. [CrossRef]
598. Baker, P.R.S.; Lin, Y.; Schopfer, F.J.; Woodcock, S.R.; Groeger, A.L.; Batthyany, C.; Sweeney, S.; Long, M.H.; Iles, K.E.; Baker, L.M.S.; et al. Fatty acid transduction of nitric oxide signaling. *J. Biol. Chem.* **2005**, *280*, 42464–42475. [CrossRef]
599. Lima, É.S.; Bonini, M.G.; Augusto, O.; Barbeiro, H.V.; Souza, H.P.; Abdalla, D.S.P. Nitrated lipids decompose to nitric oxide and lipid radicals and cause vasorelaxation. *Free Radic. Biol. Med.* **2005**, *39*, 532–539. [CrossRef]
600. Coles, B.; Bloodsworth, A.; Clark, S.R.; Lewis, M.J.; Cross, A.R.; Freeman, B.A.; O'Donnell, V.B. Nitrolinoleate inhibits superoxide generation, degranulation, and integrin expression by human neutrophils. *Circ. Res.* **2002**, *91*, 375–381. [CrossRef]
601. von Euler, U.S. On the specific vaso-dilating and plain muscle stimulating substances from accessory genital glands in man and certain animals (prostaglandin and vesiglandin). *J. Physiol.* **1936**, *88*, 213–234. [CrossRef]
602. Tang, X.; Edwards, E.M.; Holmes, B.B.; Falck, J.R.; Campbell, W.B. Role of phospholipase C and diacylglyceride lipase pathway in arachidonic acid release and acetylcholine-induced vascular relaxation in rabbit aorta. *Am. J. Physiol. Circ. Physiol.* **2006**, *290*, H37–H45. [CrossRef]
603. Abraham, R.T.; McKinney, M.M.; Forray, C.; Shipley, G.D.; Handwerger, B.S. Stimulation of arachidonic acid release and eicosanoid biosynthesis in an interleukin 2-dependent T cell Line. *Immunopharmacol. Immunotoxicol.* **1986**, *8*, 165–204. [CrossRef]
604. Yada, Y.; Higuchi, K.; Imokawa, G. Purification and biochemical characterization of membrane-bound epidermal ceramidases from Guinea pig skin. *J. Biol. Chem.* **1995**, *270*, 12677–12684. [CrossRef]
605. Imokawa, G. A possible mechanism underlying the ceramide deficiency in atopic dermatitis: Expression of a deacylase enzyme that cleaves the N-acyl linkage of sphingomyelin and glucosylceramide. *J. Dermatol. Sci.* **2009**, *55*, 1–9. [CrossRef]
606. Ishibashi, M.; Arikawa, J.; Okamoto, R.; Kawashima, M.; Takagi, Y.; Ohguchi, K.; Imokawa, G. Abnormal expression of the novel epidermal enzyme, glucosylceramide deacylase, and the accumulation of its enzymatic reaction product, glucosylsphingosine, in the skin of patients with atopic dermatitis. *Lab. Investig.* **2003**, *83*, 397–408. [CrossRef]
607. Imokawa, G.; Takagi, Y.; Higuchi, K.; Kondo, H.; Yada, Y. Sphingosylphosphorylcholine is a potent inducer of intercellular adhesion molecule-1 expression in human keratinocytes. *J. Investig. Dermatol.* **1999**, *112*, 91–96. [CrossRef]
608. Richmond, G.S.; Smith, T.K. Phospholipases A1. *Int. J. Mol. Sci.* **2011**, *12*, 588–612. [CrossRef]
609. Goñi, F.M.; Montes, L.-R.; Alonso, A. Phospholipases C and sphingomyelinases: Lipids as substrates and modulators of enzyme activity. *Prog. Lipid Res.* **2012**, *51*, 238–266. [CrossRef]
610. Yang, S.F.; Freer, S.; Benson, A.A. Transphosphatidylation by phospholipase D. *J. Biol. Chem.* **1967**, *242*, 477–484. [CrossRef]
611. Dennis, E.A.; Cao, J.; Hsu, Y.-H.; Magrioti, V.; Kokotos, G. Phospholipase A 2 enzymes: Physical structure, biological function, disease implication, chemical inhibition, and therapeutic intervention. *Chem. Rev.* **2011**, *111*, 6130–6185. [CrossRef]
612. Clark, J.D.; Lin, L.-L.; Kriz, R.W.; Ramesha, C.S.; Sultzman, L.A.; Lin, A.Y.; Milona, N.; Knopf, J.L. A novel arachidonic acid-selective cytosolic PLA2 contains a Ca^{2+}-dependent translocation domain with homology to PKC and GAP. *Cell* **1991**, *65*, 1043–1051. [CrossRef]
613. Cheon, Y.; Kim, H.-W.; Igarashi, M.; Modi, H.R.; Chang, L.; Ma, K.; Greenstein, D.; Wohltmann, M.; Turk, J.; Rapoport, S.I.; et al. Disturbed brain phospholipid and docosahexaenoic acid metabolism in calcium-independent phospholipase A2-VIA (iPLA2β)-knockout mice. *Biochim. Biophys. Acta Mol. Cell Biol. Lipids* **2012**, *1821*, 1278–1286. [CrossRef]
614. Strokin, M.; Sergeeva, M.; Reiser, G. Docosahexaenoic acid and arachidonic acid release in rat brain astrocytes is mediated by two separate isoforms of phospholipase A 2 and is differently regulated by cyclic AMP and Ca^{2+}. *Br. J. Pharmacol.* **2003**, *139*, 1014–1022. [CrossRef]

615. Murase, R.; Sato, H.; Yamamoto, K.; Ushida, A.; Nishito, Y.; Ikeda, K.; Kobayashi, T.; Yamamoto, T.; Taketomi, Y.; Murakami, M. Group X secreted phospholipase A2 releases ω3 polyunsaturated fatty acids, suppresses colitis, and promotes sperm fertility. *J. Biol. Chem.* **2016**, *291*, 6895–6911. [CrossRef]
616. Fonteh, A.N.; Bass, D.A.; Marshall, L.A.; Seeds, M.; Samet, J.M.; Chilton, F.H. Evidence that secretory phospholipase A2 plays a role in arachidonic acid release and eicosanoid biosynthesis by mast cells. *J. Immunol.* **1994**, *152*, 5438–5446. [CrossRef]
617. Wada, M.; DeLong, C.J.; Hong, Y.H.; Rieke, C.J.; Song, I.; Sidhu, R.S.; Yuan, C.; Warnock, M.; Schmaier, A.H.; Yokoyama, C.; et al. Enzymes and receptors of prostaglandin pathways with arachidonic acid-derived versus eicosapentaenoic acid-derived substrates and products. *J. Biol. Chem.* **2007**, *282*, 22254–22266. [CrossRef]
618. Chan, J.K.; McDonald, B.E.; Gerrard, J.M.; Bruce, V.M.; Weaver, B.J.; Holub, B.J. Effect of dietary α-linolenic acid and its ratio to linoleic acid on platelet and plasma fatty acids and thrombogenesis. *Lipids* **1993**, *28*, 811–817. [CrossRef]
619. Luu, W.; Sharpe, L.J.; Capell-Hattam, I.; Gelissen, I.C.; Brown, A.J. Oxysterols: Old tale, new twists. *Annu. Rev. Pharmacol. Toxicol.* **2016**, *56*, 447–467. [CrossRef]
620. Schroepfer, G.J. Oxysterols: Modulators of cholesterol metabolism and other processes. *Physiol. Rev.* **2000**, *80*, 361–554. [CrossRef]
621. Phillis, J.W.; Horrocks, L.A.; Farooqui, A.A. Cyclooxygenases, lipoxygenases, and epoxygenases in CNS: Their role and involvement in neurological disorders. *Brain Res. Rev.* **2006**, *52*, 201–243. [CrossRef]
622. Bazan, N.G.; Fletcher, B.S.; Herschman, H.R.; Mukherjee, P.K. Platelet-activating factor and retinoic acid synergistically activate the inducible prostaglandin synthase gene. *Proc. Natl. Acad. Sci. USA* **1994**, *91*, 5252–5256. [CrossRef]
623. Chandrasekharan, N.V.; Dai, H.; Roos, K.L.T.; Evanson, N.K.; Tomsik, J.; Elton, T.S.; Simmons, D.L. COX-3, a cyclooxygenase-1 variant inhibited by acetaminophen and other analgesic/antipyretic drugs: Cloning, structure, and expression. *Proc. Natl. Acad. Sci. USA* **2002**, *99*, 13926–13931. [CrossRef]
624. Simard, M.; Morin, S.; Ridha, Z.; Pouliot, R. Current knowledge of the implication of lipid mediators in psoriasis. *Front. Immunol.* **2022**, *13*, 961107. [CrossRef]
625. Alhouayek, M.; Muccioli, G.G. COX-2-derived endocannabinoid metabolites as novel inflammatory mediators. *Trends Pharmacol. Sci.* **2014**, *35*, 284–292. [CrossRef]
626. Ivanov, I.; Kuhn, H.; Heydeck, D. Structural and functional biology of arachidonic acid 15-lipoxygenase-1 (ALOX15). *Gene* **2015**, *573*, 1–32. [CrossRef]
627. Lehnert, N.; Solomon, E.I. Density-functional investigation on the mechanism of H-atom abstraction by lipoxygenase. *JBIC J. Biol. Inorg. Chem.* **2003**, *8*, 294–305. [CrossRef]
628. Dyall, S.C.; Balas, L.; Bazan, N.G.; Brenna, J.T.; Chiang, N.; da Costa Souza, F.; Dalli, J.; Durand, T.; Galano, J.-M.; Lein, P.J.; et al. Polyunsaturated fatty acids and fatty acid-derived lipid mediators: Recent advances in the understanding of their biosynthesis, structures, and functions. *Prog. Lipid Res.* **2022**, *86*, 101165. [CrossRef] [PubMed]
629. Hajeyah, A.A.; Griffiths, W.J.; Wang, Y.; Finch, A.J.; O'Donnell, V.B. The biosynthesis of enzymatically oxidized lipids. *Front. Endocrinol.* **2020**, *11*, 591819. [CrossRef] [PubMed]
630. Nebert, D.W.; Wikvall, K.; Miller, W.L. Human cytochromes P450 in health and disease. *Philos. Trans. R. Soc. B Biol. Sci.* **2013**, *368*, 20120431. [CrossRef]
631. Davis, C.M.; Liu, X.; Alkayed, N.J. Cytochrome P450 eicosanoids in cerebrovascular function and disease. *Pharmacol. Ther.* **2017**, *179*, 31–46. [CrossRef]
632. Lamb, D.C.; Waterman, M.R. Unusual properties of the cytochrome P450 superfamily. *Philos. Trans. R. Soc. B Biol. Sci.* **2013**, *368*, 20120434. [CrossRef]
633. Isin, E.M.; Guengerich, F.P. Complex reactions catalyzed by cytochrome P450 enzymes. *Biochim. Biophys. Acta Gen. Subj.* **2007**, *1770*, 314–329. [CrossRef]
634. Kumar, N.; Gupta, G.; Anilkumar, K.; Fatima, N.; Karnati, R.; Reddy, G.V.; Giri, P.V.; Reddanna, P. 15-Lipoxygenase metabolites of α-linolenic acid, [13-(S)-HPOTrE and 13-(S)-HOTrE], mediate anti-inflammatory effects by inactivating NLRP3 inflammasome. *Sci. Rep.* **2016**, *6*, 31649. [CrossRef]
635. Budowski, P.; Bartov, I.; Dror, Y.; Frankel, E.N. Lipid oxidation products and chick nutritional encephalopathy. *Lipids* **1979**, *14*, 768–772. [CrossRef]
636. Buchanan, M.R.; Haas, T.A.; Lagarde, M.; Guichardant, M. 13-Hydroxyoctadecadienoic acid is the vessel wall chemorepellant factor, LOX. *J. Biol. Chem.* **1985**, *260*, 16056–16059. [CrossRef]
637. Shearer, G.C.; Walker, R.E. An overview of the biologic effects of omega-6 oxylipins in humans. *Prostaglandins Leukot. Essent. Fat. Acids* **2018**, *137*, 26–38. [CrossRef]
638. Vangaveti, V.N.; Jansen, H.; Kennedy, R.L.; Malabu, U.H. Hydroxyoctadecadienoic acids: Oxidised derivatives of linoleic acid and their role in inflammation associated with metabolic syndrome and cancer. *Eur. J. Pharmacol.* **2016**, *785*, 70–76. [CrossRef] [PubMed]
639. Moghaddam, M.F.; Grant, D.F.; Cheek, J.M.; Greene, J.F.; Williamson, K.C.; Hammock, B.D. Bioactivation of leukotoxins to their toxic diols by epoxide hydrolase. *Nat. Med.* **1997**, *3*, 562–566. [CrossRef] [PubMed]
640. Wheeler, J.J.; Domenichiello, A.F.; Jensen, J.R.; Keyes, G.S.; Maiden, K.M.; Davis, J.M.; Ramsden, C.E.; Mishra, S.K. Endogenous derivatives of linoleic acid and their stable analogs are potential pain mediators. *JID Innov.* **2023**, *3*, 100177. [CrossRef] [PubMed]

641. Alsalem, M.; Wong, A.; Millns, P.; Arya, P.H.; Chan, M.S.L.; Bennett, A.; Barrett, D.A.; Chapman, V.; Kendall, D.A. The contribution of the endogenous TRPV1 ligands 9-HODE and 13-HODE to nociceptive processing and their role in peripheral inflammatory pain mechanisms. *Br. J. Pharmacol.* **2013**, *168*, 1961–1974. [CrossRef] [PubMed]
642. Biswas, P.; Datta, C.; Rathi, P.; Bhattacharjee, A. Fatty acids and their lipid mediators in the induction of cellular apoptosis in cancer cells. *Prostaglandins Other Lipid Mediat.* **2022**, *160*, 106637. [CrossRef]
643. Johnson, G.H.; Fritsche, K. Effect of dietary linoleic acid on markers of inflammation in healthy persons: A systematic review of randomized controlled trials. *J. Acad. Nutr. Diet.* **2012**, *112*, 1029–1041.e15. [CrossRef]
644. Takic, M.; Pokimica, B.; Petrovic-Oggiano, G.; Popovic, T. Effects of Dietary α-Linolenic Acid Treatment and the Efficiency of Its Conversion to Eicosapentaenoic and Docosahexaenoic Acids in Obesity and Related Diseases. *Molecules* **2022**, *27*, 4471. [CrossRef]
645. Arterburn, L.M.; Hall, E.B.; Oken, H. Distribution, interconversion, and dose response of n-3 fatty acids in humans. *Am. J. Clin. Nutr.* **2006**, *83*, 1467S–1476S. [CrossRef]
646. Shimizu, T. Lipid mediators in health and disease: Enzymes and receptors as therapeutic targets for the regulation of immunity and inflammation. *Annu. Rev. Pharmacol. Toxicol.* **2009**, *49*, 123–150. [CrossRef]
647. Sergeant, S.; Rahbar, E.; Chilton, F.H. Gamma-linolenic acid, Dihommo-gamma linolenic, Eicosanoids and Inflammatory Processes. *Eur. J. Pharmacol.* **2016**, *785*, 77–86. [CrossRef]
648. Miller, C.C.; McCreedy, C.A.; Jones, A.D.; Ziboh, V.A. Oxidative metabolism of dihomogammalinolenic acid by guinea pig epidermis: Evidence of generation of anti-inflammatory products. *Prostaglandins* **1988**, *35*, 917–938. [CrossRef]
649. Samuelsson, B. From studies of biochemical mechanism to novel biological mediators: Prostaglandin endoperoxides, thromboxanes, and leukotriences. *Biosci. Rep.* **1983**, *3*, 791–813. [CrossRef] [PubMed]
650. Buczynski, M.W.; Dumlao, D.S.; Dennis, E.A. Thematic review series: Proteomics. An integrated omics analysis of eicosanoid biology. *J. Lipid Res.* **2009**, *50*, 1015–1038. [CrossRef]
651. Dalvi, S.; Nguyen, H.H.; On, N.; Mitchell, R.W.; Aukema, H.M.; Miller, D.W.; Hatch, G.M. Exogenous arachidonic acid mediates permeability of human brain microvessel endothelial cells through prostaglandin E 2 activation of EP 3 and EP 4 receptors. *J. Neurochem.* **2015**, *135*, 867–879. [CrossRef]
652. Patel, P.; Cossette, C.; Anumolu, J.R.; Gravel, S.; Lesimple, A.; Mamer, O.A.; Rokach, J.; Powell, W.S. Structural requirements for activation of the 5-Oxo-6E, 8Z, 11Z, 14Z-eicosatetraenoic acid (5-Oxo-ETE) receptor: Identification of a Mead acid metabolite with potent agonist activity. *J. Pharmacol. Exp. Ther.* **2008**, *325*, 698–707. [CrossRef] [PubMed]
653. Balestrieri, B.; Di Costanzo, D.; Dwyer, D.F. Macrophage-Mediated Immune Responses: From Fatty Acids to Oxylipins. *Molecules* **2021**, *27*, 152. [CrossRef]
654. Farag, M.A.; Gad, M.Z. Omega-9 fatty acids: Potential roles in inflammation and cancer management. *J. Genet. Eng. Biotechnol.* **2022**, *20*, 48. [CrossRef]
655. Hamazaki, T.; Hamazaki, K. What Are the Physiological Roles of Mead Acid (5,8,11-Eicosatrienoic Acid)? In *Handbook of Lipids in Human Function*; Elsevier: Amsterdam, The Netherlands, 2016; pp. 483–497.
656. Duvall, M.G.; Levy, B.D. DHA- and EPA-derived resolvins, protectins, and maresins in airway inflammation. *Eur. J. Pharmacol.* **2016**, *785*, 144–155. [CrossRef]
657. Bergström, S.; Danielsson, H.; Klenberg, D.; Samuelsson, B. The enzymatic conversion of essential fatty acids into prostaglandins. *J. Biol. Chem.* **1964**, *239*, PC4006–PC4008. [CrossRef]
658. Dyerberg, J. Eicosapentaenoic acid and prevention of thrombosis and atherosclerosis? *Lancet* **1978**, *312*, 117–119. [CrossRef]
659. Levy, B.D.; Clish, C.B.; Schmidt, B.; Gronert, K.; Serhan, C.N. Lipid mediator class switching during acute inflammation: Signals in resolution. *Nat. Immunol.* **2001**, *2*, 612–619. [CrossRef] [PubMed]
660. Schunck, W.-H.; Konkel, A.; Fischer, R.; Weylandt, K.-H. Therapeutic potential of omega-3 fatty acid-derived epoxyeicosanoids in cardiovascular and inflammatory diseases. *Pharmacol. Ther.* **2018**, *183*, 177–204. [CrossRef] [PubMed]
661. Calder, P.C.; Yaqoob, P.; Thies, F.; Wallace, F.A.; Miles, E.A. Fatty acids and lymphocyte functions. *Br. J. Nutr.* **2002**, *87*, S31–S48. [CrossRef] [PubMed]
662. Mustonen, A.-M.; Nieminen, P. Dihomo-γ-linolenic acid (20:3n-6)—Metabolism, derivatives, and potential significance in chronic inflammation. *Int. J. Mol. Sci.* **2023**, *24*, 2116. [CrossRef]
663. Baker, E.J.; Valenzuela, C.A.; Dooremalen, W.T.M.; Martínez-Fernández, L.; Yaqoob, P.; Miles, E.A.; Calder, P.C. Gamma-linolenic and pinolenic acids exert anti-inflammatory effects in cultured human endothelial cells through their elongation products. *Mol. Nutr. Food Res.* **2020**, *64*, 2000382. [CrossRef]
664. Brouwers, H.; Jónasdóttir, H.S.; Kuipers, M.E.; Kwekkeboom, J.C.; Auger, J.L.; Gonzalez-Torres, M.; López-Vicario, C.; Clària, J.; Freysdottir, J.; Hardardottir, I.; et al. Anti-inflammatory and proresolving effects of the omega-6 polyunsaturated fatty acid adrenic acid. *J. Immunol.* **2020**, *205*, 2840–2849. [CrossRef]
665. Poorani, R.; Bhatt, A.; Das, U. Modulation of pro-inflammatory and pro-resolution mediators by γ-linolenic acid: An important element in radioprotection against ionizing radiation. *Arch. Med. Sci.* **2020**, *16*, 1448–1456.
666. Serhan, C.N.; Levy, B.D. Resolvins in inflammation: Emergence of the pro-resolving superfamily of mediators. *J. Clin. Investig.* **2018**, *128*, 2657–2669. [CrossRef]
667. Farooqui, A.A.; Horrocks, L.A.; Farooqui, T. Modulation of inflammation in brain: A matter of fat. *J. Neurochem.* **2006**, *101*, 577–599. [CrossRef]

668. Chiang, N.; Gronert, K.; Clish, C.B.; O'Brien, J.A.; Freeman, M.W.; Serhan, C.N. Leukotriene B4 receptor transgenic mice reveal novel protective roles for lipoxins and aspirin-triggered lipoxins in reperfusion. *J. Clin. Investig.* **1999**, *104*, 309–316. [CrossRef]
669. Clària, J.; Serhan, C.N. Aspirin triggers previously undescribed bioactive eicosanoids by human endothelial cell-leukocyte interactions. *Proc. Natl. Acad. Sci. USA* **1995**, *92*, 9475–9479. [CrossRef] [PubMed]
670. Weylandt, K.-H. Docosapentaenoic acid derived metabolites and mediators—The new world of lipid mediator medicine in a nutshell. *Eur. J. Pharmacol.* **2016**, *785*, 108–115. [CrossRef] [PubMed]
671. Serhan, C.N.; Gupta, S.S.S.K.; Perretti, M.; Godson, C.; Brennan, E.; Li, Y.; Soehnlein, O.; Shimizu, T.; Werz, O.; Chiurchiù, V.; et al. The atlas of inflammation resolution (AIR). *Mol. Asp. Med.* **2020**, *74*, 100894. [CrossRef]
672. Drouin, G.; Rioux, V.; Legrand, P. The n-3 docosapentaenoic acid (DPA): A new player in the n-3 long chain polyunsaturated fatty acid family. *Biochimie* **2019**, *159*, 36–48. [CrossRef]
673. Serhan, C.N.; Dalli, J.; Colas, R.A.; Winkler, J.W.; Chiang, N. Protectins and maresins: New pro-resolving families of mediators in acute inflammation and resolution bioactive metabolome. *Biochim. Biophys. Acta Mol. Cell Biol. Lipids* **2015**, *1851*, 397–413. [CrossRef]
674. Terranova, L.; Risé, P.; Gramegna, A.; Pinna, C.; Agostoni, C.; Syrén, M.-L.; Turolo, S.; Marchisio, P.; Amati, F.; Aliberti, S.; et al. Pro-resolving and pro-inflammatory fatty acid-derived mediators in sputum of stable state bronchiectasis patients. *Respir. Res.* **2022**, *23*, 363. [CrossRef]
675. Tang, S.; Wan, M.; Huang, W.; Stanton, R.C.; Xu, Y. Maresins: Specialized proresolving lipid mediators and their potential role in inflammatory-related diseases. *Mediat. Inflamm.* **2018**, *2018*, 2380319. [CrossRef]
676. Balas, L.; Durand, T. Dihydroxylated E,E,Z-docosatrienes. An overview of their synthesis and biological significance. *Prog. Lipid Res.* **2016**, *61*, 1–18. [CrossRef]
677. Serhan, C.N.; Chiang, N.; Van Dyke, T.E. Resolving inflammation: Dual anti-inflammatory and pro-resolution lipid mediators. *Nat. Rev. Immunol.* **2008**, *8*, 349–361. [CrossRef]
678. Serhan, C.N.; Arita, M.; Hong, S.; Gotlinger, K. Resolvins, docosatrienes, and neuroprotectins, novel omega-3-derived mediators, and their endogenous aspirin-triggered epimers. *Lipids* **2004**, *39*, 1125–1132. [CrossRef]
679. Wu, A.; Ying, Z.; Gomez-Pinilla, F. Dietary omega-3 fatty acids normalize BDNF levels, reduce oxidative damage, and counteract learning disability after traumatic brain injury in rats. *J. Neurotrauma* **2004**, *21*, 1457–1467. [CrossRef] [PubMed]
680. Hong, S.; Gronert, K.; Devchand, P.R.; Moussignac, R.-L.; Serhan, C.N. Novel docosatrienes and 17S-resolvins generated from docosahexaenoic acid in murine brain, human blood, and glial cells. *J. Biol. Chem.* **2003**, *278*, 14677–14687. [CrossRef] [PubMed]
681. Marcheselli, V.L.; Hong, S.; Lukiw, W.J.; Tian, X.H.; Gronert, K.; Musto, A.; Hardy, M.; Gimenez, J.M.; Chiang, N.; Serhan, C.N.; et al. Novel docosanoids inhibit brain ischemia-reperfusion-mediated leukocyte infiltration and pro-inflammatory gene expression. *J. Biol. Chem.* **2003**, *278*, 43807–43817. [CrossRef]
682. Serhan, C.N.; Hong, S.; Gronert, K.; Colgan, S.P.; Devchand, P.R.; Mirick, G.; Moussignac, R.-L. Resolvins. *J. Exp. Med.* **2002**, *196*, 1025–1037. [CrossRef] [PubMed]
683. Li, X.-A.; Everson, W.V.; Smart, E.J. Caveolae, lipid rafts, and vascular disease. *Trends Cardiovasc. Med.* **2005**, *15*, 92–96. [CrossRef]
684. Dalli, J.; Colas, R.A.; Serhan, C.N. Novel n-3 Immunoresolvents: Structures and Actions. *Sci. Rep.* **2013**, *3*, 1940. [CrossRef]
685. Jun, B.; Mukherjee, P.K.; Asatryan, A.; Kautzmann, M.-A.; Heap, J.; Gordon, W.C.; Bhattacharjee, S.; Yang, R.; Petasis, N.A.; Bazan, N.G. Elovanoids are novel cell-specific lipid mediators necessary for neuroprotective signaling for photoreceptor cell integrity. *Sci. Rep.* **2017**, *7*, 5279. [CrossRef]
686. Kyselová, L.; Vítová, M.; Řezanka, T. Very long chain fatty acids. *Prog. Lipid Res.* **2022**, *87*, 101180. [CrossRef]
687. Asatryan, A.; Calandria, J.M.; Kautzmann, M.-A.I.; Jun, B.; Gordon, W.C.; Do, K.V.; Bhattacharjee, S.; Pham, T.L.; Bermúdez, V.; Mateos, M.V.; et al. New retinal pigment epithelial cell model to unravel neuroprotection sensors of neurodegeneration in retinal disease. *Front. Neurosci.* **2022**, *16*, 3894. [CrossRef]
688. Bazan, N.G. Overview of how N32 and N34 elovanoids sustain sight by protecting retinal pigment epithelial cells and photoreceptors. *J. Lipid Res.* **2021**, *62*, 100058. [CrossRef]
689. Do, K.V.; Kautzmann, M.-A.I.; Jun, B.; Gordon, W.C.; Nshimiyimana, R.; Yang, R.; Petasis, N.A.; Bazan, N.G. Elovanoids counteract oligomeric β-amyloid-induced gene expression and protect photoreceptors. *Proc. Natl. Acad. Sci. USA* **2019**, *116*, 24317–24325. [CrossRef] [PubMed]
690. Bazan, N.G. Docosanoids and elovanoids from omega-3 fatty acids are pro-homeostatic modulators of inflammatory responses, cell damage and neuroprotection. *Mol. Asp. Med.* **2018**, *64*, 18–33. [CrossRef] [PubMed]
691. Bhattacharjee, S.; Jun, B.; Belayev, L.; Heap, J.; Kautzmann, M.-A.; Obenaus, A.; Menghani, H.; Marcell, S.J.; Khoutorova, L.; Yang, R.; et al. Elovanoids are a novel class of homeostatic lipid mediators that protect neural cell integrity upon injury. *Sci. Adv.* **2017**, *3*, e1700735. [CrossRef] [PubMed]

International Journal of
Molecular Sciences

MDPI

Review

The Impact of the Aryl Hydrocarbon Receptor on Antenatal Chemical Exposure-Induced Cardiovascular–Kidney–Metabolic Programming

You-Lin Tain [1,2,3] and Chien-Ning Hsu [4,5,*]

1 Division of Pediatric Nephrology, Kaohsiung Chang Gung Memorial Hospital, Kaohsiung 833, Taiwan; tainyl@cgmh.org.tw
2 Institute for Translational Research in Biomedicine, Kaohsiung Chang Gung Memorial Hospital, Kaohsiung 833, Taiwan
3 College of Medicine, Chang Gung University, Taoyuan 333, Taiwan
4 Department of Pharmacy, Kaohsiung Chang Gung Memorial Hospital, Kaohsiung 833, Taiwan
5 School of Pharmacy, Kaohsiung Medical University, Kaohsiung 807, Taiwan
* Correspondence: cnhsu@cgmh.org.tw; Tel.: +886-975-368-975

Abstract: Early life exposure lays the groundwork for the risk of developing cardiovascular–kidney–metabolic (CKM) syndrome in adulthood. Various environmental chemicals to which pregnant mothers are commonly exposed can disrupt fetal programming, leading to a wide range of CKM phenotypes. The aryl hydrocarbon receptor (AHR) has a key role as a ligand-activated transcription factor in sensing these environmental chemicals. Activating AHR through exposure to environmental chemicals has been documented for its adverse impacts on cardiovascular diseases, hypertension, diabetes, obesity, kidney disease, and non-alcoholic fatty liver disease, as evidenced by both epidemiological and animal studies. In this review, we compile current human evidence and findings from animal models that support the connection between antenatal chemical exposures and CKM programming, focusing particularly on AHR signaling. Additionally, we explore potential AHR modulators aimed at preventing CKM syndrome. As the pioneering review to present evidence advocating for the avoidance of toxic chemical exposure during pregnancy and deepening our understanding of AHR signaling, this has the potential to mitigate the global burden of CKM syndrome in the future.

Keywords: cardiovascular disease; dioxins; metabolic syndrome; chemical; chronic kidney disease; aryl hydrocarbon receptor; developmental origins of health and disease (DOHaD); prenatal exposure; hypertension

1. Introduction

Numerous epidemiological and experimental findings have established that early life exposure to adverse environmental conditions can significantly impact the likelihood of developing adult-onset diseases [1,2]. This phenomenon, now recognized as developmental programming or the developmental origins of health and disease (DOHaD), elucidates the adaptations made by a developing fetus in response to cues during early life. These adaptations result in morphological and functional adjustments that may prove detrimental in later life stages, thereby increasing the susceptibility to adult diseases.

A myriad of early life factors can trigger developmental programming, including maternal malnutrition, maternal illnesses, complications during pregnancy, substance abuse, medication usage, or exposure to chemicals during pregnancy [1–6]. Environmental chemicals possess the ability to traverse the placental barrier, and it is widely acknowledged that the prenatal stage is particularly vulnerable to chemical disruptions and subsequent health ramifications compared to later developmental stages [7]. Concurrent exposure to multiple chemicals can exacerbate health consequences either through additive or synergistic effects [8]. Notably, the scale of chemical production has expanded significantly over the last six decades [9], with nearly 8000 chemicals now being manufactured or imported in substantial quantities.

Cardiovascular–kidney–metabolic (CKM) syndrome has surfaced as a significant and pressing global public health issue [10]. The impact of CKM syndrome is thought to affect approximately 40% of the adult population in the United States [10]. In 2023, the American Heart Association initially recognized CKM syndrome as a systemic ailment defined by complex physiological interconnections among metabolic disorders, chronic kidney disease (CKD), and cardiovascular health [11]. The likelihood of adverse consequences is heightened as a result of the interplay of these factors, leading to multiorgan dysfunction [11]. CKM syndrome is categorized into four discernible stages, ranging from stage 0 to stage 4, each representing a different degree of advancement and severity across the intricate spectrum of this condition. Across the complex range of CKM syndrome, different pivotal components emerge at various stages, contributing to the nuanced evolution and severity observed. Of paramount importance to emphasize is that prioritizing early prevention, rather than solely focusing on treatment, holds the potential to improve the global burdens associated with CKM syndrome.

The DOHaD theory establishes a connection between early life programming and various recognized facets of CKM syndrome, including cardiovascular disease (CVD) [6], metabolic disease [12], hypertension [13], CKD [14], and obesity [15]. Several molecular mechanisms linked to CKM programming have been discovered, such as an aberrant renin–angiotensin system (RAS), epigenetic dysregulation, deficient nitric oxide (NO), disturbances in nutrient-sensing signals, oxidative stress, and gut microbiota dysbiosis [16–21]. Conversely, by targeting these pivotal mechanisms, there is a shift in focus from managing diseases during adulthood to intervening in disease processes before they clinically manifest, known as reprogramming, which holds promising potential as a preventive strategy.

The aryl hydrocarbon receptor (AHR) is a pivotal ligand-activated transcription factor recognized for its capacity to sense environmental chemicals and regulate various physiological processes, including fetal development [22–24]. Extensive research over the years has elucidated how activation of AHR by environmental chemicals or microbial-derived uremic toxins impacts the cardiovascular, renal, and metabolic systems, thus contributing to the development of different facets of CKM syndrome [25–27].

Although the detrimental effects of AHR activation due to adult exposure to environmental chemicals on cardiovascular–kidney–metabolic health are well-established, our understanding of its involvement in the chemical-induced programming of CKM syndrome remains limited. Therefore, this review aims to delineate the impact of prenatal chemical exposures on CKM programming by synthesizing the available epidemiological and experimental evidence, with a particular emphasis on AHR signaling. Furthermore, we discuss potential interventions targeting AHR for reprogramming purposes to mitigate the onset of CKM syndrome.

A comprehensive search of scientific databases, including MEDLINE, SCOPUS, Embase, and the Cochrane Library, was conducted to elucidate the intricate relationship between AHR, CKM syndrome, and developmental programming. This exploration encompassed various keywords and their permutations, including "cardiovascular disease", "chronic kidney disease", "obesity", "fatty liver", "metabolic syndrome", "diabetes", "hypertension", "hyperlipidemia", "pregnancy", "gestation", "lactation", "progeny", "offspring", "mother", "developmental programming", "DOHaD", "reprogramming", "aryl hydrocarbon receptor", "endocrine-disrupting chemicals", "organophosphate flame retardants", "phthalates", "microplastics", "heavy metals", "air pollution", and "$PM_{2.5}$". Additionally, supplementary investigations were selected and evaluated based on relevant references identified in eligible papers. The final search was concluded on 20 March 2024.

2. Aryl Hydrocarbon Receptor

2.1. The Structure of AHR

Belonging to the basic helix–loop–helix Per–ARNT–SIM (bHLH–PAS) family is the AHR, with its structure comprising an N-terminal bHLH domain, a central PAS domain (A and B), and a C-terminal transactivation domain [22] (Figure 1). The N-terminal bHLH domain of AHR undergoes dimerization, resulting in the formation of a four-helical bundle. This configuration serves as the DNA-binding domain and facilitates dimerization [28]. Governing DNA recognition, ligand binding, and chaperone interactions are the roles of the PAS domain [29]. Additionally, this domain, in conjunction with the bHLH domain, aids in mediating the heterodimerization of AHR with the aryl hydrocarbon receptor nuclear translocator (ARNT) [30]. Diverse ligands are accommodated by the ligand-binding domain (LBD) situated within the PAS B domain [31]. Within the C-terminal transactivation domain (TAD), the Q-rich subdomain takes precedence in the transcriptional activation of xenobiotic response elements (XRE) in the DNA [32]. Moreover, binding to coactivators during transcription exhibits a broad spectrum of diversity and tissue-specific effects [33].

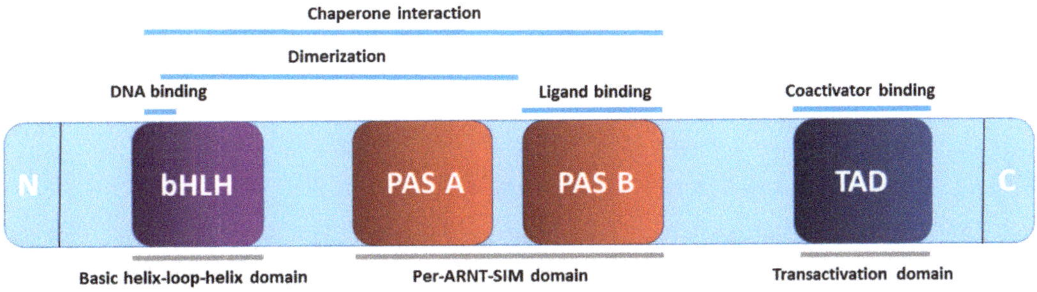

Figure 1. Schema outlining the structure of the aryl hydrocarbon receptor.

2.2. AHR Ligands

Environmental pollutants, as well as dietary- and microbiota-derived metabolites, are the major sources of AHR ligands. These ligands can be delineated as either exogenous or endogenous. The former category includes dietary compounds like polyphenols, environmental contaminants such as dioxins, pharmaceuticals like omeprazole, and a variety of synthetic compounds like SP600125. Endogenous AHR ligands include compounds synthesized within the human body (e.g., tryptamine) and those generated by the gut microbiota (e.g., indoles). The activation of AHR by these distinct classes of ligands precipitates context-dependent positive and negative consequences.

2.3. AHR Signaling

The activation of AHR and its subsequent downstream signaling transduction comprises both canonical and non-canonical pathways [34] (Figure 2). Within the canonical pathway, AHR forms complexes with molecular chaperones, remaining inert in the cytosol [35]. Included among these molecular chaperones are heat shock protein 90 (Hsp90), AHR-interacting protein (AIP), also known as ARA9 and XAP2, and p23 [36]. Upon binding with a ligand, AHR undergoes a structural alteration, leading to the release of the AHR/ligand complex from the chaperone proteins. Following this, AHR translocates to the nucleus, where it forms a heterodimer with ARNT. Subsequently, this AHR/ARNT complex binds to xenobiotic response elements (XREs) located within the regulatory regions of target genes, thus modulating their expression. Noteworthy AHR target genes encompass those encoding members of the cytochrome P450 superfamily enzymes (such as CYP1A1, CYP1A2, and CYP1B1), as well as the AHR repressor (AHRR), which have a critical role in detoxifying environmental chemicals and negatively regulating AHR-dependent gene expression, respectively [37].

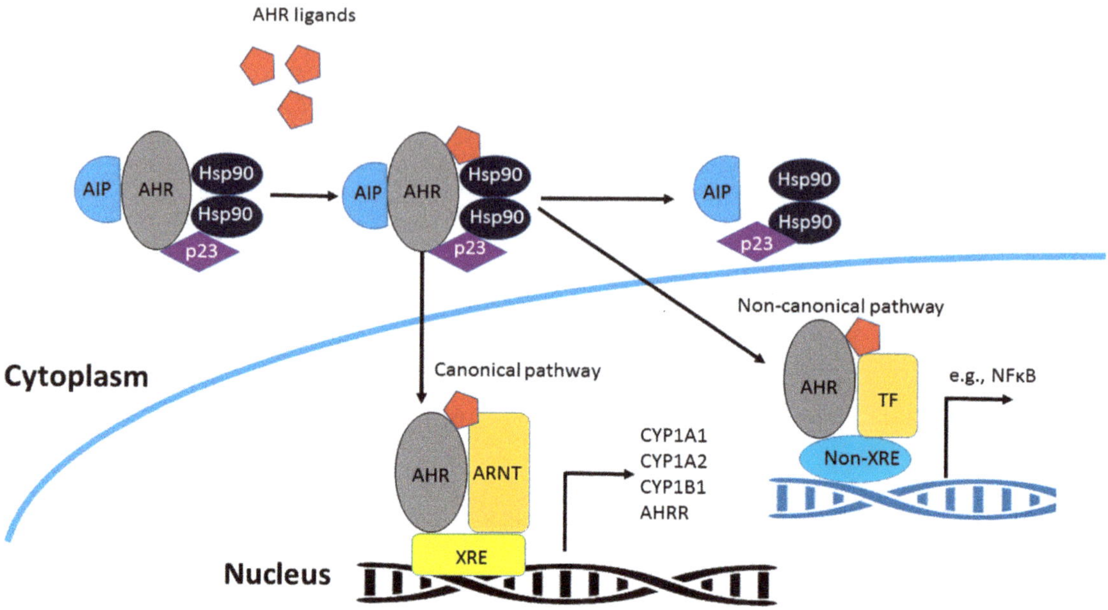

Figure 2. Classical and non-classical AHR signaling pathways.

In the non-canonical signaling pathway, once AHR is activated, it interacts with different transcription factors inside the nucleus, thereby facilitating its attachment to non-XRE DNA elements, consequently regulating the expression of target genes [38]. For instance, the interaction of AHR binding to transcription factors (e.g., NFκB) triggers specific downstream gene expression [39]. In light of the fact that the promoters of most AHR-regulated genes do not contain identifiable XREs [40], there is speculation that AHR could potentially cooperate with a multitude of other nuclear factors to control gene expression following exposure to diverse exogenous and endogenous ligands.

3. AHR and CKM Syndrome

AHR elicits varied physiological effects contingent upon its localization within distinct tissues. AhR is expressed ubiquitously, while its distribution changes significantly with age [41]. During fetal development, AhR demonstrates distinct distribution in the liver, kidneys, lungs, pancreas, thymus glands, testicles, and epithelial cells, with fairly diminished levels observed in the heart, aorta, and brain. In adulthood, AHR exhibits elevated expression in the placenta, lungs, spleen, pancreas, and liver, while displaying comparatively reduced abundance in the brain, heart, and skeletal muscles [42]. For a number of years, AHR was primarily investigated for its involvement in organ toxicity induced by environmental chemicals, given that many of these chemicals contain ligands for AHR. Increasing evidence suggests that AHR is involved in triggering pathogenesis in various components of CKM syndrome [25–27].

3.1. Cardiovascular Disease and Hypertension

Despite its low expression in the fetal heart, the AHR signaling pathway is critical for cardiac development. Genetic deficiency in AHR is associated with cardiac hypertrophy and developmental vascular defects in the heart, kidney, and liver [43,44]. Additionally, the process of cardiomyocyte differentiation is meticulously controlled by AHR signaling. Activation, inhibition, or suppression of AHR can all potentially impact the differentiation of cardiomyocytes derived from mouse embryonic stem cells [45].

AHR is also involved in the regulation of the vascular microenvironment [46]. Ischemia-induced angiogenesis was noticeably augmented in AHR knockout (KO) mice compared with that in wild-type animals, which was associated with enhanced ARNT [47]. Another study revealed that ischemic insult increases AHR expression and its transcriptional activity in neurons in vitro and in vivo, while ablation of AHR by pharmacological or genetic loss-of-function approaches leads to neuroprotection [48]. In addition, acute kynurenine administration, an AHR ligand, causes vascular dysfunction accompanied by oxidative stress [49]. Moreover, exposure to environmental chemicals containing ligands of AHR (e.g., dioxins, PAH, and benzo[a]pyrene) is reported to promote the development and progression of atherosclerosis [25].

Although the exact mechanism has yet to be fully determined, AHR is involved in the complex networks that control blood pressure (BP). AHR KO mice developed hypotension at low altitudes and hypertension at modest altitudes, which might be related to elevated plasma endothelin-1 levels [50]. In the administration of captopril (an angiotensin-converting enzyme (ACE) inhibitor) to heterozygous and homozygous AHR KO mice, it was observed that the heterozygous group exhibited a significantly greater reduction in blood pressure compared to the homozygous group [51], accompanied by higher plasma renin and ACE activity in the heterozygote AHR KO mice. These findings suggest the interplay between AHR and vasoconstrictors in the regulation of BP.

3.2. Kidney Disease

In CKD, the buildup of uremic toxins within the body poses significant risks to all tissues and organs. Among these toxins, the AHR plays a pivotal role, acting as a receptor for many uremic toxins [26]. Notably, tryptophan-derived uremic toxins such as indoxyl sulfate and indole acetic acid are known AHR ligands, contributing to kidney inflammation and the progression of CKD [26].

AHR signaling holds crucial importance in maintaining the delicate balance between regulatory T (Treg) cells and T helper type 17 (Th17) cells in CKD. However, this pathway can be dysregulated by environmental chemicals, leading to aberrant activation [52]. Depending on the specific ligand and cellular context, AHR activation can either exacerbate or mitigate inflammation. Aberrant activation of AHR signaling may trigger inflammation by promoting monocyte adhesion, increasing the expression of pro-inflammatory cytokines, and reducing the bioavailability of nitric oxide (NO) [53,54]. Conversely, AHR can also exert anti-inflammatory effects. Additionally, AHR interacts with other pathways such as Nrf2, peroxisome proliferator-activated receptor-γ (PPAR-γ), and NF-κB, contributing to the diverse responses of AHR during different stages of CKD progression [55].

Moreover, AHR competes with hypoxia-inducible factor 1-alpha (HIF-1α) in binding to ARNT, influencing pro-inflammatory responses [56]. Furthermore, AHR antagonizes transforming growth factor beta 1 (TGF-β1) signaling in fibrogenesis, suggesting the potential of targeting AHR to attenuate CKD progression [57].

3.3. Diabetes, Obesity, and NAFLD

AHR KO mice displayed low plasma insulin, imbalanced glucose homeostasis, and impaired glucose intolerance [58], indicating the significance of AHR expression in the regulation of glucose balance.

The involvement of AHR/CYP1A1 activation is suggested in the development of non-alcoholic fatty liver disease (NAFLD) and the consequent onset of diabetes [59]. AHR activation leads to decreased levels of PPARα, consequently impacting β-oxidation. This reduction is associated with diminished expression of PEPCK and G6Pase, both recognized for their roles in controlling hyperglycemia and insulin resistance [27]. Additionally, given the circadian variation in PPARα, it modulates the levels of CLOCK and BMAL1, thereby impacting glucose tolerance and disturbing the regulation of specific metabolic genes [60].

AHR plays a role in regulating adipocyte differentiation by modulating the PPAR signaling pathway, which is essential for regulating fatty acid oxidation and glucose metabolism [61]. Dioxins like TCDD bind to AHR, triggering inflammation in adipocytes and consequently leading to impairment in insulin sensitivity. Additionally, the assembly of AHR–ARNT complexes interferes with several signaling pathways. The activation of AHR by TCDD additionally disrupts lipoprotein lipase activity in adipose tissue, thus regulating adipocyte differentiation and interfering with the PPAR signaling pathway crucial for fatty acid oxidation and glucose metabolism [62]. Furthermore, the upregulation in TNF-α expression induced by TCDD exacerbates dysfunction in insulin signaling and insulin resistance.

4. Epidemiological Evidence: The Link between Chemical Exposure and CKM Syndrome

Presented in Table 1 are the principal sources and documented detrimental impacts associated with CKM syndrome in human research, attributable to various environmental chemicals encountered during routine consumer activities. Numerous adverse effects on cardiovascular–kidney–metabolic health are posed by a plethora of environmental chemicals. Subsequent sections will delve into a discussion of each of these chemicals.

Table 1. Major sources of environmental chemicals and their associated CKM syndrome in human studies.

Environmental Chemicals	Common Substances or Derivatives	Major Sources	Exposure-Associated CKM Phenotypes
Dioxins	TCDD, PCDF, PCDD, PCB	Consumption of animal products rich in fat, pesticide production, wood pulp bleaching, and the process of waste incineration	Cardiovascular disease [63], diabetes [64], metabolic syndrome [65], hypertension, and kidney disease [66,67]
Bisphenol A	BPA	Plastic containers, lenses, medical tubing, and apparatus	Cardiovascular disease [68], diabetes [69], obesity [69], NAFLD [70], hypertension, and kidney disease [66,67]
Phthalates	DEHP, DBP	Vinyl plastics, cosmetics, shampoos, medical devices, and food packaging	Cardiovascular disease [71], diabetes [72], metabolic syndrome [73], NAFLD [74], hypertension [67], and kidney disease [66,67]
Organophosphate flame retardants	DPHP, TPHP, TDCPP	Plastics, rubbers, textiles, upholstered furniture, building materials, and electronic equipment	Cardiovascular disease [75], metabolic syndrome [76], kidney disease [77], and hypertension [78]
Microplastics	MPs	Various foods, including drinking water, seafood, milk, sugar, and salt	Cardiovascular disease [79]
Per- and polyfluoroalkyl substances	PFOA, PFOS	Firefighting foams, non-stick cookware, stain-resistant fabrics, water-repellent coatings, and food packaging materials	Obesity, diabetes, NAFLD [80], hypertension, and kidney disease [66,67]
Polycyclic aromatic hydrocarbon	BaP	Vehicle exhaust, industrial processes, tobacco smoke, and grilled food	Cardiovascular disease [81], metabolic syndrome [82], NAFLD [83], and kidney disease [66]
Air pollution	PM_{10}, $PM_{2.5}$	Factories and manufacturing plants, transportation, agriculture, and waste management	Cardiovascular disease [84], diabetes [85], NAFLD [86], hypertension [67], and kidney disease [67]

Table 1. *Cont.*

Environmental Chemicals	Common Substances or Derivatives	Major Sources	Exposure-Associated CKM Phenotypes
Heavy metals	Pb, Cd, Hg	Manufacturing processes, emissions from vehicles, combustion of fossil fuels, improper disposal of electronic waste and batteries, and contaminating soil, water, and air	Cardiovascular disease [87], kidney disease [88], obesity [89], diabetes [89], and hypertension [89].

TCDD, 2,3,7,8-tetrachlorodibenzo-p-dioxin; PCDF, polychlorinated dibenzo-p-furan; PCDD, polychlorinated dibenzo-p-dioxin; PCB, dioxin-like polychlorinated biphenyl; BPA, bisphenol A; DEHP, di-2-ethylhexylphthalate; DBP, di-n-butyl phthalate; DPHP, diphenyl phosphate; TPHP, triphenyl phosphate; TDCPP, Tris-(1,3-dichloroisopropyl)phosphate; MPs, microplastics; PFOS, perfluorooctane sulfonic acid; PFOA, perfluorooctanoic acid; BaP, benzo(a)pyrene; PM_{10} (particulate matter < 10 mm in diameter), $PM_{2.5}$ (particulate matter < 2.5 mm); Pb, lead; Cd, cadmium; Hg, mercury.

4.1. Dioxins

Dioxin, the most extensively researched and toxic variant, is formally known as 2,3,7,8-tetrachlorodibenzo-pdioxin (TCDD). "Dioxins" typically refers to a group of closely related chemical compounds, including polychlorinated dibenzo-p-dioxins (PCDDs), dioxin-like polychlorinated biphenyls (PCBs), and polychlorinated dibenzofurans (PCDFs), which share similar chemical structures and properties. Dioxins are predominantly released from anthropogenic activities such as pesticide manufacturing, wood pulp bleaching, and waste incineration [90]. Accumulating in the food chain within the environment and persisting for extended periods in the body's fat tissue [91], dioxins can be encountered by pregnant mothers through the consumption of diets high in animal fat or via occupational exposure. Elevated exposure to dioxins has been linked to several facets of CKM syndrome, including cardiovascular disease [63], diabetes [64], metabolic syndrome [65], kidney disease [66,67], and hypertension [67].

4.2. Plastic Chemicals

The proliferation of plastic waste presents a significant environmental predicament, with a substantial portion of plastic being non-recyclable. Consequently, it infiltrates our surroundings, polluting oceans and disrupting ecosystems. Comprising a carbon backbone and augmented with numerous additional chemicals to form polymers, plastics harbor a plethora of toxic compounds including neurotoxicants, carcinogens, and endocrine disruptors such as Bisphenol A (BPA), di-2-ethylhexyl phthalate (DEHP)—the most prevalent phthalate, and organophosphate flame retardants (OPFRs). Furthermore, as plastics degrade, they fragment into microplastics (0.1–0.5 mm in diameter) and nanoplastics (1–1000 nm in diameter), exacerbating their detrimental impact on human health.

BPA has a characteristic structure that mimics estrogens by binding to their receptors [92], which causes it to be classified as an endocrine-disrupting chemical (EDC). Human exposure to BPA occurs primarily via the hydrolysis of polycarbonate plastics utilized in food and liquid containers, and medical devices. Several recent epidemiological studies suggest that BPA exposure is connected to the risk of developing cardiovascular disease [68], diabetes [69], obesity [69], NAFLD [70], hypertension, and kidney disease [66,67]. Importantly, evidence from mother–child cohort studies revealed that childhood obesity [93] and hypertension [94] could be related to maternal exposure to BPA.

People are constantly exposed to phthalates via plastic containers, food packaging, and medical devices. Similar to BPA, phthalates are recognized as EDCs and have been linked to all facets of CKM syndrome [66,67,71–74]. Another notable plastic compound is organophosphate flame retardant (OPFR), predominantly utilized as flame retardant plasticizers in engineering plastics [95]. OPFRs have been detected in indoor environments and are extensively employed in consumer goods like plastics, rubbers, construction materials, and electronic devices [95]. Emerging research has underscored the connections between OPFR metabolites and various components associated with CKM syndrome in humans, as outlined in Table 1 [75–78].

Microplastics (MPs) resulting from the environmental degradation of plastic waste are pervasive across diverse ecosystems, although the precise health risks to humans remain indeterminate [96]. Human exposure to MPs can range between 74,000 and 121,000 particles annually [96]. Accumulation of MPs has been observed in human blood, feces, breast milk, and certain organs [97]. A recent investigation involving 304 patients with carotid artery disease revealed that those with detectable MPs within atheroma were at heightened risk for cardiovascular events compared to those without such detection [79]. Nonetheless, there remains a dearth of information regarding the potential associations between MP exposure and other components of CKM syndrome.

4.3. Per- and Polyfluoroalkyl Substances (PFAS)

Per- and polyfluoroalkyl substances (PFAS) constitute a class of chemicals utilized in the production of fluoropolymer coatings and products engineered to resist water, heat, oil, and grease [98]. Epidemiological investigations have unveiled correlations between exposure to certain PFAS and a spectrum of CKM manifestations, encompassing obesity [80], diabetes [80], NAFLD [80], hypertension [67], and kidney disorders [99]. A longitudinal study on birth cohorts unearthed a positive relationship between prenatal PFAS exposure and subsequent obesity [100]. Additionally, maternal PFAS exposure has been associated with specific DNA methylation alterations, with these PFAS-linked CpG sites mapping to gene regions pertinent to cardiovascular health and renal function [101]. These revelations prompt consideration of the potential linkage between PFAS exposure and other components of CKM syndrome, warranting further elucidation.

4.4. Polycyclic Aromatic Hydrocarbon

With their intrinsic characteristics, polycyclic aromatic hydrocarbons (PAHs) persist as pollutants, displaying a diverse array of biological toxicities [102]. PAHs emerge during the refining of coal, crude oil, and natural gas [102]. Human studies have found a relationship between PAH exposure and cardiovascular disease [81], metabolic syndrome [82], NAFLD [83], and kidney disease [67,103]. As PAHs can pass through the placental barrier, studies have shown that exposure to PAHs during pregnancy can result in developmental toxicity [104]. Benzo(a)pyrene (BaP), a major example of PAHs, has shown epigenetic actions, such as inhibiting the activity of DNA methyltransferases and increasing histone deacetylases (HDACs) [105]. Considering the crucial role of epigenetic regulation in developmental programming [106], the interplay between PAH exposure and epigenetic regulation behind CKM programming deserves further evaluation.

4.5. Air Pollution

Air pollution, one of the greatest threats to global health, is also a risk factor for CKM syndrome. Airborne pollutants, encompassing carbon monoxide (CO), ozone (O_3), nitrogen oxides (NOx), sulfur dioxide (SO_2), volatile organic compounds (VOCs), and respirable particulate matter, exhibit variances in their chemical compositions [107]. Particulate matter is generally categorized by its mean aerodynamic diameter as PM10 (<10 μm in diameter), $PM_{2.5}$ (<2.5 μm), or ultrafine particles (UFPs, <0.1 μm). $PM_{2.5}$ and PM_{10} are frequently studied particulate matter indices, and both have been linked to various components of CKM syndrome, including cardiovascular disease [84,108], diabetes [85], NAFLD [86], kidney disease [67,109], and hypertension [108]. Highlighted in certain observational studies within exposed populations is the correlation between maternal exposure to $PM_{2.5}$ and negative outcomes in offspring, notably hypertension [110] and diabetes [111].

4.6. Heavy Metals

Considered the most significant threat to human health among all forms of pollution in drinking water and food are heavy metals, owing to their persistence in the environment and their bioavailability [112]. Epidemiological data indicate that chronic exposure to heavy metals, including cadmium (Cd), mercury (Hg), and lead (Pb), escalates the risks of cardiovascular disease [87], kidney disease [88], obesity [89], diabetes [89], and hypertension [89]. A meta-analysis comprising 13 studies demonstrates substantial links between Cd, Hg, Pb, and arsenic exposure during pregnancy and heightened risks of specific congenital heart diseases in offspring [113]. Additionally, a study revealed that elevated selenium levels are associated with an increased risk of congenital anomalies of the kidney and urinary tract (CAKUT) [114]. Reported in another investigation is an inverse relationship between maternal blood lead levels and kidney function in children aged 8–12 years who are overweight [115]. Furthermore, a study examining mother–infant pairs evaluates the impact of antenatal heavy metal exposures on childhood BP [115]. Although Cd shows no association with systolic BP overall, the inverse correlation between manganese and systolic BP is more pronounced at higher Cd levels [116].

5. Evidence from Animal Models: The Role of AHR in CKM Programming

While epidemiological observations suggest a correlation between environmental chemical exposures and CKM syndrome, there remains a scarcity of comprehensive information regarding antenatal chemical exposure and the manifestation of CKM syndrome in adulthood. It is important to note that these observational studies alone cannot definitively establish a causal relationship between antenatal chemical exposure and adult CKM syndrome. Moreover, these human studies fail to elucidate the molecular mechanisms underlying the development of CKM syndrome or provide strategies for reprogramming.

To delve into the role of the AHR in antenatal chemical exposure-induced CKM programming, animal models serve as invaluable tools. They facilitate the understanding of mechanisms and aid in the development of preventive strategies. Table 2 outlines animal studies that demonstrate the association between maternal chemical exposure and subsequent CKM syndrome in offspring, with a particular focus on AHR signaling. This review exclusively focuses on chemical exposures occurring during pregnancy and/or lactation, with an emphasis on reporting offspring outcomes commencing from childhood onwards.

Table 2. Overview of animal models of antenatal chemical-induced programmed CKM syndrome related to AHR signaling.

Chemical	Exposure Dose and Period	Species	Age at Evaluation (Weeks)	CKM Phenotypes	Refs.
TCDD	200 ng/kg orally on gestational days 14 and 21 and days 7 and 14 after birth	SD rats/M	12	Hypertension	[117]
TCDD	200 ng/kg in four once-weekly oral doses throughout pregnancy and lactation	SD rats/M	12	Hypertension	[118,119]
TCDD	200 ng/kg in four weekly oral doses throughout pregnancy and lactation	SD rats/M	16	Hypertension	[120]
TCDD	6.0 μg/g orally on gestational day 14.5	C57BL/6N mice/M	12	Cardiovascular dysfunction and kidney malformation	[121]
BPA	10 or 100 mg/kg/day throughout gestational days 9–16	OF1 mice/ M and F	5	Kidney dysfunction	[122]
BPA	50 μg/kg/day throughout gestation and lactation	SD rats/F	7	Obesity	[123]

Table 2. *Cont.*

Chemical	Exposure Dose and Period	Species	Age at Evaluation (Weeks)	CKM Phenotypes	Refs.
BPA	50 μg/kg/day throughout gestation and lactation	SD rats/F	7	NAFLD	[124]
BPA	50 mg/kg/day throughout gestation and lactation	SD rats/M	16	Hypertension	[125]
BPA	10 or 100 μg/kg/day throughout gestational days 9–16	SD rats/F	16	Insulin resistance and hyperlipidemia	[126]
DEHP	0.25 or 6.25 mg/kg/day throughout pregnancy	Wistar rats/ M and F	21	Kidney dysfunction and hypertension	[127]
DEHP	10 mg/kg/day throughout pregnancy and lactation	SD rats/M	12	Hypertension	[128]
DEHP	0.2, 2, or 20 mg/kg/day throughout pregnancy and lactation	ICR mice/M	12	Abnormal adipogenesis and glucose metabolism	[129]
DBP	850 mg/kg/day throughout gestational days 14–18	SD rats/M	8	Kidney dysfunction and renal fibrosis	[130]
DBP	33, 66, or 132 mg/kg/day from gestational day 7 throughout postnatal day 21	SD rats/F	12	Obesity	[131]
PFOS	50 μg/mL from gestational day 4 until delivery	SD rats/ M and F	16	Hypertension	[132]
BaP	600 or 1200 mg/kg/day throughout gestational days 14–17	LEH rats/ M and F	8	Hypertension	[133]
Cd	Cd chloride 2.0 or 2.5 mg/kg/day on gestational days 8, 10, 12, and 14	SD rats/M	7	Kidney injury	[134]
Cd	Cd chloride 0.5 mg/kg/day throughout pregnancy	Wistar rats/ M and F	8	Kidney dysfunction	[135]
Cd	500 ppb $CdCl_2$ in drinking water throughout pregnancy to postnatal day 10	CD-1 mice/ M and F	17	Obesity, hyperlipidemia, insulin resistance, and steatosis in F	[136]
$PM_{2.5}$	$PM_{2.5}$ exposure for 16 weeks prior to delivery	C57BL/6N mice/M and F	12	Hypertension	[137]
$PM_{2.5}$	$PM_{2.5}$ exposure 300 μg/m^3 for 2 h/day throughout	C57BL/6N mice/M and F	12	Cardiac hypertrophy	[138]
$PM_{2.5}$	Oropharyngeal drip of $PM_{2.5}$ (1.0 mg/kg) on gestational days 8, 10, and 12	SD rats/M	14	Hypertension	[139]
$PM_{2.5}$	Concentrated ambient $PM_{2.5}$ exposure throughout gestation and lactation	C57BL/6N mice/M and F	22	Obesity	[140]
$PM_{2.5}$	Diesel exhaust $PM_{2.5}$ 8.6 μg/day intratracheal instillation throughout pregnancy and lactation	C57BL/6N mice/M and F	22	Glucose intolerance and pancreatic islet dysfunction	[141]

TCDD, 2,3,7,8-tetrachlorodibenzo-p-dioxin; BPA, bisphenol A; DEHP, di-2-ethylhexylphthalate; DBP, di-n-butyl phthalate; PFOS, perfluorooctane sulfonic acid; BaP, benzo(a)pyrene; Cd, cadmium; $PM_{2.5}$ (particulate matter < 2.5 mm); SD, Sprague–Dawley rat; LEH, Long–Evans Hooded; OF1, Oncins France 1; M, male; F, female.

Table 2 illustrates that rodents are the predominant animal species utilized, with large animals not currently employed for studying similar exposures. The programming effects of environmental chemicals have been documented in rats aged between 7 and 21 weeks, corresponding to human ages from childhood to young adulthood [142].

The earliest AHR agonists identified were typically constituents of environmental chemicals including dioxins, BPA, phthalates, and PFOS, as well as polycyclic aromatic hydrocarbons [143]. Various chemicals have been assessed, including TCDD [117–121], BPA [122–126], DEHP [127–129], DBP [130,131], perfluorooctane sulfonic acid (PFOS) [132], BaP [133], Cd [134–136], and $PM_{2.5}$ [137,141]. Maternal exposure to the AHR ligand TCDD induces hypertension in offspring, correlated with AHR/CYP1A1 induction and TH17-mediated renal inflammation [118]. Additionally, cardiovascular dysfunction and kidney malformations have been observed in rat offspring prenatally exposed to TCDD [121].

Similarly to TCDD, BPA acts as an AHR ligand [35]. Exposure during pregnancy and lactation induces various components of CKM syndrome in rats, including kidney disease, obesity, NAFLD, hypertension, insulin resistance, and hyperlipidemia [122–126]. In a rat model of maternal BPA exposure, adult offspring developed hypertension alongside increased protein levels of AHR and mRNA expression of AHRR, CYP1A1, and ARNT in the offspring kidneys [125].

DEHP and DBP, two widely used phthalates acting as endocrine disruptors and AHR ligands, exhibit detrimental effects on offspring when maternally exposed. These effects include kidney dysfunction, hypertension, abnormal adipogenesis, and glucose metabolism alterations [127–131]. Maternal exposure to DBP is also implicated in offspring exhibiting kidney dysfunction, renal fibrosis, and obesity [130,131].

PFOS, another investigated environmental chemical, induces hypertension in both male and female rat offspring at 16 weeks of age when the dams are exposed during gestation [132]. Despite being known to activate AHR, PFOS's mechanism in this study remains unclear [144]. BaP, a polycyclic aromatic hydrocarbon, contributes to cardiovascular disease via AHR activation [145], and gestational exposure leads to offspring hypertension [133].

Maternal heavy metal exposure studies indicate Cd is the primary cause of adverse cardiovascular–kidney–metabolic outcomes programmed by early life exposure [134–136]. Prenatal Cd exposure in rats leads to kidney disease features in some studies, and obesity, hyperlipidemia, insulin resistance, and steatosis in a sex-specific manner [134–136].

Moreover, prenatal exposure to $PM_{2.5}$ has been linked to hypertension, cardiac hypertrophy, obesity, glucose intolerance, and pancreatic islet dysfunction in rodents [137–141]. Notably, AHR is implicated in $PM_{2.5}$'s prooxidative and pro-inflammatory effects [146].

6. Reprogramming Strategies Targeting AHR Signaling

In CKM programming, the pivotal roles played by the AHR underscore its significance as a potential therapeutic target. Indeed, the pharmacotherapy of numerous diseases has explored the targeting of the AHR, as extensively reviewed elsewhere [25,147–149]. Functioning as a ligand-driven receptor, the AHR exhibits complex pharmacology wherein its activation is contingent upon the type and concentration of ligands [147]. Ligands for the AHR can be categorized into three distinct groups: full agonists, partial agonists, and antagonists. Both agonists and antagonists bind to the receptor, yet only agonists induce a response. Partial agonists, on the other hand, stimulate a sub-maximal response despite occupying all the receptor sites; when paired with a full agonist, partial agonists act as functional antagonists [150]. Moreover, the activity of the AHR can be modulated by mechanisms independent of ligand binding [148].

Significantly, AHR facilitates developmental programming not only during gestation but also in the early stages of postnatal ontogenesis. This is crucial as interventions targeting AHR can be employed as reprogramming strategies during both pregnancy and the early postnatal period. Presently, various compounds that interact with the AHR have been identified as potential interventions for reprogramming to prevent CKM syndrome, including tryptophan metabolites, resveratrol, and butyrate. Each of these will be discussed sequentially.

6.1. Tryptophan Metabolites

Tryptophan undergoes conversion into a series of metabolites, many of which have been identified as ligands for the AHR (e.g., indole and tryptamine) [143]. In the gut, tryptophan metabolism traverses three principal pathways, encompassing the kynurenine pathway, the indole pathway, and the serotonin pathway [151]. Within a maternal CKD-induced hypertension model, the therapeutic efficacy of tryptophan in lowering BP is associated with its modulation of the AHR signaling pathway [152].

So far, the focus of research on microbial-derived metabolites involved in AHR modulation has primarily centered on tryptophan metabolites. While numerous microbial-derived tryptophan metabolites have demonstrated the ability to bind to and regulate AHR activity, only a few have been thoroughly investigated for their potential to reprogram AHR-related inflammatory responses and mitigate CKM syndrome. Nonetheless, additional studies are warranted to further elucidate their mechanisms and potential therapeutic applications.

Kynurenine, an AHR ligand, emerges through the degradation of tryptophan catalyzed by the enzyme indoleamine 2,3-dioxygenase (IDO). In the context of CKD, elevated kynurenine levels indicate activation of the IDO–kynurenine pathway [153]. IDO exhibits the capacity to foster the differentiation of Treg cells while impeding the differentiation of TH17 cells. TH17 cells, known for their production of interleukin 17 (IL-17), play a role in inflammation and tissue damage. Given kynurenine's potential to incite AHR-mediated inflammation, inhibitors of IDO present themselves as promising targets for the treatment of cardiovascular disease [154]. Considering the regulatory influence of AHR on both Treg and TH17 cells [52], and the participation of several microbial tryptophan catabolites as AHR ligands in the developmental programming of kidney disease and hypertension [155], further investigation is warranted to comprehensively understand the protective role of tryptophan metabolites and IDO inhibitors in modulating CKM syndrome of developmental origin.

6.2. Resveratrol

Studied for their AHR modulatory potential, polyphenols have garnered attention as compounds capable of reaching cells and potentially influencing AHR activity across the gut and other organs [149]. Among these polyphenols, several have been identified as AHR ligands, exhibiting either agonistic or antagonistic properties [143]. Despite numerous published studies highlighting the anti-inflammatory effects of various polyphenol types in the prevention and treatment of diverse diseases [156], investigations into the beneficial actions of resveratrol specifically in AHR-related inflammation in various animal models of CKM programming remain limited.

Resveratrol, a natural polyphenol abundant in grapes, is renowned for its antioxidant, anti-inflammatory, and prebiotic properties, and its ability to modulate AHR [157]. Resveratrol has been characterized as an antagonist of the AHR, capable of inhibiting the activation of members of the CYP1 family by impeding the recruitment of the transcription factors AHR and ARNT to XREs within the enhancer regions of CYP1 family genes [158,159].

It has been proposed as a reprogramming strategy to forestall cardiovascular disease, kidney disease, and metabolic syndrome [160,161]. Previous research has shown that TCDD-induced hypertension correlates with AHR activation and TH17-induced renal inflammation [118]. Conversely, supplementation with resveratrol during gestation and lactation can counteract TCDD-induced AHR signaling activation and TH17 responses. Similarly, perinatal resveratrol therapy restores maternal BPA exposure-induced increases in AHR protein levels and mRNA expression of AHRR, CYPA1A1, and ARNT [125]. Furthermore, resveratrol has been documented to function as an antagonist of the AHR, showing efficacy in mitigating offspring hypertension in alternative models of the developmental origins of hypertension [120,162].

Despite its advantages, the challenge of translating basic scientific findings into clinical practice is posed by the limited bioavailability of resveratrol [163]. To address this hurdle, previous efforts have centered on esterifying resveratrol with butyrate, thereby producing resveratrol butyrate esters (RBEs) with the aim of enhancing efficacy [164]. An improvement in hyperlipidemia and obesity in female progeny and hepatic steatosis in male offspring, both induced by maternal exposure to BPA, have been demonstrated in studies through the administration of low-dose RBEs (30 mg/L) [123,124]. In a maternal DEHP exposure model, low-dose RBE treatment significantly shielded adult rat offspring against hypertension, accompanied by a reduction in renal mRNA expression of CYP1A1 and ARNT [128].

6.3. Butyrate

Butyrate, a prevalent short-chain fatty acid (SCFA) derived from gut microbiota, exerts its effects through various mechanisms, including acting as a histone deacetylase (HDAC) inhibitor, signaling via SCFA receptors, or functioning as a postbiotic [165,166]. Reports indicate that butyrate can activate the AHR and enhance the functions of AHR activated by ligands [148]. While butyrate itself does not directly bind to AHR, it can induce the nuclear translocation of AHR and activate AHR independently of its HDAC activity and SCFA receptors [167]. Given that perinatal supplementation with butyrate has demonstrated protective effects against offspring hypertension and metabolic dysfunction in various developmental programming models [168–170], further investigation is warranted to elucidate the potential role of the AHR signaling pathway in these protective actions.

6.4. Others

AHR is a pivotal player in orchestrating epigenetic regulations impacting transcriptome alterations, chromatin architecture adjustments, and involvement in crucial signaling pathways [171]. Epigenetic mechanisms such as aberrant DNA methylation, histone modification, and microRNAs can drive changes in gene expression, potentially leading to developmental programming [172]. Chemical exposure-related developmental programming implicates various molecular pathways including oxidative stress, dysregulated nutrient-sensing signals, aberrant RAS, reduced nephron numbers, and dysbiotic gut microbiota, all intertwined with epigenetic programming [172–177]. Given the interconnectedness of these mechanisms with AHR signaling, there exists significant potential for cross-talk among them in the context of CKM programming. Consequently, targeting AHR-related epigenetic modifications within these pathways could offer promising avenues for intervention.

Despite the identification of numerous AHR ligands across diverse chemical structural classes [143,147,148], only a subset of them has been scrutinized for their effects on developmental programming. Therefore, further investigation is warranted to comprehensively understand the impact of AHR agonists/antagonists on key molecular pathways. This deeper understanding is essential for developing AHR-targeted reprogramming therapies aimed at mitigating CKM syndrome.

Furthermore, previous research has indicated that genetic variability in response to AHR ligands may highlight specific genes that could be under the regulation of AHR [178,179]. Identifying these genes has the potential to enhance comprehension of the AHR's involvement in developmental programming and could potentially pave the way for therapeutic interventions targeting adverse AHR-mediated CKM phenotypes.

7. Conclusions and Perspectives

The reviewed literature collectively suggests the impact of early developmental exposure to environmental chemicals on offspring's cardiovascular–kidney–metabolic health, culminating in CKM syndrome. The emerging concept from animal studies posits the intimate involvement of the AHR signaling pathway in CKM programming, induced by antenatal chemical exposure. Despite indications from animal models proposing that early interventions targeting AHR modulation could avert CKM syndrome, there persists a scarcity of human trials exploring their clinical translation. The relationship between antenatal chemical exposure and CKM programming and reprogramming via AHR links is depicted in Figure 3.

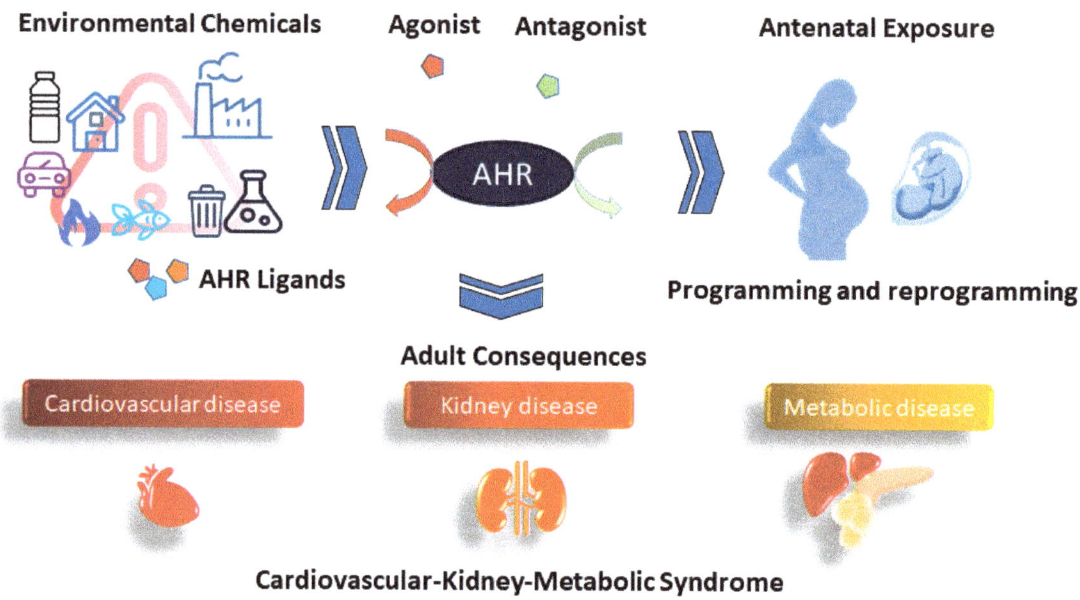

Figure 3. The AHR connects antenatal chemical exposure to CKM programming and reprogramming.

The impact of chemical exposure across various developmental stages can exhibit notable variations. Exposure during the initial phases of fetal development has the potential to interfere with organogenesis, while exposure later in development could impact the growth and functionality of organs that have already formed. By taking into account the timing of exposure and the vulnerability of organs, distinct effects on CKM programming can arise. Further exploration through animal studies is imperative to gain deeper insights into how exposure to different chemicals at various developmental stages can contribute to specific components of CKM syndrome later in life and to what extent. Another unresolved aspect pertains to the predominant use of mother–child cohorts in epidemiological studies, which pose challenges in extending observations into adulthood. There is a pressing need for additional long-term follow-up studies to comprehensively elucidate the chronic effects of antenatal chemical exposure mediated through AHR.

There exists an urgent imperative for multidisciplinary endeavors to undertake investigations discerning hazardous chemicals within the environment. Throughout pregnancy and early childhood, it is crucial to prioritize avoiding exposure to harmful substances and toxins in various settings such as the home, workplace, and recreational activities, as this is essential for promoting cardiovascular–kidney–metabolic well-being. While numerous environmental chemicals have been identified thus far, proactive efforts should persist in uncovering additional potentially harmful chemicals.

The growing spectrum of individual compounds that bind to and influence AHR-mediated responses and genes is continuously broadening [35,180,181]. This includes a wide range of structurally diverse synthetic chemicals, microbial metabolites, phytochemicals, and endogenous biochemicals. The potential exists within some of these compounds to serve as selective modulators of AHR for reprogramming strategies [147]. However, a significant caveat in the development of selective AHR modulators is the inability to readily predict their response selectivity as agonists or antagonists. The double-edged sword effects of AHR necessitate comprehensive testing to identify the optimal ligand for specific clinical applications.

In conclusion, antenatal exposure to environmental chemicals serves as a significant pathogenetic link in the developmental programming of CKM syndrome. With an enhanced comprehension of AHR's role in CKM programming, AHR modulators show promise in maximizing benefits without exacerbating toxicity, thereby averting maternal chemical-induced CKM syndrome.

Author Contributions: Conceptualization, Y.-L.T. and C.-N.H.; funding acquisition, Y.-L.T. and C.-N.H.; data curation, Y.-L.T. and C.-N.H.; writing—original draft, Y.-L.T. and C.-N.H.; writing—review and editing, Y.-L.T. and C.-N.H. All authors have read and agreed to the published version of the manuscript.

Funding: This work was supported by Kaohsiung Chang Gung Memorial Hospital, Kaohsiung, Taiwan, under grants CMRPG8M0752, CMRPG8M0721, CMRPG8M0722, CORPG8P0031, CMRPG8P0081, and CORPG8P0011.

Data Availability Statement: Data are contained within the article.

Conflicts of Interest: The authors declare no conflicts of interest.

References

1. Hanson, M.; Gluckman, P. Developmental origins of noncommunicable disease: Population and public health implications. *Am. J. Clin. Nutr.* **2011**, *94*, 1754S–1758S. [CrossRef] [PubMed]
2. Hanson, M.A.; Gluckman, P.D. Early developmental conditioning of later health and disease: Physiology or pathophysiology? *Physiol. Rev.* **2014**, *94*, 1027–1076. [CrossRef] [PubMed]
3. Fleming, T.P.; Velazquez, M.A.; Eckert, J.J. Embryos, DOHaD and David Barker. *J. Dev. Orig. Health Dis.* **2015**, *6*, 377–383. [CrossRef] [PubMed]
4. Alves, J.G.B.; Alves, L.V. Early-life nutrition and adult-life outcomes. *J. Pediatr.* **2024**, *100*, S4–S9. [CrossRef] [PubMed]
5. Lapehn, S.; Paquette, A.G. The Placental Epigenome as a Molecular Link between Prenatal Exposures and Fetal Health Outcomes through the DOHaD Hypothesis. *Curr. Environ. Health Rep.* **2022**, *9*, 490–501. [CrossRef] [PubMed]
6. Arima, Y.; Fukuoka, H. Developmental origins of health and disease theory in cardiology. *J. Cardiol.* **2020**, *76*, 14–17. [CrossRef] [PubMed]
7. American College of Obstetricians and Gynecologists. Exposure to toxic environmental agents. *Obstet. Gynecol.* **2013**, *122*, 931–935. [CrossRef] [PubMed]
8. National Research Council (US) Committee on Improving Risk Analysis Approaches Used by the U.S. EPA. *Science and Decisions: Advancing Risk Assessment*; National Academies Press (US): Washington, DC, USA, 2009.
9. Centers For Disease Control and Prevention. *Fourth National Report on Human Exposure to Environmental Chemicals, Updated Tables, January 2019*; US Department of Health and Human Services, Centers for Disease Control and Prevention: Atlanta, GA, USA, 2019.
10. Jaradat, J.H.; Nashwan, A.J. Cardiovascular-kidney-metabolic syndrome: Understanding the interconnections and the need for holistic intervention. *J. Med. Surg. Public Health* **2023**, *1*, 100028. [CrossRef]
11. Ndumele, C.E.; Rangaswami, J.; Chow, S.L.; Neeland, I.J.; Tuttle, K.R.; Khan, S.S.; Coresh, J.; Mathew, R.O.; Baker-Smith, C.M.; Carnethon, M.R.; et al. American Heart Association. Cardiovascular-Kidney-Metabolic Health: A Presidential Advisory from the American Heart Association. *Circulation* **2023**, *148*, 1606–1635. [CrossRef]
12. Hoffman, D.J.; Powell, T.L.; Barrett, E.S.; Hardy, D.B. Developmental origins of metabolic diseases. *Physiol. Rev.* **2021**, *101*, 739–795. [CrossRef]
13. Iturzaeta, A.; Sáenz Tejeira, M.M. Early programming of hypertension. *Arch. Argent. Pediatr.* **2022**, *120*, e8–e16. [PubMed]
14. Chevalier, R.L. Evolution, kidney development, and chronic kidney disease. *Semin. Cell Dev. Biol.* **2019**, *91*, 119–131. [CrossRef] [PubMed]
15. Saavedra, L.P.J.; Piovan, S.; Moreira, V.M.; Gonçalves, G.D.; Ferreira, A.R.O.; Ribeiro, M.V.G.; Peres, M.N.C.; Almeida, D.L.; Raposo, S.R.; da Silva, M.C.; et al. Epigenetic programming for obesity and noncommunicable disease: From womb to tomb. *Rev. Endocr. Metab. Disord.* **2023**, *25*, 309–324. [CrossRef]
16. Paauw, N.D.; van Rijn, B.B.; Lely, A.T.; Joles, J.A. Pregnancy as a critical window for blood pressure regulation in mother and child: Programming and reprogramming. *Acta Physiol.* **2017**, *219*, 241–259. [CrossRef] [PubMed]
17. Tain, Y.L.; Hsu, C.N. Interplay between oxidative stress and nutrient sensing signaling in the developmental origins of cardiovascular disease. *Int. J. Mol. Sci.* **2017**, *18*, 841. [CrossRef]
18. Kett, M.M.; Denton, K.M. Renal programming: Cause for concern? *Am. J. Physiol. Regul. Integr. Comp. Physiol.* **2011**, *300*, R791–R803. [CrossRef] [PubMed]
19. Goyal, D.; Limesand, S.W.; Goyal, R. Epigenetic responses and the developmental origins of health and disease. *J. Endocrinol.* **2019**, *242*, T105–T119. [CrossRef] [PubMed]
20. Sarkar, A.; Yoo, J.Y.; Valeria Ozorio Dutra, S.; Morgan, K.H.; Groer, M. The Association between Early-Life Gut Microbiota and Long-Term Health and Diseases. *J. Clin. Med.* **2021**, *10*, 459. [CrossRef] [PubMed]
21. Tain, Y.L.; Hsu, C.N. The Renin Angiotensin System and the Cardiovascular-Kidney-Metabolic Syndrome: Focus on Early-Life Programming. *Int. J. Mol. Sci.* **2024**, *25*, 3298. [CrossRef] [PubMed]
22. Kou, Z.; Dai, W. Aryl hydrocarbon receptor: Its roles in physiology. *Biochem. Pharmacol.* **2021**, *185*, 114428. [CrossRef]

23. Vogel, C.F.A.; Van Winkle, L.S.; Esser, C.; Haarmann-Stemmann, T. The aryl hydrocarbon receptor as a target of environmental stressors—Implications for pollution mediated stress and inflammatory responses. *Redox Biol.* **2020**, *34*, 101530. [CrossRef]
24. Zablon, H.A.; Ko, C.I.; Puga, A. Converging Roles of the Aryl Hydrocarbon Receptor in Early Embryonic Development, Maintenance of Stemness, and Tissue Repair. *Toxicol. Sci.* **2021**, *182*, 1–9. [CrossRef]
25. Yi, T.; Wang, J.; Zhu, K.; Tang, Y.; Huang, S.; Shui, X.; Ding, Y.; Chen, C.; Lei, W. Aryl Hydrocarbon Receptor: A New Player of Pathogenesis and Therapy in Cardiovascular Diseases. *BioMed Res. Int.* **2018**, *2018*, 6058784. [CrossRef]
26. Brito, J.S.; Borges, N.A.; Esgalhado, M.; Magliano, D.C.; Soulage, C.O.; Mafra, D. Aryl Hydrocarbon Receptor Activation in Chronic Kidney Disease: Role of Uremic Toxins. *Nephron* **2017**, *137*, 1–7. [CrossRef]
27. Sayed, T.S.; Maayah, Z.H.; Zeidan, H.A.; Agouni, A.; Korashy, H.M. Insight into the physiological and pathological roles of the aryl hydrocarbon receptor pathway in glucose homeostasis, insulin resistance, and diabetes development. *Cell. Mol. Biol. Lett.* **2022**, *27*, 103. [CrossRef]
28. Jones, S. An overview of the basic helix-loop-helix proteins. *Genome Biol.* **2004**, *5*, 226. [CrossRef]
29. Busbee, P.B.; Rouse, M.; Nagarkatti, M.; Nagarkatti, P.S. Use of natural AhR ligands as potential therapeutic modalities against inflammatory disorders. *Nutr. Rev.* **2013**, *71*, 353–369. [CrossRef]
30. Fukunaga, B.N.; Probst, M.R.; Reisz Porszasz, S.; Hankinson, O. Identification of functional domains of the aryl hydrocarbon receptor. *J. Biol. Chem.* **1995**, *270*, 29270–29278. [CrossRef]
31. Ho, P.P.; Steinman, L. The aryl hydrocarbon receptor: A regulator of Th17 and Treg cell development in disease. *Cell Res.* **2008**, *18*, 605–608. [CrossRef]
32. Kumar, M.B.; Ramadoss, P.; Reen, R.K.; Perdew, G.H. The Q-rich subdomain of the human Ah receptor transactivation domain is required for dioxin-mediated transcriptional activity. *J. Biol. Chem.* **2001**, *276*, 42302–42310. [CrossRef]
33. Hankinson, O. Role of coactivators in transcriptional activation by the aryl hydrocarbon receptor. *Arch. Biochem. Biophys.* **2005**, *433*, 379–386. [CrossRef]
34. Wright, E.J.; Pereira De Castro, J.; Joshi, A.D.; Elferink, C.J. Canonical and non-canonical aryl hydrocarbon receptor signaling pathways. *Curr. Opin. Toxicol.* **2017**, *2*, 87–92. [CrossRef]
35. Avilla, M.N.; Malecki, K.M.C.; Hahn, M.E.; Wilson, R.H.; Bradfield, C.A. The Ah receptor: Adaptive metabolism, ligand diversity, and the xenokine model. *Chem. Res. Toxicol.* **2020**, *33*, 860–879. [CrossRef]
36. Meyer, B.K.; Petrulis, J.R.; Perdew, G.H. Aryl hydrocarbon (Ah) receptor levels are selectively modulated by hsp90-associated immunophilin homolog XAP2. *Cell Stress Chaperones* **2000**, *5*, 243–254. [CrossRef]
37. Larigot, L.; Juricek, L.; Dairou, J.; Coumoul, X. AhR signaling pathways and regulatory functions. *Biochim. Open* **2018**, *7*, 1–9. [CrossRef]
38. Jackson, D.P.; Joshi, A.D.; Elferink, C.J. Ah receptor pathway intricacies; signaling through diverse protein partners and DNA-motifs. *Toxicol. Res.* **2015**, *4*, 1143–1158. [CrossRef]
39. Vogel, C.F.; Matsumura, F. A new cross-talk between the aryl hydrocarbon receptor and RelB, a member of the NF-kappaB family. *Biochem. Pharmacol.* **2009**, *77*, 734–745. [CrossRef]
40. Huang, G.; Elferink, C.J. A novel nonconsensus xenobiotic response element capable of mediating aryl hydrocarbon receptor-dependent gene expression. *Mol. Pharmacol.* **2012**, *81*, 338–347. [CrossRef]
41. Esser, C.; Rannug, A. The aryl hydrocarbon receptor in barrier organ physiology, immunology, and toxicology. *Pharmacol. Rev.* **2015**, *67*, 259–279. [CrossRef]
42. Jiang, Y.Z.; Wang, K.; Fang, R.; Zheng, J. Expression of aryl hydrocarbon receptor in human placentas and fetal tissues. *J. Histochem. Cytochem.* **2010**, *58*, 679–685. [CrossRef]
43. Lund, A.K.; Goens, M.B.; Nuñez, B.A.; Walker, M.K. Characterizing the role of endothelin-1 in the progression of cardiac hypertrophy in aryl hydrocarbon receptor (AhR) null mice. *Toxicol. Appl. Pharmacol.* **2006**, *212*, 127–135. [CrossRef]
44. Thackaberry, E.A.; Gabaldon, D.M.; Walker, M.K.; Smith, S.M. Aryl hydrocarbon receptor null mice develop cardiac hypertrophy and increased hypoxia-inducible factor-1alpha in the absence of cardiac hypoxia. *Cardiovasc. Toxicol.* **2002**, *2*, 263–274. [CrossRef]
45. Ko, C.I.; Fan, Y.; de Gannes, M.; Wang, Q.; Xia, Y.; Puga, A. Repression of the Aryl Hydrocarbon Receptor Is Required to Maintain Mitotic Progression and Prevent Loss of Pluripotency of Embryonic Stem Cells. *Stem Cells* **2016**, *34*, 2825–2839. [CrossRef]
46. Lahvis, G.P.; Lindell, S.L.; Thomas, R.S.; McCuskey, R.S.; Murphy, C.; Glover, E.; Bentz, M.; Southard, J.; Bradfield, C.A. Portosystemic shunting and persistent fetal vascular structures in aryl hydrocarbon receptor-deficient mice. *Proc. Natl. Acad. Sci. USA* **2000**, *97*, 10442–10447. [CrossRef]
47. Ichihara, S.; Yamada, Y.; Ichihara, G.; Nakajima, T.; Li, P.; Kondo, T.; Gonzalez, F.J.; Murohara, T. A role for the aryl hydrocarbon receptor in regulation of ischemia-induced angiogenesis. *Arterioscler. Thromb. Vasc. Biol.* **2007**, *27*, 1297–1304. [CrossRef]
48. Cuartero, M.I.; Ballesteros, I.; de la Parra, J.; Harkin, A.L.; Abautret-Daly, A.; Sherwin, E.; Fernández-Salguero, P.; Corbí, A.L.; Lizasoain, I.; Moro, M.A. L-kynurenine/aryl hydrocarbon receptor pathway mediates brain damage after experimental stroke. *Circulation* **2014**, *130*, 2040–2051. [CrossRef]
49. Nakagawa, K.; Kobayashi, F.; Kamei, Y.; Tawa, M.; Ohkita, M. Acute Kynurenine Exposure of Rat Thoracic Aorta Induces Vascular Dysfunction via Superoxide Anion Production. *Biol. Pharm. Bull.* **2022**, *45*, 522–527. [CrossRef]
50. Lund, A.K.; Agbor, L.N.; Zhang, N.; Baker, A.; Zhao, H.; Fink, G.D.; Kanagy, N.L.; Walker, M.K. Loss of the Aryl Hydrocarbon Receptor Induces Hypoxemia, Endothelin-1, and Systemic Hypertension at Modest Altitude. *Hypertension* **2008**, *51*, 803–809. [CrossRef]

51. Zhang, N.; Agbor, L.N.; Scott, J.A.; Zalobowski, T.; Elased, K.M.; Trujillo, A.; Duke, M.S.; Wolf, V.; Walsh, M.T.; Born, J.L.; et al. An Activated Renin-Angiotensin System Maintains Normal Blood Pressure in Aryl Hydrocarbon Receptor Heterozygous Mice but Not in Null Mice. *Biochem. Pharmacol.* **2010**, *80*, 197–204. [CrossRef]
52. Stevens, E.A.; Mezrich, J.D.; Bradfield, C.A. The aryl hydrocarbon receptor: A perspective on potential roles in the immune system. *Immunology* **2009**, *127*, 299–311. [CrossRef]
53. Sallée, M.; Dou, L.; Cerini, C.; Poitevin, S.; Brunet, P.; Burtey, S. The aryl hydrocarbon receptor-activating effect of uremic toxins from tryptophan metabolism: A new concept to understand cardiovascular complications of chronic kidney disease. *Toxins* **2014**, *6*, 934–949. [CrossRef]
54. Neavin, D.R.; Liu, D.; Ray, B.; Weinshilboum, R.M. The Role of the Aryl Hydrocarbon Receptor (AHR) in Immune and Inflammatory Diseases. *Int. J. Mol. Sci.* **2018**, *19*, 3851. [CrossRef]
55. Curran, C.S.; Kopp, J.B. Aryl Hydrocarbon Receptor Mechanisms Affecting Chronic Kidney Disease. *Front. Pharmacol.* **2022**, *13*, 782199. [CrossRef]
56. Ding, M.; Coward, R.J.; Jeansson, M.; Kim, W.; Quaggin, S.E. Regulation of Hypoxia-Inducible Factor 2-a Is Essential for Integrity of the Glomerular Barrier. *Am. J. Physiol. Ren. Physiol.* **2013**, *304*, F120–F126. [CrossRef]
57. Nakano, N.; Sakata, N.; Katsu, Y.; Nochise, D.; Sato, E.; Takahashi, Y.; Yamaguchi, S.; Haga, Y.; Ikeno, S.; Motizuki, M.; et al. Dissociation of the AhR/ARNT complex by TGF-β/Smad signaling represses CYP1A1 gene expression and inhibits benze[a]pyrene-mediated cytotoxicity. *J. Biol. Chem.* **2020**, *295*, 9033–9051. [CrossRef]
58. Thackaberry, E.A.; Bedrick, E.J.; Goens, M.B.; Danielson, L.; Lund, A.K.; Gabaldon, D.; Smith, S.M.; Walker, M.K. Insulin regulation in AhR-null Mice: Embryonic cardiac enlargement, neonatal macrosomia, and altered insulin regulation and response in pregnant and aging AhR-null females. *Toxicol. Sci.* **2003**, *76*, 407–417. [CrossRef]
59. Xia, H.; Zhu, X.; Zhang, X.; Jiang, H.; Li, B.; Wang, Z.; Li, D.; Jin, Y. Alpha-naphthoflavone attenuates non-alcoholic fatty liver disease in oleic acid-treated HepG2 hepatocytes and in high fat diet-fed mice. *Biomed. Pharmacother.* **2019**, *118*, 109287. [CrossRef]
60. Tseng, H.L.; Yang, S.C.; Yang, S.H.; Shieh, K.R. Hepatic circadian-clock system altered by insulin resistance, diabetes and insulin sensitizer in mice. *PLoS ONE* **2015**, *10*, e0120380. [CrossRef]
61. Dou, H.; Duan, Y.; Zhang, X.; Yu, Q.; Di, Q.; Song, Y.; Li, P.; Gong, Y. Aryl hydrocarbon receptor (AhR) regulates adipocyte differentiation by assembling CRL4B ubiquitin ligase to target PPARγ for proteasomal degradation. *J. Biol. Chem.* **2019**, *294*, 18504–18515. [CrossRef]
62. Kern, P.A.; Dicker-Brown, A.; Said, S.T.; Kennedy, R.; Fonseca, V.A. The stimulation of tumor necrosis factor and inhibition of glucose transport and lipoprotein lipase in adipose cells by 2,3,7,8-tetrachlorodibenzo-p-dioxin. *Metabolism* **2002**, *51*, 65–68. [CrossRef]
63. Ha, M.H.; Lee, D.H.; Jacobs, D.R., Jr. Association between serum concentrations of persistent organic pollutants and self-reported cardiovascular disease prevalence: Results from the National Health and Nutrition Examination Survey, 1999–2002. *Environ. Health Perspect.* **2007**, *115*, 1204–1209. [CrossRef]
64. Magliano, D.J.; Loh, V.H.Y.; Harding, J.L.; Botton, J.; Shaw, J.E. Persistent organic pollutants and diabetes: A review of the epidemiological evidence. *Diabetes Metab.* **2014**, *40*, 1–14. [CrossRef]
65. Gao, J.; Xu, Y.; Zhong, T.; Yu, X.; Wang, L.; Xiao, Y.; Peng, Y.; Sun, Q. A review of food contaminant 2,3,7,8-tetrachlorodibenzo-p-dioxin and its toxicity associated with metabolic disorders. *Curr. Res. Food Sci.* **2023**, *7*, 100617. [CrossRef]
66. Kataria, A.; Trasande, L.; Trachtmanm, H. The effects of environmental chemicals on renal function. *Nat. Rev. Nephrol.* **2015**, *11*, 610–625. [CrossRef]
67. Hsu, C.N.; Tain, Y.L. Adverse Impact of Environmental Chemicals on Developmental Origins of Kidney Disease and Hypertension. *Front. Endocrinol.* **2021**, *12*, 745716. [CrossRef]
68. Chen, M.; Yang, Y.; Baral, K.; Fu, Y.; Meng, Y.; Zhang, Y.; Sun, F.; Zhao, M. Relationship between bisphenol A and the cardiovascular disease metabolic risk factors in American adults: A population-based study. *Chemosphere* **2023**, *324*, 138289. [CrossRef]
69. Pérez-Bermejo, M.; Mas-Pérez, I.; Murillo-Llorente, M.T. The Role of the Bisphenol A in Diabetes and Obesity. *Biomedicines* **2021**, *9*, 666. [CrossRef]
70. Dallio, M.; Masarone, M.; Errico, S.; Gravina, A.G.; Nicolucci, C.; Di Sarno, R.; Gionti, L.; Tuccillo, C.; Persico, M.; Stiuso, P.; et al. Role of bisphenol A as environmental factor in the promotion of non-alcoholic fatty liver disease: In vitro and clinical study. *Aliment. Pharmacol. Ther.* **2018**, *47*, 826–837. [CrossRef]
71. Mariana, M.; Cairrao, E. Phthalates Implications in the Cardiovascular System. *J. Cardiovasc. Dev. Dis.* **2020**, *7*, 26. [CrossRef]
72. Mariana, M.; Cairrao, E. The Relationship between Phthalates and Diabetes: A Review. *Metabolites* **2023**, *13*, 746. [CrossRef]
73. Mérida, D.M.; Moreno-Franco, B.; Marquès, M.; León-Latre, M.; Laclaustra, M.; Guallar-Castillón, P. Phthalate exposure and the metabolic syndrome: A systematic review and meta-analysis. *Environ. Pollut.* **2023**, *333*, 121957. [CrossRef]
74. Cai, S.; Fan, J.; Ye, J.; Rao, X.; Li, Y. Phthalates exposure is associated with non-alcoholic fatty liver disease among US adults. *Ecotoxicol. Environ. Saf.* **2021**, *224*, 112665. [CrossRef]
75. Guo, X.; Wu, B.; Xia, W.; Gao, J.; Xie, P.; Feng, L.; Sun, C.; Liang, M.; Ding, X.; Zhao, D.; et al. Association of organophosphate ester exposure with cardiovascular disease among US adults: Cross-sectional findings from the 2011–2018 National Health and Nutrition Examination Survey. *Chemosphere* **2022**, *308*, 136428. [CrossRef] [PubMed]
76. Luo, K.; Zhang, R.; Aimuzi, R.; Wang, Y.; Nian, M.; Zhang, J. Exposure to Organophosphate esters and metabolic syndrome in adults. *Environ. Int.* **2020**, *143*, 105941. [CrossRef] [PubMed]

77. Tsai, K.F.; Cheng, F.J.; Huang, W.T.; Kung, C.T.; Lee, C.T.; Cheng, B.C.; Chen, J.B.; Li, S.H.; Wang, C.C.; Wang, L.J.; et al. The associations between renal disease severity and exposure to organophosphate flame retardants in patients with chronic kidney disease. *Environ. Int.* **2022**, *170*, 107573. [CrossRef]
78. Guo, X.; Ke, Y.; Wu, B.; Song, Q.; Sun, C.; Li, Y.; Wang, H.; Su, W.; Liang, Q.; Lowe, S.; et al. Exploratory analysis of the association between organophosphate ester mixtures with high blood pressure of children and adolescents aged 8–17 years: Cross-sectional findings from the National Health and Nutrition Examination Survey. *Environ. Sci. Pollut. Res. Int.* **2023**, *30*, 22900–22912. [CrossRef]
79. Marfella, R.; Prattichizzo, F.; Sardu, C.; Fulgenzi, G.; Graciotti, L.; Spadoni, T.; D'Onofrio, N.; Scisciola, L.; La Grotta, R.; Frigé, C.; et al. Microplastics and Nanoplastics in Atheromas and Cardiovascular Events. *N. Engl. J. Med.* **2024**, *390*, 900–910. [CrossRef]
80. Qi, W.; Clark, J.M.; Timme-Laragy, A.R.; Park, Y. Per- and Polyfluoroalkyl Substances and Obesity, Type 2 Diabetes and Non-alcoholic Fatty Liver Disease: A Review of Epidemiologic Findings. *Toxicol. Environ. Chem.* **2020**, *102*, 1–36. [CrossRef] [PubMed]
81. Mallah, M.A.; Changxing, L.; Mallah, M.A.; Naveed, M.; Liu, Y.; Noreen, S.; Xi, H.; Wang, W.; Feng, F.; Zhang, Q. Association of urinary polycyclic aromatic hydrocarbon metabolites and cardiovascular disease among US population: A cross-sectional study. *Environ. Res.* **2022**, *209*, 112775. [CrossRef] [PubMed]
82. Yang, X.; Xue, Q.; Wen, Y.; Huang, Y.; Wang, Y.; Mahai, G.; Yan, T.; Liu, Y.; Rong, T.; Wang, Y.; et al. Environmental polycyclic aromatic hydrocarbon exposure in relation to metabolic syndrome in US adults. *Sci. Total Environ.* **2022**, *840*, 156673. [CrossRef]
83. Choi, Y.H.; Lee, J.Y.; Moon, K.W. Exposure to volatile organic compounds and polycyclic aromatic hydrocarbons is associated with the risk of non-alcoholic fatty liver disease in Korean adolescents: Korea National Environmental Health Survey (KoNEHS) 2015–2017. *Ecotoxicol. Environ. Saf.* **2023**, *251*, 114508. [CrossRef] [PubMed]
84. Krittanawong, C.; Qadeer, Y.K.; Hayes, R.B.; Wang, Z.; Virani, S.; Thurston, G.D.; Lavie, C.J. PM2.5 and Cardiovascular Health Risks. *Curr. Probl. Cardiol.* **2023**, *48*, 101670. [CrossRef] [PubMed]
85. Li, X.; Wang, M.; Song, Y.; Ma, H.; Zhou, T.; Liang, Z.; Qi, L. Obesity and the relation between joint exposure to ambient air pollutants and incident type 2 diabetes: A cohort study in UK Biobank. *PLoS Med.* **2021**, *18*, e1003767. [CrossRef] [PubMed]
86. Chen, J.; Wu, L.; Yang, G.; Zhang, C.; Liu, X.; Sun, X.; Chen, X.; Wang, N. The influence of PM2.5 exposure on non-alcoholic fatty liver disease. *Life Sci.* **2021**, *270*, 119135. [CrossRef] [PubMed]
87. Duan, W.; Xu, C.; Liu, Q.; Xu, J.; Weng, Z.; Zhang, X.; Basnet, T.B.; Dahal, M.; Gu, A. Levels of a mixture of heavy metals in blood and urine and all-cause, cardiovascular disease and cancer mortality: A population-based cohort study. *Environ. Pollut.* **2020**, *263*, 114630. [CrossRef] [PubMed]
88. Tsai, H.J.; Hung, C.H.; Wang, C.W.; Tu, H.P.; Li, C.H.; Tsai, C.C.; Lin, W.Y.; Chen, S.C.; Kuo, C.H. Associations among Heavy Metals and Proteinuria and Chronic Kidney Disease. *Diagnostics* **2021**, *11*, 282. [CrossRef] [PubMed]
89. Wang, X.; Mukherjee, B.; Park, S.K. Associations of cumulative exposure to heavy metal mixtures with obesity and its comorbidities among U.S. adults in NHANES 2003–2014. *Environ. Int.* **2018**, *121*, 683–694. [CrossRef] [PubMed]
90. Dopico, M.; Gómez, A. Review of the Current State and Main Sources of Dioxins around the World. *J. Air Waste Manag. Assoc.* **2015**, *65*, 1033–1049. [CrossRef] [PubMed]
91. Milbrath, M.O.; Wenger, Y.; Chang, C.W.; Emond, C.; Garabrant, D.; Gillespie, B.W.; Jolliet, O. Apparent half-lives of dioxins, furans, and polychlorinated biphenyls as a function of age, body fat, smoking status, and breast-feeding. *Environ. Health Perspect.* **2009**, *117*, 417–425. [CrossRef]
92. Alonso-Magdalena, P.; Ropero, A.B.; Soriano, S.; García-Arévalo, M.; Ripoll, C.; Fuentes, E.; Quesada, I.; Nadal, A. Bisphenol-A acts as a potent estrogen via non-classical estrogen triggered pathways. *Mol. Cell. Endocrinol.* **2012**, *355*, 201–207. [CrossRef]
93. Agay-Shay, K.; Martinez, D.; Valvi, D.; Garcia-Esteban, R.; Basagaña, X.; Robinson, O.; Casas, M.; Sunyer, J.; Vrijheid, M. Exposure to Endocrine-Disrupting Chemicals during Pregnancy and Weight at 7 Years of Age: A Multi-pollutant Approach. *Environ. Health Perspect.* **2015**, *123*, 1030–1037. [CrossRef] [PubMed]
94. Bae, S.; Lim, Y.H.; Lee, Y.A.; Shin, C.H.; Oh, S.Y.; Hong, Y.C. Maternal Urinary Bisphenol a Concentration during Midterm Pregnancy and Children's Blood Pressure at Age 4. *Hypertension* **2017**, *69*, 367–374. [CrossRef] [PubMed]
95. Kung, H.C.; Hsieh, Y.K.; Huang, B.W.; Cheruiyot, N.K.; Chang-Chien, G.P. An Overview: Organophosphate Flame Retardants in the Atmosphere. *Aerosol Air Qual. Res.* **2022**, *22*, 220148. [CrossRef]
96. Cox, K.D.; Covernton, G.A.; Davies, H.L.; Dower, J.F.; Juanes, F.; Dudas, S.E. Human Consumption of Microplastics. *Environ. Sci. Technol.* **2019**, *53*, 7068–7074. [CrossRef] [PubMed]
97. Kutralam-Muniasamy, G.; Shruti, V.C.; Pérez-Guevara, F.; Roy, P.D. Microplastic diagnostics in humans: "The 3Ps" Progress, problems, and prospects. *Sci. Total Environ.* **2023**, *856*, 159164. [CrossRef] [PubMed]
98. Sunderland, E.M.; Hu, X.C.; Dassuncao, C.; Tokranov, A.K.; Wagner, C.C.; Allen, J.G. A Review of the Pathways of Human Exposure to Poly- and Perfluoroalkyl Substances (Pfass) and Present Understanding of Health Effects. *J. Expo. Sci. Environ. Epidemiol.* **2019**, *29*, 131–147. [CrossRef] [PubMed]
99. Shankar, A.; Xiao, J.; Ducatman, A. Perfluoroalkyl Chemicals and Chronic Kidney Disease in US Adults. *Am. J. Epidemiol.* **2011**, *174*, 893–900. [CrossRef] [PubMed]
100. Starling, A.P.; Adgate, J.L.; Hamman, R.F.; Kechris, K.; Calafat, A.M.; Dabelea, D. Prenatal exposure to per- and polyfluoroalkyl substances and infant growth and adiposity: The Healthy Start Study. *Environ. Int.* **2019**, *131*, 104983. [CrossRef] [PubMed]

101. Liu, Y.; Eliot, M.N.; Papandonatos, G.D.; Kelsey, K.T.; Fore, R.; Langevin, S.; Buckley, J.; Chen, A.; Lanphear, B.P.; Cecil, K.M.; et al. Gestational Perfluoroalkyl Substance Exposure and DNA Methylation at Birth and 12 Years of Age: A Longitudinal Epigenome-Wide Association Study. *Environ. Health Perspect.* **2022**, *130*, 37005. [CrossRef]
102. Patel, A.B.; Shaikh, S.; Jain, K.R.; Desai, C.; Madamwar, D. Polycyclic Aromatic Hydrocarbons: Sources, Toxicity, and Remediation Approaches. *Front. Microbiol.* **2020**, *11*, 562813. [CrossRef]
103. Rahman, H.H.; Niemann, D.; Munson-McGee, S.H. Association of chronic kidney disease with exposure to polycyclic aromatic hydrocarbons in the US population. *Environ. Sci. Pollut. Res. Int.* **2022**, *29*, 24024–24034. [CrossRef]
104. Drwal, E.; Rak, A.; Gregoraszczuk, E.L. Review: Polycyclic Aromatic Hydrocarbons (Pahs)-Action on Placental Function and Health Risks in Future Life of Newborns. *Toxicology* **2019**, *411*, 133–142. [CrossRef]
105. Bukowska, B.; Sicińska, P. Influence of Benzo(a)pyrene on Different Epigenetic Processes. *Int. J. Mol. Sci.* **2021**, *22*, 13453. [CrossRef]
106. Bianco-Miotto, T.; Craig, J.M.; Gasser, Y.P.; van Dijk, S.J.; Ozanne, S.E. Epigenetics and DOHaD: From basics to birth and beyond. *J. Dev. Orig. Health Dis.* **2017**, *8*, 513–519. [CrossRef]
107. Kampa, M.; Castanas, E. Human health effects of air pollution. *Environ. Pollut.* **2008**, *151*, 362–367. [CrossRef]
108. Zhang, S.; Qian, Z.M.; Chen, L.; Zhao, X.; Cai, M.; Wang, C.; Zou, H.; Wu, Y.; Zhang, Z.; Li, H.; et al. Exposure to Air Pollution during Pre-Hypertension and Subsequent Hypertension, Cardiovascular Disease, and Death: A Trajectory Analysis of the UK Biobank Cohort. *Environ. Health Perspect.* **2023**, *131*, 17008. [CrossRef]
109. An, Y.; Liu, Z.H. Air Pollution and Kidney Diseases: PM2.5 as an Emerging Culprit. *Contrib. Nephrol.* **2021**, *199*, 274–284.
110. Zhang, M.; Mueller, N.T.; Wang, H.; Hong, X.; Appel, L.J.; Wang, X. Maternal Exposure to Ambient Particulate Matter $\leq$ 2.5 μm During Pregnancy and the Risk for High Blood Pressure in Childhood. *Hypertension* **2018**, *72*, 194–201. [CrossRef]
111. Elten, M.; Donelle, J.; Lima, I.; Burnett, R.T.; Weichenthal, S.; Stieb, D.M.; Hystad, P.; van Donkelaar, A.; Chen, H.; Paul, L.A.; et al. Ambient air pollution and incidence of early-onset paediatric type 1 diabetes: A retrospective population-based cohort study. *Environ. Res.* **2020**, *184*, 109291. [CrossRef]
112. Rehman, K.; Fatima, F.; Waheed, I.; Akash, M.S.H. Prevalence of exposure of heavy metals and their impact on health consequences. *J. Cell. Biochem.* **2018**, *119*, 157–184. [CrossRef]
113. Li, S.; Wang, Q.; Luo, W.; Jia, S.; Liu, D.; Ma, W.; Gu, H.; Wei, X.; He, Y.; Cao, S.; et al. Relationship between maternal heavy metal exposure and congenital heart defects: A systematic review and meta-analysis. *Environ. Sci. Pollut. Res. Int.* **2022**, *29*, 55348–55366. [CrossRef]
114. Iwaya, Y.; Sanefuji, M.; Nishiyama, K.; Sonoda, Y.; Hamada, N.; Suga, R.; Ochiai, M.; Shimono, M.; Kusuhara, K.; Ohga, S.; et al. Prenatal metal levels and congenital anomalies of the kidney and urinary tract: The Japan Environment and Children's Study. *Sci. Total Environ.* **2023**, *890*, 164356. [CrossRef]
115. Saylor, C.; Tamayo-Ortiz, M.; Pantic, I.; Amarasiriwardena, C.; McRae, N.; Estrada-Gutierrez, G.; Parra-Hernandez, S.; Tolentino, M.C.; Baccarelli, A.A.; Fadrowski, J.J.; et al. Prenatal blood lead levels and reduced preadolescent glomerular filtration rate: Modification by body mass index. *Environ. Int.* **2021**, *154*, 106414. [CrossRef]
116. Zhang, M.; Liu, T.; Wang, G.; Buckley, J.P.; Guallar, E.; Hong, X.; Wang, M.C.; Wills-Karp, M.; Wang, X.; Mueller, N.T. In Utero Exposure to Heavy Metals and Trace Elements and Childhood Blood Pressure in a U.S. Urban, Low-Income, Minority Birth Cohort. *Environ. Health Perspect.* **2021**, *129*, 67005. [CrossRef]
117. Hsu, C.N.; Chan, J.Y.H.; Yu, H.R.; Lee, W.C.; Wu, K.L.H.; Chang-Chien, G.P.; Lin, S.; Hou, C.Y.; Tain, Y.L. Targeting on Gut Microbiota-Derived Metabolite Trimethylamine to Protect Adult Male Rat Offspring against Hypertension Programmed by Combined Maternal High-Fructose Intake and Dioxin Exposure. *Int. J. Mol. Sci.* **2020**, *21*, 5488. [CrossRef]
118. Hsu, C.N.; Hung, C.H.; Hou, C.Y.; Chang, C.I.; Tain, Y.L. Perinatal Resveratrol Therapy to Dioxin-Exposed Dams Prevents the Programming of Hypertension in Adult Rat Offspring. *Antioxidants* **2021**, *10*, 1393. [CrossRef]
119. Hsu, C.N.; Hou, C.Y.; Lee, C.T.; Chang-Chien, G.P.; Lin, S.; Tain, Y.L. Maternal 3,3-Dimethyl-1-Butanol Therapy Protects Adult Male Rat Offspring against Hypertension Programmed by Perinatal TCDD Exposure. *Nutrients* **2021**, *13*, 3041. [CrossRef]
120. Hsu, C.N.; Lin, Y.J.; Lu, P.C.; Tain, Y.L. Maternal Resveratrol Therapy Protects Male Rat Offspring Against Programmed Hypertension Induced by TCDD and Dexamethasone Exposures: Is it Relevant to Aryl Hydrocarbon Receptor? *Int. J. Mol. Sci.* **2018**, *19*, 2459. [CrossRef]
121. Aragon, A.C.; Kopf, P.G.; Campen, M.J.; Huwe, J.K.; Walker, M.K. In Utero and Lactational 2,3,7,8-Tetrachlorodibenzo-P-Dioxin Exposure: Effects on Fetal and Adult Cardiac Gene Expression and Adult Cardiac and Renal Morphology. *Toxicol. Sci.* **2008**, *101*, 321–330. [CrossRef]
122. Nuñez, P.; Fernandez, T.; García-Arévalo, M.; Alonso-Magdalena, P.; Nadal, A.; Perillan, C. Effects of Bisphenol a Treatment During Pregnancy on Kidney Development in Mice: A Stereological and Histopathological Study. *J. Dev. Orig. Health Dis.* **2018**, *9*, 208–214. [CrossRef]
123. Shih, M.K.; Tain, Y.L.; Chen, Y.W.; Hsu, W.H.; Yeh, Y.T.; Chang, S.K.C.; Liao, J.X.; Hou, C.Y. Resveratrol Butyrate Esters Inhibit Obesity Caused by Perinatal Exposure to Bisphenol A in Female Offspring Rats. *Molecules* **2021**, *26*, 4010. [CrossRef]
124. Liao, J.X.; Chen, Y.W.; Shih, M.K.; Tain, Y.L.; Yeh, Y.T.; Chiu, M.H.; Chang, S.K.C.; Hou, C.Y. Resveratrol Butyrate Esters Inhibit BPA-Induced Liver Damage in Male Offspring Rats by Modulating Antioxidant Capacity and Gut Microbiota. *Int. J. Mol. Sci.* **2021**, *22*, 5273. [CrossRef]

125. Hsu, C.N.; Lin, Y.J.; Tain, Y.L. Maternal Exposure to Bisphenol a Combined with High-Fat Diet-Induced Programmed Hypertension in Adult Male Rat Offspring: Effects of Resveratrol. *Int. J. Mol. Sci.* **2019**, *20*, 4382. [CrossRef]
126. Alonso-Magdalena, P.; Vieira, E.; Soriano, S.; Menes, L.; Burks, D.; Quesada, I.; Nadal, A. Bisphenol A exposure during pregnancy disrupts glucose homeostasis in mothers and adult male offspring. *Environ. Health Perspect.* **2010**, *118*, 1243–1250. [CrossRef]
127. Wei, Z.; Song, L.; Wei, J.; Chen, T.; Chen, J.; Lin, Y.; Xia, W.; Xu, B.; Li, X.; Chen, X.; et al. Maternal Exposure to Di-(2-Ethylhexyl)Phthalate Alters Kidney Development through the Renin-Angiotensin System in Offspring. *Toxicol. Lett.* **2012**, *212*, 212–221. [CrossRef]
128. Tain, Y.L.; Hou, C.Y.; Chang-Chien, G.P.; Lin, S.; Hsu, C.N. Resveratrol Butyrate Ester Supplementation Blunts the Development of Offspring Hypertension in a Maternal Di-2-ethylhexyl Phthalate Exposure Rat Model. *Nutrients* **2023**, *15*, 697. [CrossRef]
129. Fan, Y.; Qin, Y.; Chen, M.; Li, X.; Wang, R.; Huang, Z.; Xu, Q.; Yu, M.; Zhang, Y.; Han, X.; et al. Prenatal low-dose DEHP exposure induces metabolic adaptation and obesity: Role of hepatic thiamine metabolism. *J. Hazard. Mater.* **2020**, *385*, 121534. [CrossRef]
130. Ye, Q.; Zhao, S.; Zhang, Y.; Su, Y.M.; Chen, M.; Zhao, J.; Jia, G.Z.; Han, B.M.; Jiang, J.T. Activation of the Rhoa/ROCK Pathway Contributes to Renal Fibrosis in Offspring Rats Induced by Maternal Exposure to Di-N-Butyl Phthalate. *Toxicology* **2020**, *443*, 152573. [CrossRef]
131. Zhou, K.; Cheng, R.; Zhu, M.; Yang, M.; Shen, X.; Luo, X.; Ma, L.; Xu, L.; Zhang, J. The influence of perinatal maternal exposure to dibutyl phthalate on glucolipid metabolism in adult female offspring. *Obes. Res. Clin. Pract.* **2022**, *16*, 500–506. [CrossRef]
132. Dangudubiyyam, S.V.; Mishra, J.S.; Zhao, H.; Kumar, S. Perfluorooctane sulfonic acid (PFOS) exposure during pregnancy increases blood pressure and impairs vascular relaxation mechanisms in the adult offspring. *Reprod. Toxicol.* **2020**, *98*, 165–173. [CrossRef]
133. Jules, G.E.; Pratap, S.; Ramesh, A.; Hood, D.B. In Utero Exposure to Benzo(a)Pyrene Predisposes Offspring to Cardiovascular Dysfunction in Later-Life. *Toxicology* **2012**, *295*, 56–67. [CrossRef] [PubMed]
134. Saillenfait, A.M.; Payan, J.P.; Brondeau, M.T.; Zissu, D.; de Ceaurriz, J. Changes in Urinary Proximal Tubule Parameters in Neonatal Rats Exposed to Cadmium Chloride during Pregnancy. *J. Appl. Toxicol.* **1991**, *11*, 23–27. [CrossRef] [PubMed]
135. Jacquillet, G.; Barbier, O.; Rubera, I.; Tauc, M.; Borderie, A.; Namorado, M.C.; Martin, D.; Sierra, G.; Reyes, J.L.; Poujeol, P.; et al. Cadmium Causes Delayed Effects on Renal Function in the Offspring of Cadmium-Contaminated Pregnant Female Rats. *Am. J. Physiol. Renal Physiol.* **2007**, *293*, F1450–F1560. [CrossRef]
136. Jackson, T.W.; Ryherd, G.L.; Scheibly, C.M.; Sasser, A.L.; Guillette, T.C.; Belcher, S.M. Gestational Cd Exposure in the CD-1 Mouse Induces Sex-Specific Hepatic Insulin Insensitivity, Obesity, and Metabolic Syndrome in Adult Female Offspring. *Toxicol. Sci.* **2020**, *178*, 264–280. [CrossRef]
137. Pan, K.; Jiang, S.; Du, X.; Zeng, X.; Zhang, J.; Song, L.; Lei, L.; Zhou, J.; Kan, H.; Sun, Q. Parental PM2.5 Exposure Changes Th17/Treg Cells in Offspring, Is Associated with the Elevation of Blood Pressure. *Environ. Toxicol.* **2021**, *36*, 1152–1161. [CrossRef] [PubMed]
138. Wu, X.; Pan, B.; Liu, L.; Zhao, W.; Zhu, J.; Huang, X.; Tian, J. In utero exposure to PM2.5 during gestation caused adult cardiac hypertrophy through histone acetylation modification. *J. Cell. Biochem.* **2019**, *120*, 4375–4384. [CrossRef] [PubMed]
139. Ye, Z.; Lu, X.; Deng, Y.; Wang, X.; Zheng, S.; Ren, H.; Zhang, M.; Chen, T.; Jose, P.A.; Yang, J.; et al. In Utero Exposure to Fine Particulate Matter Causes Hypertension due to Impaired Renal Dopamine D1 Receptor in Offspring. *Cell. Physiol. Biochem.* **2018**, *46*, 148–159. [CrossRef]
140. Chen, M.; Wang, X.; Hu, Z.; Zhou, H.; Xu, Y.; Qiu, L.; Qin, X.; Zhang, Y.; Ying, Z. Programming of mouse obesity by maternal exposure to concentrated ambient fine particles. *Part. Fibre Toxicol.* **2017**, *14*, 20. [CrossRef] [PubMed]
141. Chen, M.; Liang, S.; Qin, X.; Zhang, L.; Qiu, L.; Chen, S.; Hu, Z.; Xu, Y.; Wang, W.; Zhang, Y.; et al. Prenatal exposure to diesel exhaust $PM_{2.5}$ causes offspring β cell dysfunction in adulthood. *Am. J. Physiol. Endocrinol. Metab.* **2018**, *315*, E72–E80. [CrossRef]
142. Sengupta, P. The Laboratory Rat: Relating Its Age with Human's. *Int. J. Prev. Med.* **2013**, *4*, 624–630.
143. Lin, L.; Dai, Y.; Xia, Y. An overview of aryl hydrocarbon receptor ligands in the Last two decades (2002–2022): A medicinal chemistry perspective. *Eur. J. Med. Chem.* **2022**, *244*, 114845. [CrossRef] [PubMed]
144. Fang, C.; Wu, X.; Huang, Q.; Liao, Y.; Liu, L.; Qiu, L.; Shen, H.; Dong, S. PFOS elicits transcriptional responses of the ER, AHR and PPAR pathways in *Oryzias melastigma* in a stage-specific manner. *Aquat. Toxicol.* **2012**, *106–107*, 9–19. [CrossRef] [PubMed]
145. Fu, C.; Li, Y.; Xi, H.; Niu, Z.; Chen, N.; Wang, R.; Yan, Y.; Gan, X.; Wang, M.; Zhang, W.; et al. Benzo(a)pyrene and cardiovascular diseases: An overview of pre-clinical studies focused on the underlying molecular mechanism. *Front. Nutr.* **2022**, *9*, 978475. [CrossRef]
146. Feng, S.; Duan, E.; Shi, X.; Zhang, H.; Li, H.; Zhao, Y.; Chao, L.; Zhong, X.; Zhang, W.; Li, R.; et al. Hydrogen ameliorates lung injury in a rat model of subacute exposure to concentrated ambient PM2.5 via Aryl hydrocarbon receptor. *Int. Immunopharmacol.* **2019**, *77*, 105939. [CrossRef]
147. Safe, S.; Jin, U.H.; Park, H.; Chapkin, R.S.; Jayaraman, A. Aryl Hydrocarbon Receptor (AHR) Ligands as Selective AHR Modulators (SAhRMs). *Int. J. Mol. Sci.* **2020**, *21*, 6654. [CrossRef]
148. Sládeková, L.; Mani, S.; Dvořák, Z. Ligands and agonists of the aryl hydrocarbon receptor AhR: Facts and myths. *Biochem. Pharmacol.* **2023**, *213*, 115626. [CrossRef]
149. Pinto, C.J.G.; Ávila-Gálvez, M.Á.; Lian, Y.; Moura-Alves, P.; Nunes Dos Santos, C. Targeting the aryl hydrocarbon receptor by gut phenolic metabolites: A strategy towards gut inflammation. *Redox Biol.* **2023**, *61*, 102622. [CrossRef] [PubMed]
150. Stephenson, R.P. A modification of receptor theory. *Br. J. Pharmacol. Chemother.* **1956**, *11*, 379–393. [CrossRef]

151. Agus, A.; Planchais, J.; Sokol, H. Gut Microbiota Regulation of Tryptophan Metabolism in Health and Disease. *Cell Host Microbe* **2018**, *23*, 716–724. [CrossRef]
152. Hsu, C.N.; Lin, I.C.; Yu, H.R.; Huang, L.T.; Tiao, M.M.; Tain, Y.L. Maternal Tryptophan Supplementation Protects Adult Rat Offspring against Hypertension Programmed by Maternal Chronic Kidney Disease: Implication of Tryptophan-Metabolizing Microbiome and Aryl Hydrocarbon Receptor. *Int. J. Mol. Sci.* **2020**, *21*, 4552. [CrossRef]
153. Zakrocka, I.; Załuska, W. Kynurenine pathway in kidney diseases. *Pharmacol. Rep.* **2022**, *74*, 27–39. [CrossRef] [PubMed]
154. Campesato, L.F.; Budhu, S.; Tchaicha, J.; Weng, C.H.; Gigoux, M.; Cohen, I.J.; Redmond, D.; Mangarin, L.; Pourpe, S.; Liu, C.; et al. Blockade of the AHR restricts a Treg-macrophage suppressive axis induced by L-Kynurenine. *Nat. Commun.* **2020**, *11*, 4011. [CrossRef] [PubMed]
155. Hus, C.N.; Tain, Y.L. Developmental Programming and Reprogramming of Hypertension and Kidney Disease: Impact of Tryptophan Metabolism. *Int. J. Mol. Sci.* **2020**, *21*, 8705. [CrossRef]
156. Hussain, T.; Tan, B.; Yin, Y.; Blachier, F.; Tossou, M.C.; Rahu, N. Oxidative Stress and Inflammation: What Polyphenols Can Do for Us? *Oxid. Med. Cell. Longev.* **2016**, *2016*, 7432797. [CrossRef] [PubMed]
157. Kursvietiene, L.; Staneviciene, I.; Mongirdiene, A.; Bernatoniene, J. Multiplicity of effects and health benefits of resveratrol. *Medicina* **2016**, *52*, 148–155. [CrossRef] [PubMed]
158. Mohammadi-Bardbori, A.; Bengtsson, J.; Rannug, U.; Rannug, A.; Wincent, E. Quercetin, resveratrol, and curcumin are indirect activators of the aryl hydrocarbon receptor (AHR). *Chem. Res. Toxicol.* **2012**, *25*, 1878–1884. [CrossRef] [PubMed]
159. Ciolino, H.P.; Yeh, G.C. Inhibition of aryl hydrocarbon-induced cytochrome P-450 1A1 enzyme activity and CYP1A1 expression by resveratrol. *Mol. Pharmacol.* **1999**, *56*, 760–767.
160. Hsu, C.N.; Hou, C.Y.; Tain, Y.L. Preventive Aspects of Early Resveratrol Supplementation in Cardiovascular and Kidney Disease of Developmental Origins. *Int. J. Mol. Sci.* **2021**, *22*, 4210. [CrossRef]
161. Tain, Y.L.; Hsu, C.N. Developmental Programming of the Metabolic Syndrome: Can We Reprogram with Resveratrol? *Int. J. Mol. Sci.* **2018**, *19*, 2584. [CrossRef] [PubMed]
162. Hsu, C.N.; Hou, C.Y.; Chang-Chien, G.P.; Lin, S.; Chan, J.Y.H.; Lee, C.T.; Tain, Y.L. Maternal resveratrol therapy protected adult rat offspring against hypertension programmed by combined exposures to asymmetric dimethylarginine and trimethylamine-N oxide. *J. Nutr. Biochem.* **2021**, *93*, 108630. [CrossRef]
163. Walle, T.; Hsieh, F.; DeLegge, M.H.; Oatis, J.E., Jr.; Walle, U.K. High absorption but very low bioavailability of oral resveratrol in humans. *Drug Metab. Dispos.* **2004**, *32*, 1377–1382. [CrossRef] [PubMed]
164. Tain, Y.L.; Chang, S.K.C.; Liao, J.X.; Chen, Y.W.; Huang, H.T.; Li, Y.L.; Hou, C.Y. Synthesis of Short-Chain-Fatty-Acid Resveratrol Esters and Their Antioxidant Properties. *Antioxidants* **2021**, *10*, 420. [CrossRef] [PubMed]
165. Liu, H.; Wang, J.; He, T.; Becker, S.; Zhang, G.; Li, D.; Ma, X. Butyrate: A Double-Edged Sword for Health? *Adv. Nutr.* **2018**, *9*, 21–29. [CrossRef] [PubMed]
166. Pluznick, J.L. Microbial short-chain fatty acids and blood pressure regulation. *Curr. Hypertens. Rep.* **2017**, *19*, 25. [CrossRef]
167. Marinelli, L.; Martin-Gallausiaux, C.; Bourhis, J.M.; Béguet-Crespel, F.; Blottière, H.M.; Lapaque, N. Identification of the novel role of butyrate as AhR ligand in human intestinal epithelial cells. *Sci. Rep.* **2019**, *9*, 643. [CrossRef] [PubMed]
168. Hsu, C.N.; Yu, H.R.; Lin, I.C.; Tiao, M.M.; Huang, L.T.; Hou, C.Y.; Chang-Chien, G.P.; Lin, S.; Tain, Y.L. Sodium butyrate modulates blood pressure and gut microbiota in maternal tryptophan-free diet-induced hypertension rat offspring. *J. Nutr. Biochem.* **2022**, *108*, 109090. [CrossRef] [PubMed]
169. Tain, Y.L.; Hou, C.Y.; Chang-Chien, G.P.; Lin, S.; Tzeng, H.T.; Lee, W.C.; Wu, K.L.H.; Yu, H.R.; Chan, J.Y.H.; Hsu, C.N. Reprogramming Effects of Postbiotic Butyrate and Propionate on Maternal High-Fructose Diet-Induced Offspring Hypertension. *Nutrients* **2023**, *15*, 1682. [CrossRef]
170. Wu, K.L.H.; Liu, W.C.; Wu, C.W.; Fu, M.H.; Huang, H.M.; Tain, Y.L.; Liang, C.K.; Hung, C.Y.; Chen, I.C.; Hung, P.L.; et al. Butyrate reduction and HDAC4 increase underlie maternal high fructose-induced metabolic dysfunction in hippocampal astrocytes in female rats. *J. Nutr. Biochem.* **2024**, *126*, 109571. [CrossRef] [PubMed]
171. Rejano-Gordillo, C.M.; Marín-Díaz, B.; Ordiales-Talavero, A.; Merino, J.M.; González-Rico, F.J.; Fernández-Salguero, P.M. From Nucleus to Organs: Insights of Aryl Hydrocarbon Receptor Molecular Mechanisms. *Int. J. Mol. Sci.* **2022**, *23*, 14919. [CrossRef]
172. Tain, Y.L.; Hsu, C.N. Interplay between maternal nutrition and epigenetic programming on offspring hypertension. *J. Nutr. Biochem.* **2024**, *127*, 109604. [CrossRef]
173. Tain, Y.L.; Huang, L.T.; Chan, J.Y.; Lee, C.T. Transcriptome analysis in rat kidneys: Importance of genes involved in programmed hypertension. *Int. J. Mol. Sci.* **2015**, *16*, 4744–4758. [CrossRef] [PubMed]
174. Anatskaya, O.V.; Runov, A.L.; Ponomartsev, S.V.; Vonsky, M.S.; Elmuratov, A.U.; Vinogradov, A.E. Long-Term Transcriptomic Changes and Cardiomyocyte Hyperpolyploidy after Lactose Intolerance in Neonatal Rats. *Int. J. Mol. Sci.* **2023**, *24*, 7063. [CrossRef] [PubMed]
175. Rosenfeld, C.S. Gut Dysbiosis in Animals due to Environmental Chemical Exposures. *Front. Cell. Infect. Microbiol.* **2017**, *7*, 396. [CrossRef] [PubMed]
176. Tain, Y.L.; Hsu, C.N. Metabolic Syndrome Programming and Reprogramming: Mechanistic Aspects of Oxidative Stress. *Antioxidants* **2022**, *11*, 2108. [CrossRef] [PubMed]
177. Hsu, C.N.; Tain, Y.L. Targeting the Renin-Angiotensin-Aldosterone System to Prevent Hypertension and Kidney Disease of Developmental Origins. *Int. J. Mol. Sci.* **2021**, *22*, 2298. [CrossRef] [PubMed]

178. Dornbos, P.; Warren, M.; Crawford, R.B.; Kaminski, N.E.; Threadgill, D.W.; LaPres, J.J. Characterizing Serpinb2 as a Modulator of TCDD-Induced Suppression of the B Cell. *Chem. Res. Toxicol.* **2018**, *31*, 1248–1259. [CrossRef]
179. Jurgelewicz, A.; Dornbos, P.; Warren, M.; Nault, R.; Arkatkar, A.; Lin, H.; Threadgill, D.W.; Zacharewski, T.; LaPres, J.J. Genetics-Based Approach to Identify Novel Genes Regulated by the Aryl Hydrocarbon Receptor in Mouse Liver. *Toxicol. Sci.* **2021**, *181*, 285–294. [CrossRef] [PubMed]
180. Denison, M.S.; Faber, S.C. And Now for Something Completely Different: Diversity in Ligand-Dependent Activation of Ah Receptor Responses. *Curr. Opin. Toxicol.* **2017**, *2*, 124–131. [CrossRef]
181. Dolciami, D.; Ballarotto, M.; Gargaro, M.; Lopez-Cara, L.C.; Fallarino, F.; Macchiarulo, A. Targeting Aryl hydrocarbon receptor for next-generation immunotherapies: Selective modulators (SAhRMs) versus rapidly metabolized ligands (RMAhRLs). *Eur. J. Med. Chem.* **2020**, *185*, 111842. [CrossRef]

International Journal of
Molecular Sciences

MDPI

Article

Maternal Exposure to Low-Dose BDE-47 Induced Weight Gain and Impaired Insulin Sensitivity in the Offspring

Sandra Strunz [1,2], Rebecca Strachan [1], Mario Bauer [1], Ana C. Zenclussen [1], Beate Leppert [1], Kristin M. Junge [1,3] and Tobias Polte [1,2,*]

[1] Helmholtz Centre for Environmental Research—UFZ, Department of Environmental Immunology, 04318 Leipzig, Germany; mario.bauer@ufz.de (M.B.); ana.zenclussen@ufz.de (A.C.Z.); kristin.junge@akad.de (K.M.J.)

[2] Department of Dermatology, Venerology and Allergology, Leipzig University Medical Center, University of Leipzig, 04318 Leipzig, Germany

[3] School of Health and Social Sciences, AKAD University Stuttgart, 70191 Stuttgart, Germany

* Correspondence: tobias.polte@ufz.de

Abstract: Polybrominated diphenyl ethers (PBDEs), commonly used as synthetic flame retardants, are present in a variety of consumer products, including electronics, polyurethane foams, textiles, and building materials. Initial evidence from epidemiological and experimental studies suggests that maternal PBDE exposure may be associated with a higher BMI in children, with disturbance of energy metabolism and an increased risk of Type 2 diabetes. However, the causality between early exposure to real-life PBDE concentrations and increased weight as well as mechanisms underlying impaired metabolic pathways in the offspring remain elusive. Here, using a mouse model we examined the effect of maternal exposure to 2,2′,4,4′-tetrabrominated diphenyl ether (BDE-47), the most abundant congener in human samples, on offspring weight gain and energy homeostasis using a mouse model. Maternal exposure to BDE-47 at low dose resulted in weight gain in female offspring together with an impaired glucose and insulin tolerance in both female and male mice. In vitro and in vivo data suggest increased adipogenesis induced by BDE-47, possibly mediated by DNA hypermethylation. Furthermore, mRNA data suggest that neuronal dysregulation of energy homeostasis, driven via a disturbed leptin signaling may contribute to the observed weight gain as well as impaired insulin and glucose tolerance.

Keywords: BDE-47; overweight; metabolism; leptin; pregnancy; sex-specific effect; epigenetics

1. Introduction

Childhood overweight and obesity has reached pandemic proportions in most developed countries and continues to increase worldwide. Recent reports suggest that overweight and obesity affect up to a third of the child population in Europe and North America [1,2]. In addition to genetic predisposition and high-caloric food intake, which is often associated with predominantly sedentary behavior and physical inactivity, environmental factors were found to contribute to and/or worsen the onset of overweight and obesity [3–5]. It has been shown that environmental influences like chemical exposures during critical time windows, such as fetal development, leads to disruptions in physiological, endocrine, and metabolic signaling, resulting in long-lasting health effects [6,7]. Epigenetic changes such as altered DNA methylation have been described to play a critical role as mediators between exposure and early developmental programming of disease [8,9]. Synthetic chemicals that disrupt the endocrine system are endocrine disrupting chemicals (EDCs), some of which, known as obesogens, have already been shown to contribute to developmental programming towards obesity following exposure in the perinatal period [9,10]. EDCs can be found in a variety of everyday products, to which they have been added, for example as plasticizers, preservatives, or stabilizers [11,12]. Exposure to EDCs is ubiquitous and inevitable. These chemicals can enter the body through food and water intake, skin absorption, or inhalation [4]. One group of chemicals with suspected obesogenic effects are polybrominated diphenyl ethers (PBDEs), which are widely used as synthetic flame retardants in consumer products, including electronics, polyurethane foams, textiles, and building materials [13], and are absorbed through, for example, contaminated food and drinking water [14].

There is initial evidence from epidemiological studies that maternal PBDE exposure may be associated with a higher BMI in children [15], with disturbance of energy metabolism and an increased risk of Type 2 diabetes [16]. Experimental studies in rat models have also shown that exposure to PBDEs may be associated with weight gain in offspring [17,18]. However, the mechanisms of early exposure to PBDEs leading to increased weight and impaired energy metabolism in offspring remain elusive.

In the present study, we examined the effects of maternal exposure to 2,2′,4,4′-tetrabromodiphenyl ether (BDE-47), the dominant congener in human samples, at a very low concentration relevant to daily life, on weight gain and metabolic parameters in the offspring. Using an in vivo mouse model, we hereby demonstrate that maternal exposure to BDE-47 resulted in weight gain in females and impaired glucose and insulin tolerance in both female and male offspring. The weight gain induced by BDE-47 was associated with increased adipogenesis mediated by DNA hypermethylation. Furthermore, our data suggest that neuronal dysregulation of energy homeostasis, potentially via disturbed leptin signaling, may contribute to the observed weight gain and altered insulin/glucose tolerance.

2. Results

2.1. Maternal BDE-47 Exposure Resulted in Sex-Specific Weight Gain and Impaired Insulin and Glucose Tolerance in the Offspring

Using a cross-generational mouse model (Supplementary Figure S1), female offspring from dams exposed to BDE-47 showed a significantly higher weight than control animals over the entire observation period (Figure 1A, left column). The increased body weight became noticeable shortly after birth and was associated with higher fat and lower lean mass, as measured by whole-body composition analysis using nuclear magnetic resonance technology (Figure 1B). In contrast, the weight of the male offspring of BDE-47-exposed dams was not significantly affected by maternal exposure to BDE-47 and no difference in lean muscle and fat mass was observed (Figure 1A,B, right column). However, both female and male offspring of dams exposed to BDE-47 developed an impaired insulin and glucose tolerance; showing significant differences in ITT and GTT compared to control mice (Figure 1C,D). Since the generationally mediated weight-increasing effect of BDE-47 was mainly observed in the female offspring, gene expression analyses were performed in adipose tissue of female mice at the end of the observation period of 12-weeks. The results revealed an increased mRNA expression of peroxisome proliferator-activated receptor gamma (ppar-γ), a transcription factor important for lipid uptake and adipo-genesis, caveolin-1 (cav-1), and perilipin-1 (plin-1), both of which regulate the metabolism of lipid droplets and glucose transporter 4 (glut4, slc2a4) in the offspring of BDE-47-exposed dams (Figure 2A). In addition, transcript of sterol regulatory element-binding transcription factor 1 (scrbf1), which plays a key role in lipogenesis, is augmented in adipose tissue and liver. In the liver, expression of insulin receptor 1 (insr-1) and glucose transporter 2 (slc2a2) was also increased compared to the offspring of control mice (Figure 2B). While serum leptin levels were significantly increased in the female offspring of BDE-47-exposed dams when compared to the controls, adiponectin, resistin, and ghrelin concentrations were not affected (Figure 2C). Quite remarkable is that a direct exposure of adult female mice to low-dose BDE-47 had no effects on weight, insulin, and glucose tolerance, or serum leptin levels (Figure 3A–C). These data demonstrate the importance of the developmental period as a particularly vulnerable time for chemical exposure, which affects disease development later in life.

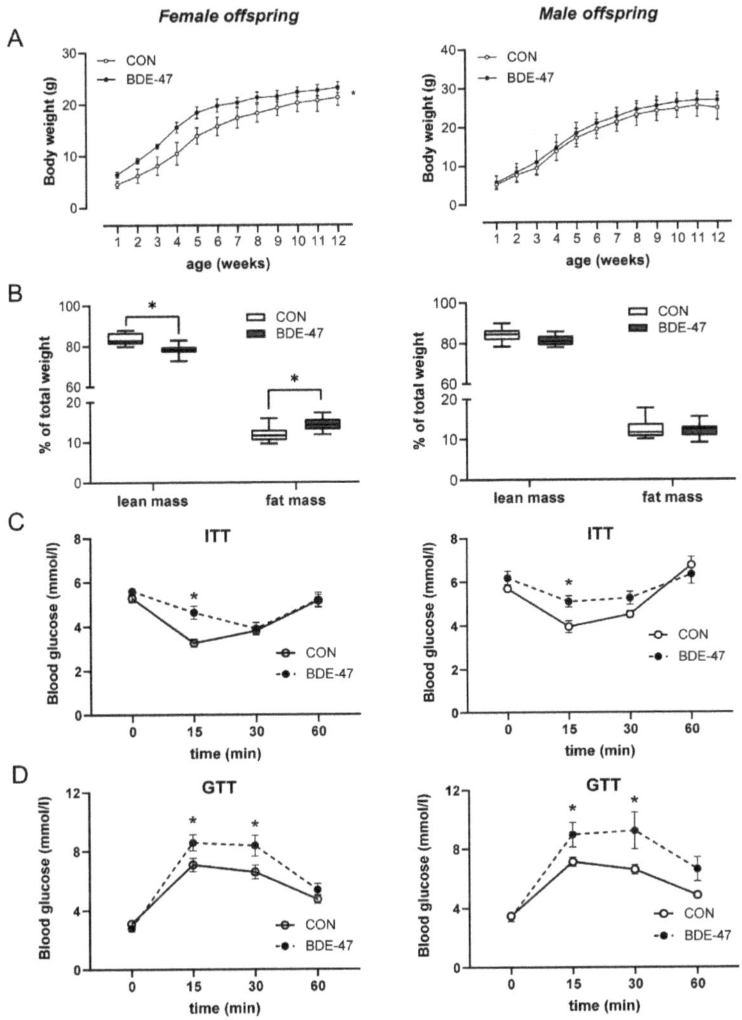

Figure 1. Maternal exposure to BDE-47 and weight development and glycose metabolism in the offspring. (**A**) Bodyweight development and (**B**) body composition are shown for female (**left**) and male offspring (**right**) from BDE-47-exposed dams, (**C**) insulin tolerance test (ITT), and (**D**) glucose tolerance test (GTT) were performed in 9-week- and 10-week-old offspring, respectively. Data are expressed as mean ± SEM, $n \geq 12$. Significant differences were derived by two-way ANOVA (**A**) or Mann–Whitney test (**B–D**) with * $p < 0.05$.

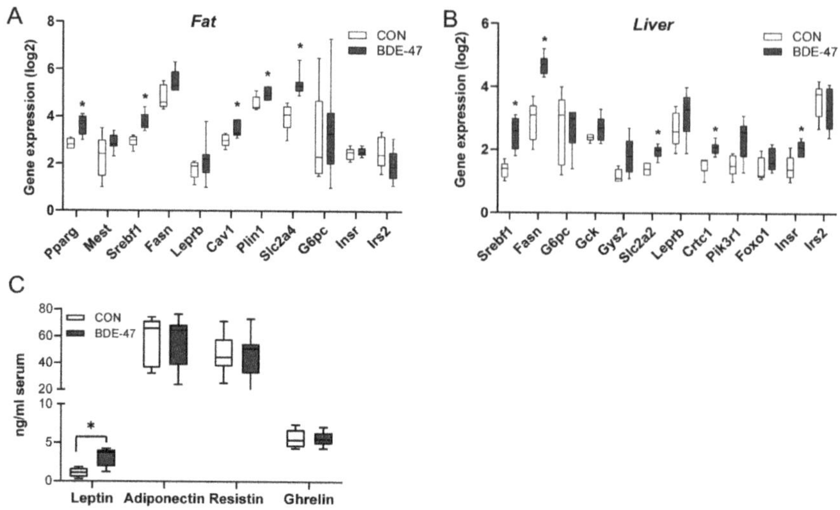

Figure 2. Maternal BDE-47 exposure and expression of key genes in adipose tissue and liver and serum adipokines in the female offspring. (**A**) The mRNA expression levels of selected target genes investigated in adipose tissue and (**B**) liver of 12-week-old female offspring. (**C**) Levels of leptin, adiponectin, resistin, and ghrelin measured in serum of female offspring. Data are expressed as mean ± SEM, $n \geq 5$. Significant differences were derived by Mann–Whitney test with * $p < 0.05$.

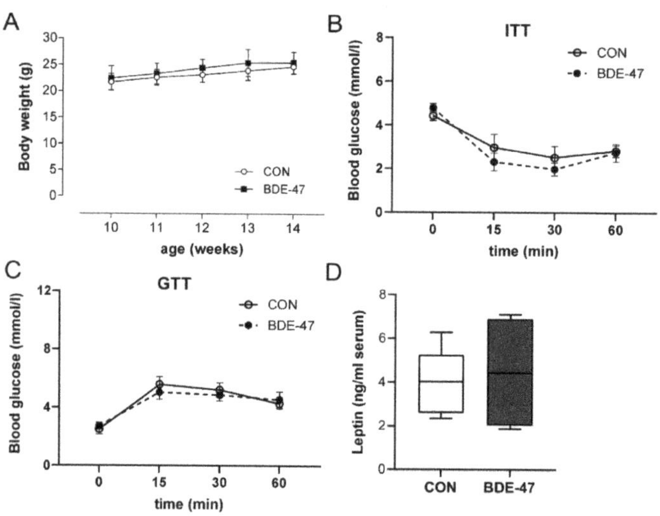

Figure 3. Effect of BDE-47 exposure to adult mice on weight development, glycose metabolism and leptin serum levels. (**A**) Bodyweight development, (**B**) ITT, (**C**) GTT and (**D**) leptin serum levels are shown from adult female mice exposed to BDE-47. Data are expressed as mean ± SEM, $n \geq 10$.

2.2. Impact of BDE-47 Exposure on Murine and Human Adipocyte Development In Vitro

To better understand the mechanistic pathways underlying the observed obesogenic effects, we next evaluated the potential effect of BDE-47 on adipocyte differentiation. For this, we employed 3T3-L1 mouse preadipocytes and treated them with different BDE-47 concentrations from 10 pM to 1 µM during culture. Interestingly, both the very low and the higher BDE concentrations showed an increased amount of triglyceride storage, while the medium concentrations had no effect (Figure 4A,B). These results indicate a U-shaped influence of BDE-47 on adipogenesis.

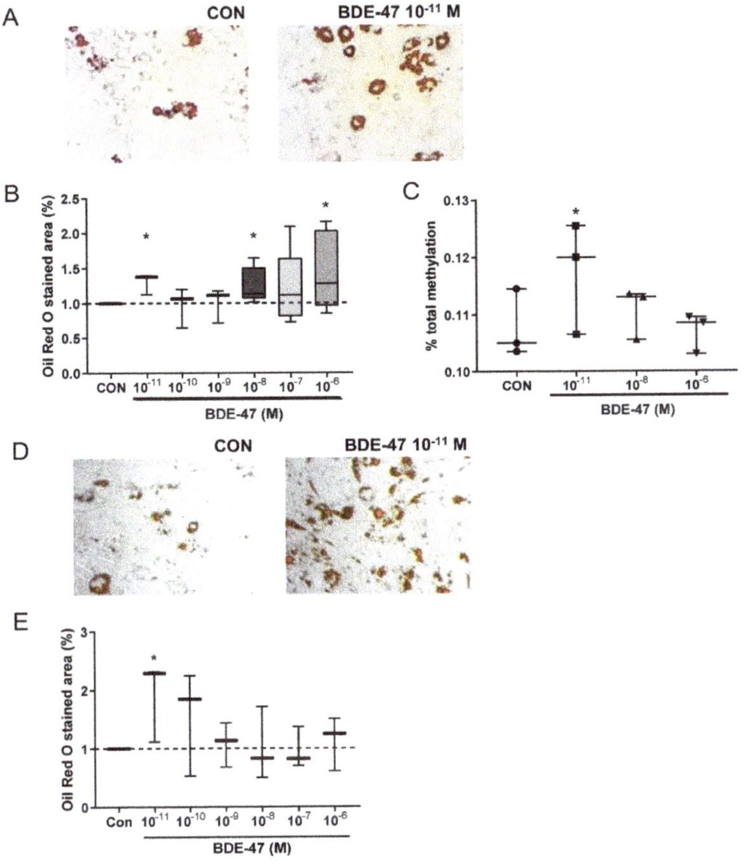

Figure 4. Effect of BDE-47 exposure on adipocyte differentiation. (**A**) Representative Oil Red O-stained pictures after differentiation of 3T3L1 mouse preadipocytes in the presence of BDE-47 (scale bar: 100 µm). (**B**) Triglyceride storage of mouse adipocytes assessed via Oil Red O staining. (**C**) Representative Oil Red O-stained pictures after differentiation from human MSCs in the presence of BDE-47 (**D**) Representative Oil Red O-stained pictures after adipocyte differentiation from MSC in the presence of BDE-47 (scale bar: 100 µm). (**E**) Triglyceride storage of human adipocytes assessed via Oil Red O staining. Data are expressed as mean ± SEM ofn = 3 experiments. Significant differences were derived by ANOVA with * $p < 0.05$.

To evaluate a possible involvement of epigenetic modifications in mediating the adipogenic effect of BDE-47 a global DNA methylation assay based on detection and quantitation of 5-methylcytosine (5-mC) was used in the 3T3L1 cells. Here, BDE-47 treatment at a very low concentration (10 pM) induced an increased DNA methylation (Figure 4C).

In order to translate the murine results of BDE-47 in adipocyte differentiation to the situation in humans, an established differentiation assay for human mesenchymal stem cells (MSC) was applied [9]. Similar to the results observed for the murine cells, BDE-47 treatment already at low concentration of 10 pM increased adipocyte differentiation of human MSC, assessed by the amount of triglyceride storage (Figure 4D). These results underline the potential obesogenic causality of BDE-47 driven effects.

2.3. Maternal BDE-47 Exposure Induced a Disturbed Regulation of Energy Homeostasis in Female Offspring

To additionally investigate whether BDE-47 exposure during the prenatal and lactational period also affects the central regulation of satiety and hunger in the offspring immediately after weaning, an analysis of the genes involved in this regulation in the hypothalamus was carried out. Data show that melanocortin type 3 and 4 receptor (mc3r, mc4r) expression was significantly downregulated in 4-week old offspring from BDE-47-exposed dams, as was forkhead transcription factor-1 (foxo1) and neuropeptide Y (npy, Figure 5A). Other genes, such as agouti-related neuropeptide gene (agrp), proopiomelanocortin (pomc), insulin receptor substrate 2 (irs2), and the leptin receptor (lepr) also seemed to have a reduced expression, but this was not significant. Investigating the feeding behaviour directly after weaning revealed a higher food intake in the offspring from BDE-47-exposed dams compared to control mice from unexposed dams (Figure 5B). It should be noted that gene expression in fat and liver remained unaffected in the offspring of BDE-47-exposed dams directly after weaning (Supplementary Figure S3A,B). These results suggest that early weight gain is due to impaired central satiety regulation, followed by increased adipogenesis manifesting at a later stage.

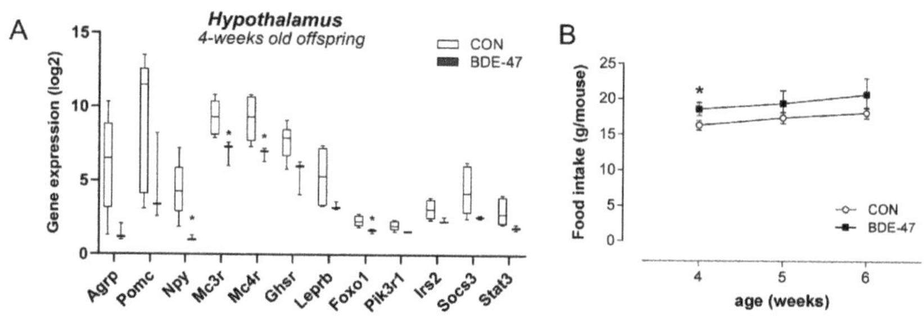

Figure 5. Maternal BDE-47 exposure and expression of key genes in hypothalamus and food intake of 4-week-old female offspring. (**A**) The mRNA expression levels of selected target genes investigated in hypothalamus of 4-week-old female offspring. (**B**) Food intake of 4–6-week-old female offspring from BDE-47-exposed dams compared to control mice. Data are expressed as mean ± SEM, $n \geq 5$. Significant differences were derived by Mann–Whitney test with * $p < 0.05$.

2.4. Offspring Weight Gain Caused by Maternal BDE-47 Exposure Is Associated with DNA Hypermethylation

The results obtained with BDE-47 in the in vitro differentiation of adipocytes suggest the involvement of DNA hypermethylation. To assess whether this also applies in vivo with regard to maternal BDE-47 exposure and the development of overweight in the offspring, one-week-old pups from BDE-47-exposed dams were treated with the DNA methyltransferase inhibitor 5-aza-2′-deoxycytidine (Aza) for two weeks until weaning [9,19]. Treatment of the offspring with Aza reduced the body weight and increased the lean mass (Figure 6A,B). Although Aza treatment had an increasing effect on fat mass in control animals, Aza also reduced fat mass in the offspring of BDE-47-exposed dams, but not significantly (Figure 6B). Leptin levels after Aza treatment were diminished but without reaching significance (Figure 6C). Furthermore, it is very interesting that Aza treatment had no effect on the altered insulin and glucose tolerance in the offspring of dams exposed to BDE-47 (Figure 6D). These results suggest an important role of the DNA hypermethylation in the observed BDE-47-induced weight gain, but not in the altered insulin and glucose tolerance.

3. Discussion

This study applied a cross-generational mouse model as well as murine und human in vitro adipocyte differentiation assays to investigate the effects of maternal exposure to the flame-retardant BDE-47 on offspring weight development. In vivo, maternal exposure of mice to BDE-47 resulted in weight gain in the female offspring together with an altered glucose and insulin tolerance. Impaired glucose and insulin tolerance was also observed in males; however, this was not accompanied by weight gain. Furthermore, it is difficult to assess the extent to which delayed glucose metabolism has a physiological relevance. The observed weight gain appears to be due to increased adipogenesis induced by BDE-47, additionally driven via an impaired satiety regulation in the hypothalamus of the offspring.

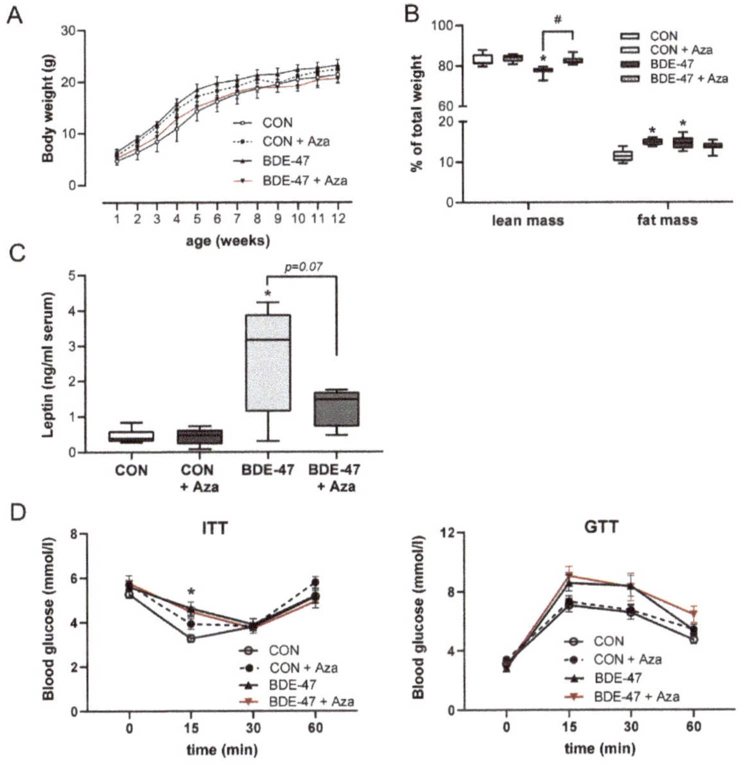

Figure 6. Offspring weight gain caused by maternal BDE-47 exposure is reversed by treatment with a DNA methyltransferase inhibitor. After treatment of F1 mice with the DNA methyltransferase inhibitor Aza, weight development (**A**), body composition (**B**), leptin serum levels (**C**), and ITT and GTT (**D**) were evaluated. Data are expressed as mean ± SEM, $n \geq 7$. (Significant differences were derived by two-way ANOVA or Mann–Whitney test with * $p < 0.05$, BDE-47 or CON + Aza vs. CON, # $p < 0.05$, BDE-47 vs. BDE-47 + Aza.

Since there is evidence from some epidemiological studies that maternal PBDE exposure may be associated with higher BMI in children [15] and may also lead to impaired energy metabolism and an increased risk of type 2 diabetes [16], it is particularly important to verify and deepen these results using in vivo disease models to proof a causal relationship between exposure and disease. To do so, the use of relevant exposure concentrations in the experimental model is a crucial point. The dose of BDE-47, the most abundant PBDE in both the environment and human bodies [20], was determined from an acceptable daily human intake of 0.02 μg kg^{-1} body weight based on no observed adverse effect levels (NOAELs) (0.002 mg kg^{-1} body weight/day) reported in intrauterine exposure studies [17,21]. In previous work, where we examined the health effects of other EDCs, such as phthalates or parabens, we found that exposure to ADI concentrations in mice resulted in comparable serum concentrations to those measured in highly exposed mothers included in our prospective mother–child cohort study LINA [9,19]. Notably, the used BDE-47 dose had no direct impact on weight or other metabolic parameters in adult female mice, confirming that only exposure during the critical period of pregnancy and lactation resulted in weight gain in offspring. This highlights the importance of developmental exposures in disease risk later in life. Interestingly, the effect on weight was sex-specific, as it was only observed in females. This is in contrast to another experimental study investigating the effect of perigestational BDE-47 exposure in Sprague–Dawley rats. Here, the authors describe a weight gain in male but not in female offspring [18]. However, the BDE-47 concentration in that approach was many times higher than the one used in the present study. Different concentrations of hormonally active chemicals can probably not only lead to different sex-specific effects but can also be mediated via other mechanisms. This is particularly interesting because effects of EDCs often follow non-monotonic dose–response, such as a U-shaped curve [22,23]. We also obtained such a u-shaped curve in our in vitro experiments in which broad ranges of different concentrations were used during murine adipocyte differentiation. Both very low concentrations (10 pM) and higher concentrations (1 μM) showed an impact on triglyceride storage, while at the intermediate concentrations there was no effect. While BDE-47 has previously been shown to promote adipocyte differentiation at higher concentrations [24,25], an effect of ultra-low concentrations, as presented in the current project, has not been documented before. Furthermore, we were even able to confirm an adipogenic effect of low-dose BDE-47 in both mice and humans, which suggests that our results could indeed be of general (patho)physiological relevance. Another interesting finding is that altered DNA methylation was only observed at the low BDE-47 concentration, although higher concentrations had a similar effect on triglyceride storage. This could suggest different underlying mechanisms of the concentration ranges applied. An epigenetic regulation of adipogenesis has already been described in previous work [26], as has modulation of DNA methylation by BDE-47 [27].

The increased triglyceride storage induced by BDE-47, as we observed during adipocyte differentiation, is consistent with the increased expression of transcripts in adipose tissue of the female offspring of BDE-47-exposed dams involved in adipogenesis and lipid metabolism (Pparγ, Cav-1, Plin-1, Scrbf1). In addition, there is enhanced expression of glucose transporters in adipose tissue (Slc2a4) and liver (Slc2a2), which may be a (so far unsuccessful) compensating response to the elevated blood glucose levels: glucose levels in the offspring of BDE-47-exposed dams are elevated in both ITT and GTT compared to control animals, with no detected differences in basal glucose levels. These findings indicate an impaired insulin sensitivity that may point to an early development of type 2 diabetes. In fact, there are initial indications from epidemiological studies that exposure to BDE-47 [28] or other PBDEs [16] may be associated with increased diabetes prevalence. In contrast to the weight increase in the female offspring of BDE-47-exposed dams, the impaired insulin sensitivity in GTT and ITT affects both sexes. Why this imbalance in the glucose/insulin metabolism is not associated with weight gain in the male offspring requires further research but suggests different initial molecular events of BDE-47, in terms of weight and metabolism, and reinforces the need to include both biological sexes in risk assessment of chemicals. As expected, serum leptin levels are only increased in the weight-/fat-accumulating female offspring. However, the expected satiety signal of leptin via the following signal transduction in the hypothalamus obviously was not sufficient due to the higher food intake already shown of the young female offspring. Potentially a leptin resistance could contribute to the impaired weight gain and insulin sensitivity, as such an association has been previously considered [29]. This hypothesis was supported by our mRNA expression analyses of the hypothalamus of 4-week-old female offspring of mothers exposed to BDE-47. Here we found significantly reduced expression of Mc3r and Mc4r compared to control mice. Both receptors are indeed linked with leptin resistance [30]. In general, the leptin-associated central regulation of satiety and hunger appears to be affected by BDE-47 with a downregulation of orexigenic genes such as Foxo1 or Npy. Foxo1 was shown to be involved in the hypothalamic control of energy homeostasis, as its ablation in hypothalamic Pomc neurons reduced food intake and body weight. Furthermore, Foxo1 was shown to interfere with both adipogenesis and glucose/insulin metabolism [31]. However, although Foxo1 was downregulated, exposure to BDE-47 was rather associated with an increased weight and fat mass, together with an early increased food intake, underlining the disturbance of the central regulation in this context. It should be noted that all RNA expression data shown in the current study still needs to be verified at the protein level. However, due to the limited organic material available per animal (particularly for the hypothalamus) we chose gene expression analyses for broader hypothesis generation rather than confirmatory protein levels when conceptualizing the experimental setup and the analytical protocol. Therefore, our interpretation remains hypothesis-driven and partly speculative. We must also acknowledge the small sample size as a limitation for the hypothalamic data as well as the lack of analysis of specific regions of the hypothalamus. Further experiments and analysis are required to better evaluate the hypothesis proposed in the present study. Nevertheless, the reduced expression of genes in the hypothalamus in our study is consistent with gene expression data from female offspring of dams exposed to a PBDE mixture that also contained BDE-47 [32]. In line, the authors saw a downregulation of genes such as Agrp, NPY, or Mc4r. However, they did not observe any weight gain in this study [32]. This could be due to an interaction of the different PBDEs in the mixture.

Early life exposure and the above-mentioned gene expression changes might be mediated via epigenetic modifications [9]. Although several studies have rather shown a BDE-47-induced hypomethylation [25,27], results from our in vitro differentiation of mouse adipocytes as well as results from other studies [33,34], suggest the involvement of an increased DNA methylation in mediating the BDE-47-induced effects. Interestingly, hypermethylation was only observed at very low BDE-47 concentrations, suggesting different mechanisms that led to increased adipocyte differentiation at low or at higher concentrations. In the present analyses, treating pups with the DNA methyltransferase inhibitor Aza reduced the BDE-47-induced weight gain and leptin serum levels and increased the lean mass. However, it is noteworthy that Aza-induced hypomethylation alone resulted in increased weight and fat mass in unexposed control animals. This could be because hypomethylation can lead to increased expression of genes involved in adipogenesis, such as PPARg [25]. Based on that observation, it is even more surprising that Aza can still reverse the BDE-47-driven effect on weight development, probably by increasing the expression of anti-adipogenic genes, which still have to be identified. It is also interesting that Aza has no effect on impaired insulin sensitivity in the offspring of BDE-47-exposed dams. As indicated above, there seem to be different underlying mechanisms/molecular initial events of BDE-47 in terms of weight and metabolism. Anyhow, further investigations are needed to clarify the epigenetic alterations more specifically. We also cannot exclude that additional epigenetic regulations, like histone modifications, as shown, e.g., for Mc4r [35] or other pathways, might be involved in mediating the BDE-47-induced changes in weight development and energy homeostasis.

In summary, our study results show that maternal exposure to very low concentrations of BDE-47 results in overweight development in the female offspring. The effect is mediated by an enhanced adipogenesis and might be promoted by a disturbed neuronal regulation of food intake. Perinatal BDE-47 exposure also led to an impaired insulin sensitivity in both sexes that might be associated with an observed disturbed leptin signaling. Our results draw attention to the great importance of exposure to EDCs, even at very low concentrations, during pregnancy and breastfeeding, for disease susceptibility in later life.

4. Materials and Methods

4.1. Mice

Balb/cByJ mice (6–8 weeks of age) were purchased from the Elevage Janvier Laboratory (Le Genest St Isle, France) with a 7-day adaption period before the start of experiments. Animals were maintained in groups of 3–6 mice per cage in the animal facility at the University of Leipzig (Germany) under conventional conditions with 21.5–23 °C room temperature, an average of 55% humidity, and a 12-h day/night rhythm. Exposed and control dams as well as the offspring of exposed and control mice were housed separately. All mice were kept in multiple sealed cages with HEPA filters by Sealsafe® and bedded with LIGNOCEL® bedding material. Dams and pups received a phytoestrogen-free diet (C1077 from Altromin, Lage, Germany) and water ad libitum. All animal experiments were performed at least 2 times with at least 3 dams per group with a maximum of 4 pups per sex per dam. All animal experiments were conducted in accordance with institutional and state guidelines. Animal protocols used in this study were approved by the Committee on Animal Welfare of Saxony/Leipzig (Permit Number: TVV01/15, 14/18).

4.2. Low-Dose Exposure to BDE-47

To evaluate the impact of an exposure during the prenatal and lactational period on weight development in the offspring we exposed female Balb/c mice to BDE-47 (0.02 µg/kg body weight/day) orally administered by gavage in 200 µL corn oil twice per week. The intervention lasted from one week before mating with BALB/c males until weaning of the pups at 4 weeks. Female mice received a BDE-47 concentration estimated from a calculated human acceptable intake based on earlier intrauterine exposure studies [21]. Control dams received the vehicle by gavage. To investigate a possible involvement of epigenetic alterations, offspring were treated i.p. with Aza (160 µg/kg body weight, Sigma-Aldrich, Munich, Germany) dissolved in PBS 3 times per week starting 1 week after birth until weaning [9,19]. Food intake was monitored after weaning until week 6.

4.3. Weight Assessment, Insulin and Glucose Tolerance Test and Metabolic Serum Proteins

Body weight of the pups was measured twice a week, and a mean weight per week was calculated for each mouse. At the end of the observation period (12 weeks), whole-body composition (fat mass and lean mass) was determined in awake mice based on nuclear magnetic resonance technology using an EchoMRI700™ instrument (Echo Medical Systems, Houston, TX, USA) in the offspring of control and BDE-47 exposed dams. An insulin tolerance test (ITT) was performed in the offspring 9 weeks after birth. Insulin (0.75 U/kg body weight) was injected intraperitoneally (i.p.). For glucose measurements blood from the tail vein was taken at four time points at 0, 15, 30 and 60 min after insulin injection. For the glucose tolerance test (GTT), glucose (2 g/kg body weight) was injected i.p. into fasting mice (10 weeks after birth), and the glucose measurement was performed equally to ITT. Adiponectin, leptin, resistin, and acetylated ghrelin serum concentrations were determined by ELISA using mouse standards according to the manufacturer's guidelines (mouse adiponectin, leptin, resistin ELISA; R&D Systems, Minneapolis, MN, USA), (mouse/rat acylated ghrelin ELISA; BioVendor, Karasek, Czech Republic).

4.4. In Vitro Adipocyte Differentiation

For adipocyte differentiation, 3T3L1 mouse preadipocytes were used according to Ruiz-Ojeda et al. [35]. In brief, for differentiation, 3T3-L1 cells were passaged onto 12-well plates at a concentration of 6000 cells/cm2 and were allowed to confluent for 48 h in basal media (BM; 90% DMEM, 10% FBS, 1% Ala/Gln, 0.1% Pen/Strep). After another 48 h in BM, cells were placed into differentiation medium (BM + 1 µM dexamethasone, 0.5 mM IBMX, 1 µg/mL Insulin) along with 50 µL of exposure solution (BDE-47: 10^{-11}–10^{-6} M in maximal 0.05% ethanol or respective solvent control) for another 48 h. Finally, they were placed in maintenance medium (BM + 1 µg/mL Insulin) with the same 50 µL of exposure solution. The medium was changed every 2–3 days for another 8 days until they were harvested and processed for further analyses. For Oil Red O staining, cells were stained with Oil Red O for triglyceride depots and quantified via absorbance measurement at 510 nm.

Human adipose-derived mesenchymal stem cells derived from a female donor (MSC; ATCC®, PCS-500-011; #59753760) and appropriate culture media were purchased from LGC Standards (Wesel, Germany). Cells were cultured under standardized conditions at 37 °C, 5% CO_2, and 95% humidity according to the manufacturer's instructions, as described previously [10]. In brief, for adipocyte differentiation, MSC passages 3–5 were seeded at 9.000 cells/cm^2 in a 96-well plate and grown to 70% confluence. For initiation of differentiation, cells were fed with adipocyte differentiation initiation medium (ADIM; ATCC Adipocyte Differentiation Toolkit PCS-500-050). ADIM was changed to adipocyte maintenance medium (ADMM; ATCC® Adipocyte Differentiation Toolkit PCS-500-050) after 4 days and changed regularly. Cells were treated with BDE-47 at different concentrations during the entire differentiation period. After a total of 16 days, cells were stained with Oil Red O for triglyceride depots.

4.5. RNA Extraction, cDNA Synthesis, and qPCR

Dissection of the hypothalamus was conducted from the ventral side of the brain. The optic chiasm was removed away from the anterior portion of the hypothalamus. The mammillary nuclei were dissected from the posterior of the hypothalamus. The entire hypothalamus was prepared, including the arcuate, ventromedial, dorsomedial, and paraventricular nuclei. Total RNA was extracted from adipocytes of humans, 3T3-L1 cells, visceral adipose tissue, liver, and hypothalamus of mice by using QIAzol Lysis Reagent (QIAGEN, Hilden, Germany) and RNAeasy Plus Mini Kit (QIAGEN) following the manufacture's instructions; 200 ng were used for cDNA synthesis by RevertAid™ H Minus Reverse Transcriptase (Thermo Fisher Scientific, Waltham, MA, USA). Primers (Supplementary Table S1) were designed using the web-based Primer3Plus package (www.primer3plus.com assessed on 3 April 2024). The semi-quantitative PCR was performed on the Biomark HD system (Standard BioTools, San Francisco, CA, USA) using EvaGreen DNA binding dye with BioMark™ 48.48 Dynamic Array Integrated Fluidic Circuits according to the manufacture's recommendations. Gene expression was determined via the $2^{\Delta\Delta CT}$ method [36,37]. Data were normalized to geometric mean of the four reference genes ATP synthase subunit b (*atp5f1*), eukaryotic elongation factor 2 (*eef2*), retention in endoplasmic reticulum sorting receptor 1 (*rer1*), and ribosomal protein l13a (*rpl13a*). Stability of reference genes was checked with the geNorm algorithm implemented in the open-source qbase+ data analysis software (www.qbaseplus.com assessed on 6 May 2024).

4.6. Statistical Analysis

Experimental data sets from in vivo mouse studies and in vitro experiments were processed and analyzed in GraphPad PRISM 7.02 for Windows (GraphPad Software, Inc., Boston, MA, USA). Data were expressed as mean ± SEM or min to max, and p values of less than 0.05 were considered significant by Wilcoxon–Mann–Whitney test or ANOVA.

Supplementary Materials: The following supporting information can be downloaded at: https://www.mdpi.com/article/10.3390/ijms25168620/s1.

Author Contributions: Conceptualization, T.P. and K.M.J.; methodology, T.P. and K.M.J.; formal analysis, S.S., R.S. and T.P.; investigation, S.S. and R.S.; resources, T.P.; data curation, M.B.; writing—original draft preparation, T.P.; writing—review and editing, S.S., R.S., B.L., A.C.Z., M.B., K.M.J. and T.P.; visualization, T.P.; supervision, T.P. and B.L.; project administration, T.P. All authors have read and agreed to the published version of the manuscript.

Funding: This work was supported by institutional funding from the Helmholtz Centre for Environmental Research—UFZ. The graphical abstract was created with BioRender.com.

Institutional Review Board Statement: Animal protocols used in this study were approved by the Committee on Animal Welfare of Saxony/Leipzig (Permit Number: TVV01/15, 14/18).

Data Availability Statement: The data generated during this study is available upon request.

Acknowledgments: We thank Marita Wagner, Elena Elter, and Beate Fink for their excellent technical assistance.

Conflicts of Interest: The authors declare no conflicts of interest. The funders had no role in the design of the study; in the collection, analyses, or interpretation of data; in the writing of the manuscript; or in the decision to publish the results.

References

1. Ng, M.; Fleming, T.; Robinson, M.; Thomson, B.; Graetz, N.; Margono, C.; Mullany, E.C.; Biryukov, S.; Abbafati, C.; Abera, S.F.; et al. Global, regional, and national prevalence of overweight and obesity in children and adults during 1980–2013: A systematic analysis for the Global Burden of Disease Study 2013. *Lancet* **2014**, *384*, 766–781. [CrossRef] [PubMed]
2. OECD. OECD Obesity Update. 2017. Available online: www.oecd.org/els/health-systems/Obesity-Update-2017.pdf (accessed on 20 May 2024).
3. Heindel, J.J.; vom Saal, F.S. Role of nutrition and environmental endocrine disrupting chemicals during the perinatal period on the aetiology of obesity. *Mol. Cell. Endocrinol.* **2009**, *304*, 90–96. [CrossRef] [PubMed]
4. Heindel, J.J.; Vom Saal, F.S.; Blumberg, B.; Bovolin, P.; Calamandrei, G.; Ceresini, G.; Cohn, B.A.; Fabbri, E.; Gioiosa, L.; Kassotis, C.; et al. Parma consensus statement on metabolic disruptors. *Environ. Health A Glob. Access Sci. Source* **2015**, *14*, 54. [CrossRef] [PubMed]
5. Gillman, M.W.; Ludwig, D.S. How early should obesity prevention start? *N. Engl. J. Med.* **2013**, *369*, 2173–2175. [CrossRef] [PubMed]
6. Rubin, B.S.; Soto, A.M. Bisphenol A: Perinatal exposure and body weight. *Mol. Cell. Endocrinol.* **2009**, *304*, 55–62. [CrossRef] [PubMed]
7. Janesick, A.; Blumberg, B. Obesogens, stem cells and the developmental programming of obesity. *Int. J. Androl.* **2012**, *35*, 437–448. [CrossRef]
8. Martino, D.; Prescott, S. Epigenetics and prenatal influences on asthma and allergic airways disease. *Chest* **2011**, *139*, 640–647. [CrossRef] [PubMed]
9. Leppert, B.; Strunz, S.; Seiwert, B.; Schlittenbauer, L.; Schlichting, R.; Pfeiffer, C.; Roder, S.; Bauer, M.; Borte, M.; Stangl, G.I.; et al. Maternal paraben exposure triggers childhood overweight development. *Nat. Commun.* **2020**, *11*, 561. [CrossRef] [PubMed]
10. Junge, K.M.; Leppert, B.; Jahreis, S.; Wissenbach, D.K.; Feltens, R.; Grutzmann, K.; Thurmann, L.; Bauer, T.; Ishaque, N.; Schick, M.; et al. MEST mediates the impact of prenatal bisphenol A exposure on long-term body weight development. *Clin. Epigenetics* **2018**, *10*, 58. [CrossRef]
11. Dodson, R.E.; Nishioka, M.; Standley, L.J.; Perovich, L.J.; Brody, J.G.; Rudel, R.A. Endocrine disruptors and asthma-associated chemicals in consumer products. *Environ. Health Perspect.* **2012**, *120*, 935–943. [CrossRef]
12. Bond, G.G.; Dietrich, D.R. Human cost burden of exposure to endocrine disrupting chemicals. A critical review. *Arch. Toxicol.* **2017**, *91*, 2745–2762. [CrossRef] [PubMed]
13. Malliari, E.; Kalantzi, O.I. Children's exposure to brominated flame retardants in indoor environments—A review. *Environ. Int.* **2017**, *108*, 146–169. [CrossRef]
14. Li, J.; Zhao, L.; Letcher, R.J.; Zhang, Y.; Jian, K.; Zhang, J.; Su, G. A review on organophosphate Ester (OPE) flame retardants and plasticizers in foodstuffs: Levels, distribution, human dietary exposure, and future directions. *Environ. Int.* **2019**, *127*, 35–51. [CrossRef] [PubMed]
15. Erkin-Cakmak, A.; Harley, K.G.; Chevrier, J.; Bradman, A.; Kogut, K.; Huen, K.; Eskenazi, B. In utero and childhood polybrominated diphenyl ether exposures and body mass at age 7 years: The CHAMACOS study. *Environ. Health Perspect.* **2015**, *123*, 636–642. [CrossRef] [PubMed]
16. Lim, J.S.; Lee, D.H.; Jacobs, D.R., Jr. Association of brominated flame retardants with diabetes and metabolic syndrome in the U.S. population, 2003–2004. *Diabetes Care* **2008**, *31*, 1802–1807. [CrossRef] [PubMed]
17. Suvorov, A.; Battista, M.C.; Takser, L. Perinatal exposure to low-dose 2,2′,4,4′-tetrabromodiphenyl ether affects growth in rat offspring: What is the role of IGF-1? *Toxicology* **2009**, *260*, 126–131. [CrossRef] [PubMed]
18. Gao, H.; Li, P.; Liu, L.; Yang, K.; Xiao, B.; Zhou, G.; Tian, Z.; Luo, C.; Xia, T.; Dong, L.; et al. Perigestational low-dose BDE-47 exposure alters maternal serum metabolome and results in sex-specific weight gain in adult offspring. *Chemosphere* **2019**, *233*, 174–182. [CrossRef]
19. Jahreis, S.; Trump, S.; Bauer, M.; Bauer, T.; Thurmann, L.; Feltens, R.; Wang, Q.; Gu, L.; Grutzmann, K.; Roder, S.; et al. Maternal phthalate exposure promotes allergic airway inflammation over 2 generations through epigenetic modifications. *J. Allergy Clin. Immunol.* **2018**, *141*, 741–753. [CrossRef] [PubMed]
20. Birnbaum, L.S.; Staskal, D.F. Brominated flame retardants: Cause for concern? *Environ. Health Perspect.* **2004**, *112*, 9–17. [CrossRef]
21. Schrenk, D.; Bignami, M.; Bodin, L.; Chipman, J.K.; del Mazo, J.; Grasl-Kraupp, B.; Hogstrand, C.; Hoogenboom, L.R.; Leblanc, J.-C.; Nebbia, C.S.; et al. Update of the risk assessment of polybrominated diphenyl ethers (PBDEs) in food. *EFSA J.* **2024**, *22*, e8497.
22. Hill, C.E.; Myers, J.P.; Vandenberg, L.N. Nonmonotonic Dose-Response Curves Occur in Dose Ranges That Are Relevant to Regulatory Decision-Making. *Dose Response* **2018**, *16*, 1559325818798282. [CrossRef] [PubMed]
23. Vandenberg, L.N. Non-monotonic dose responses in studies of endocrine disrupting chemicals: Bisphenol a as a case study. *Dose Response* **2014**, *12*, 259–276. [CrossRef] [PubMed]
24. Yang, C.; Wong, C.M.; Wei, J.; Chung, A.C.K.; Cai, Z. The brominated flame retardant BDE 47 upregulates purine metabolism and mitochondrial respiration to promote adipocyte differentiation. *Sci. Total Environ.* **2018**, *644*, 1312–1322. [CrossRef]
25. Kamstra, J.H.; Hruba, E.; Blumberg, B.; Janesick, A.; Mandrup, S.; Hamers, T.; Legler, J. Transcriptional and epigenetic mechanisms underlying enhanced in vitro adipocyte differentiation by the brominated flame retardant BDE-47. *Environ. Sci. Technol.* **2014**, *48*, 4110–4119. [CrossRef] [PubMed]

26. Pant, R.; Firmal, P.; Shah, V.K.; Alam, A.; Chattopadhyay, S. Epigenetic Regulation of Adipogenesis in Development of Metabolic Syndrome. *Front. Cell Dev. Biol.* **2020**, *8*, 619888. [CrossRef]
27. Byun, H.M.; Benachour, N.; Zalko, D.; Frisardi, M.C.; Colicino, E.; Takser, L.; Baccarelli, A.A. Epigenetic effects of low perinatal doses of flame retardant BDE-47 on mitochondrial and nuclear genes in rat offspring. *Toxicology* **2015**, *328*, 152–159. [CrossRef]
28. Zhang, Z.; Li, S.; Liu, L.; Wang, L.; Xiao, X.; Sun, Z.; Wang, X.; Wang, C.; Wang, M.; Li, L.; et al. Environmental exposure to BDE47 is associated with increased diabetes prevalence: Evidence from community-based case-control studies and an animal experiment. *Sci. Rep.* **2016**, *6*, 27854. [CrossRef]
29. Amitani, M.; Asakawa, A.; Amitani, H.; Inui, A. The role of leptin in the control of insulin-glucose axis. *Front. Neurosci.* **2013**, *7*, 51. [CrossRef]
30. Morris, D.L.; Rui, L. Recent advances in understanding leptin signaling and leptin resistance. *Am. J. Physiol. Endocrinol. Metab.* **2009**, *297*, E1247–E1259. [CrossRef]
31. Doan, K.V.; Kinyua, A.W.; Yang, D.J.; Ko, C.M.; Moh, S.H.; Shong, K.E.; Kim, H.; Park, S.K.; Kim, D.H.; Kim, I.; et al. FoxO1 in dopaminergic neurons regulates energy homeostasis and targets tyrosine hydroxylase. *Nat. Commun.* **2016**, *7*, 12733. [CrossRef]
32. Kozlova, E.V.; Denys, M.E.; Benedum, J.; Valdez, M.C.; Enriquez, D.; Bishay, A.E.; Chinthirla, B.D.; Truong, E.; Krum, J.M.; DiPatrizio, N.V.; et al. Developmental exposure to indoor flame retardants and hypothalamic molecular signatures: Sex-dependent reprogramming of lipid homeostasis. *Front. Endocrinol.* **2022**, *13*, 997304. [CrossRef] [PubMed]
33. Robinson, J.F.; Kapidzic, M.; Hamilton, E.G.; Chen, H.; Puckett, K.W.; Zhou, Y.; Ona, K.; Parry, E.; Wang, Y.; Park, J.S.; et al. Genomic Profiling of BDE-47 Effects on Human Placental Cytotrophoblasts. *Toxicol. Sci.* **2019**, *167*, 211–226. [CrossRef] [PubMed]
34. Chen, H.; Seifikar, H.; Larocque, N.; Kim, Y.; Khatib, I.; Fernandez, C.J.; Abello, N.; Robinson, J.F. Using a Multi-Stage hESC Model to Characterize BDE-47 Toxicity during Neurogenesis. *Toxicol. Sci.* **2019**, *171*, 221–234. [CrossRef] [PubMed]
35. Tabachnik, T.; Kisliouk, T.; Marco, A.; Meiri, N.; Weller, A. Thyroid Hormone-Dependent Epigenetic Regulation of Melanocortin 4 Receptor Levels in Female Offspring of Obese Rats. *Endocrinology* **2017**, *158*, 842–851. [CrossRef] [PubMed]
36. Ruiz-Ojeda, F.J.; Ruperez, A.I.; Gomez-Llorente, C.; Gil, A.; Aguilera, C.M. Cell Models and Their Application for Studying Adipogenic Differentiation in Relation to Obesity: A Review. *Int. J. Mol. Sci.* **2016**, *17*, 1040. [CrossRef]
37. Livak, K.J.; Schmittgen, T.D. Analysis of relative gene expression data using real-time quantitative PCR and the 2(-Delta Delta C(T)) Method. *Methods* **2001**, *25*, 402–408. [CrossRef]

International Journal of
Molecular Sciences

MDPI

Article

Modulation of the *ETV6::RUNX1* Gene Fusion Prevalence in Newborns by Corticosteroid Use During Pregnancy

Leticia Benítez [1,2,*], Ute Fischer [3,4,5], Fàtima Crispi [1,2], Sara Castro-Barquero [1,6,7], Francesca Crovetto [1,8], Marta Larroya [1,2], Lina Youssef [1,2,9], Ersen Kameri [3,5], Helena Castillo [1], Clara Bueno [9,10,11], Rosa Casas [12,13,14], Roger Borras [14], Eduard Vieta [15,16], Ramon Estruch [10,11,12], Pablo Menéndez [9,10,11], Arndt Borkhardt [3,4,*] and Eduard Gratacós [1,2,8]

1. BCNatal | Fetal Medicine Research Center, Hospital Clínic and Hospital Sant Joan de Deu, University of Barcelona, 08950 Barcelona, Spain; fcrispi@clinic.cat (F.C.); sacastro@clinic.cat (S.C.-B.); francesca.crovetto@sjd.es (F.C.); larroya@clinic.cat (M.L.); lyoussef@recerca.clinic.cat (L.Y.); hcastillo@clinic.cat (H.C.); egratacos@ub.edu (E.G.)
2. Institut d'Investigacions Biomèdiques August Pi i Sunyer, 08036 Barcelona, Spain
3. Department of Pediatric Oncology, Hematology and Clinical Immunology, University Children's Hospital, Medical Faculty, Heinrich Heine University, 40225 Düsseldorf, Germany; ute.fischer@med.uni-duesseldorf.de (U.F.); ersen.kameri@med.uni-duesseldorf.de (E.K.)
4. German Cancer Consortium, Partner Site Essen-Düsseldorf, 40225 Düsseldorf, Germany
5. Deutsches Krebsforschungszentrum (DKFZ), 69120 Heidelberg, Germany
6. Red Española de Terapias Avanzadas—Instituto de Salud Carlos III (ISCII), 28029 Madrid, Spain
7. Centro de Investigación Biomédica en Red Cáncer(CIBER-ONC), Instituto de Salud Carlos III (ISCIII), 28029 Madrid, Spain
8. Institut de Recerca Sant Joan de Déu, 08950 Barcelona, Spain
9. Josep Carreras Leukemia Research Institute, 08916 Barcelona, Spain; cbueno@carrerasresearch.org (C.B.); pmenendez@carrerasresearch.org (P.M.)
10. Department of Biomedicine, School of Medicine, University of Barcelona, 08014 Barcelona, Spain; restruch@clinic.cat
11. Institució Catalana de Recerca i Estudis Avançats, 08010 Barcelona, Spain
12. Centro de Investigación Biomédica en Red de Fisiopatología de la Obesidad y Nutrición, 28029 Madrid, Spain; rcasas1@recerca.clinic.cat
13. Institut de Recerca en Nutrició i Seguretat Alimentaria, University of Barcelona, 08014 Barcelona, Spain
14. Department of Internal Medicine, Hospital Clinic, University of Barcelona, 08014 Barcelona, Spain; rborras@recerca.clinic.cat
15. Centro de Investigación Biomédica en Red y Salud Mental, CIBERSAM, Instituto de Salud Carlos III, 28029 Madrid, Spain; evieta@clinic.cat
16. Department of Psychiatry and Psychology, Hospital Clinic, Neuroscience Institute, IDIBAPS, University of Barcelona, CIBERSAM, 08014 Barcelona, Spain

* Correspondence: lbenitez@clinic.cat (L.B.); arndt.borkhardt@med.uni-duesseldorf.de (A.B.)

Abstract: *ETV6::RUNX1*-positive pediatric acute lymphoblastic leukemia frequently has a prenatal origin and follows a two-hit model: a first somatic alteration leads to the formation of the oncogenic fusion gene *ETV6::RUNX1* and the generation of a preleukemic clone in utero. Secondary hits after birth are necessary to convert the preleukemic clone into clinically overt leukemia. However, prenatal factors triggering the first hit have not yet been determined. Here, we explore the influence of maternal factors during pregnancy on the prevalence of the *ETV6::RUNX1* fusion. To this end, we employed a nested interventional cohort study (IMPACT-BCN trial), including 1221 pregnancies (randomized into usual care, a Mediterranean diet, or mindfulness-based stress reduction) and determined the prevalence of the fusion gene in the DNA of cord blood samples at delivery (n = 741) using the state-of-the-art GIPFEL (genomic inverse PCR for exploration of ligated breakpoints) technique. A total of 6.5% (n = 48 of 741) of healthy newborns tested positive for *ETV6::RUNX1*. Our multiple regression analyses showed a trend toward lower *ETV6::RUNX1* prevalence in offspring of the high-adherence intervention groups. Strikingly, corticosteroid use for lung maturation during pregnancy was significantly

https://doi.org/10.3390/ijms26072971

associated with *ETV6::RUNX1* (adjusted OR 3.9, 95% CI 1.6–9.8) in 39 neonates, particularly if applied before 26 weeks of gestation (OR 7.7, 95% CI 1.08–50) or if betamethasone (OR 4.0, 95% CI 1.4–11.3) was used. Prenatal exposure to corticosteroids within a critical time window may therefore increase the risk of developing *ETV6::RUNX1+* preleukemic clones and potentially leukemia after birth. Taken together, this study indicates that *ETV6::RUNX1* preleukemia prevalence may be modulated and potentially prevented.

Keywords: *ETV6::RUNX1*; childhood leukemia; cord blood; prenatal; corticosteroids

1. Introduction

Childhood B-cell acute lymphoblastic leukemia (B-ALL) is the most frequent pediatric cancer and one of the leading causes of childhood mortality in developed countries, with increasing incidence in the past years [1–3]. Although advances in molecular risk-stratification, diagnosis, and targeted treatments have significantly improved survival rates, B-ALL remains a health priority that fits the standards for primary prevention [4–6].

The most prevalent subtypes of B-ALL frequently have a prenatal origin, following a two-hit model: (1) Somatic acquisition of oncogenic fusion genes (such as *ETV6::RUNX1* or *TCF3::PBX1*) or aberrant chromosome numbers (e.g., high hyperdiploidy) occurs in hematopoietic stem or progenitor cells as a first genetic alteration arising during prenatal life. (2) Secondary oncogenic hits after birth convert the preleukemic clone into clinically overt leukemia [7–9]. Retrospective analyses of Guthrie cards from children who subsequently developed leukemia demonstrated the presence of such first-hit molecular lesions at birth [10,11]. Such molecular lesions have subsequently been detected in healthy newborns [12–14].

The reported prevalence of the *ETV6::RUNX1* fusion was about 100× higher than the actual leukemia incidence, indicating that the fusion gene is only mildly oncogeneic.

In twins with concordant *ETV6::RUNX1*-positive leukemia, ALL develops at different times, and postnatal latency can be protracted [15]. Most frequently, *ETV6::RUNX1+* B-ALL harbors deletions of the second, non-translocated ETV6 allele [16], implying that the initial gene fusion predisposes individuals to leukemia, while subsequent ETV6 deletion acts as a promoting factor. While the postnatal triggers of leukemia have been extensively investigated [8], few studies have focused on the association between prenatal factors and preleukemic lesions. Parental age, ethnicity, maternal diet, folate intake, alcohol consumption, X-ray exposure, pesticides, perinatal infections, and fetal growth may play a role [17].

The investigation of prenatal determinants of childhood leukemia is limited by the technical challenge of evaluating a wide spectrum of preleukemic lesions in large sample sizes [18]. Cord blood (CB) biobanks from well-characterized large cohorts would be the most obvious source for this research, but sample sizes are commonly small and rarely suitable for this type of investigation. First, the reported frequency of cells in healthy newborns with preleukemic lesions at birth is estimated to range from 1 in 10^3 to 10^4, making them undetectable by most available laboratory methods. Moreover, the detection of these lesions has been optimized for RNA rather than DNA owing to patients' unique genomic DNA breakpoints, often occurring in extensive genomic regions not targetable by conventional techniques. Unfortunately, RNA samples are very rarely available in CB biobanks, are very unstable, and are susceptible to contamination, making them non-amenable for these studies.

Fueller et al. recently designed a new laboratory technique called "Genomic inverse PCR for exploration of ligated breakpoints" (GIPFEL) [19], which allows for the sensitive detection of recurrent chromosomal translocations. This technique relies on PCR-based quantification of unique DNA sequences that are created by circular ligation of restricted genomic DNA from translocation-bearing cells. Therefore, prior knowledge of the individual-specific interchromosomal breakpoints is not required. Recently, GIPFEL analysis allowed the assessment of the prevalence of preleukemic lesions in a Danish cohort of 1000 CBs from healthy newborns, and yielded a prevalence of 5% for *ETV6::RUNX1* and 0.6% for *TCF3::PBX1* [20,21].

Previous studies evaluating prenatal risk factors for childhood leukemia were mostly retrospective case-control studies, and CB samples were rarely available [17]. However, in the IMPACT-BCN (Improving Mothers for a better PrenAtal Care Trial BarCeloNa) randomized trial, CB samples were collected at delivery, and CB-derived mononuclear cells were stored as part of the study design. The study evaluated the impact of lifestyle interventions (Mediterranean diet and stress reduction) in 1221 pregnant women and demonstrated benefits on several perinatal outcomes [22].

In the present study, we evaluated the association between prenatal interventions, maternal and perinatal factors, and the emergence of neonatal preleukemic *ETV6::RUNX1* lesions in the well-characterized cohort of CB samples from the IMPACT-BCN trial.

2. Results

2.1. High Adherence to Mediterranean Diet and Stress Reduction Intervention Is Associated with a Trend Toward Lower Prevalence of the ETV6::RUNX1 Fusion

The *ETV6::RUNX1* fusion gene was screened using GIPFEL in $n = 741$ cord blood samples (Figure 1) in a blinded fashion ($n = 245$ without intervention, $n = 250$ Mediterranean diet intervention, and $n = 246$ stress reduction intervention). The *ETV6::RUNX1* fusion was detected in 48 cord blood samples (6.5%) of the study population, as determined by GIPFEL analysis (Table S1).

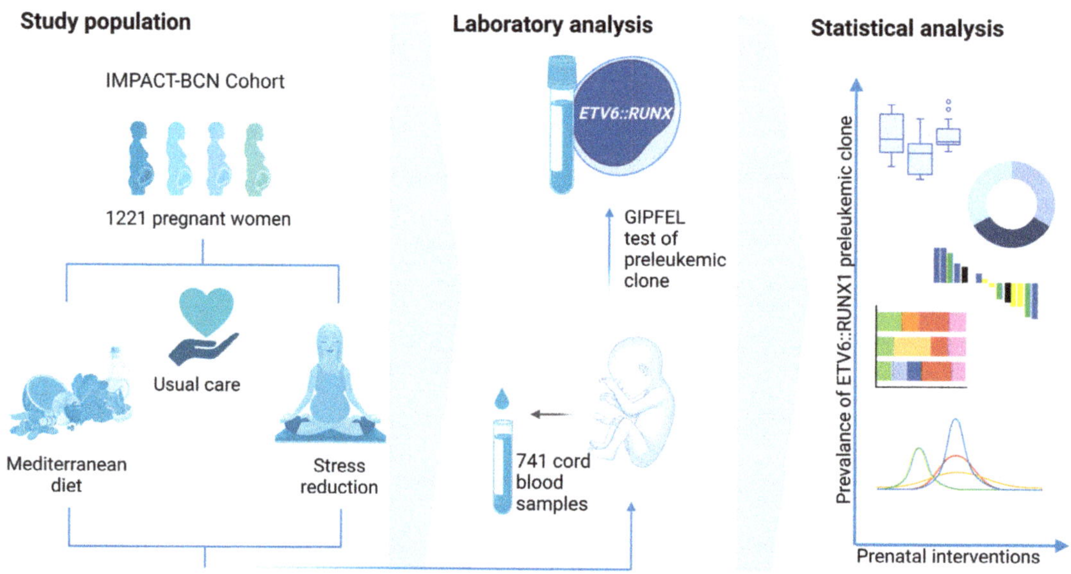

Figure 1. Study design depicting the study population, the intervention arms, and performed laboratory and statistical analyses. In the nested interventional cohort study (IMPACT-BCN trial).

High adherence to the Mediterranean diet intervention was defined as an improvement of at least 3 points in the final score of the 17-item dietary screener compared with the baseline score. Adherence to the stress reduction intervention was considered high if at least 6 of 9 stress reduction sessions were attended.

Regarding the intervention group, there were no differences in the prevalence of neonatal *ETV6::RUNX1* fusion gene among the three groups. However, when only considering participants who demonstrated a high adherence to the intervention, a non-significant trend toward lower neonatal ETV6::RUNX1 prevalence was detected in both interventions (prevalence of 7.0% for usual care versus 5.9% for stress reduction (OR 0.8; 95% CI 0.3–2.0) versus 5.3% for Mediterranean diet (OR 0.7; 95% CI 0.3–1.7) (Figure 2A, Table 1).

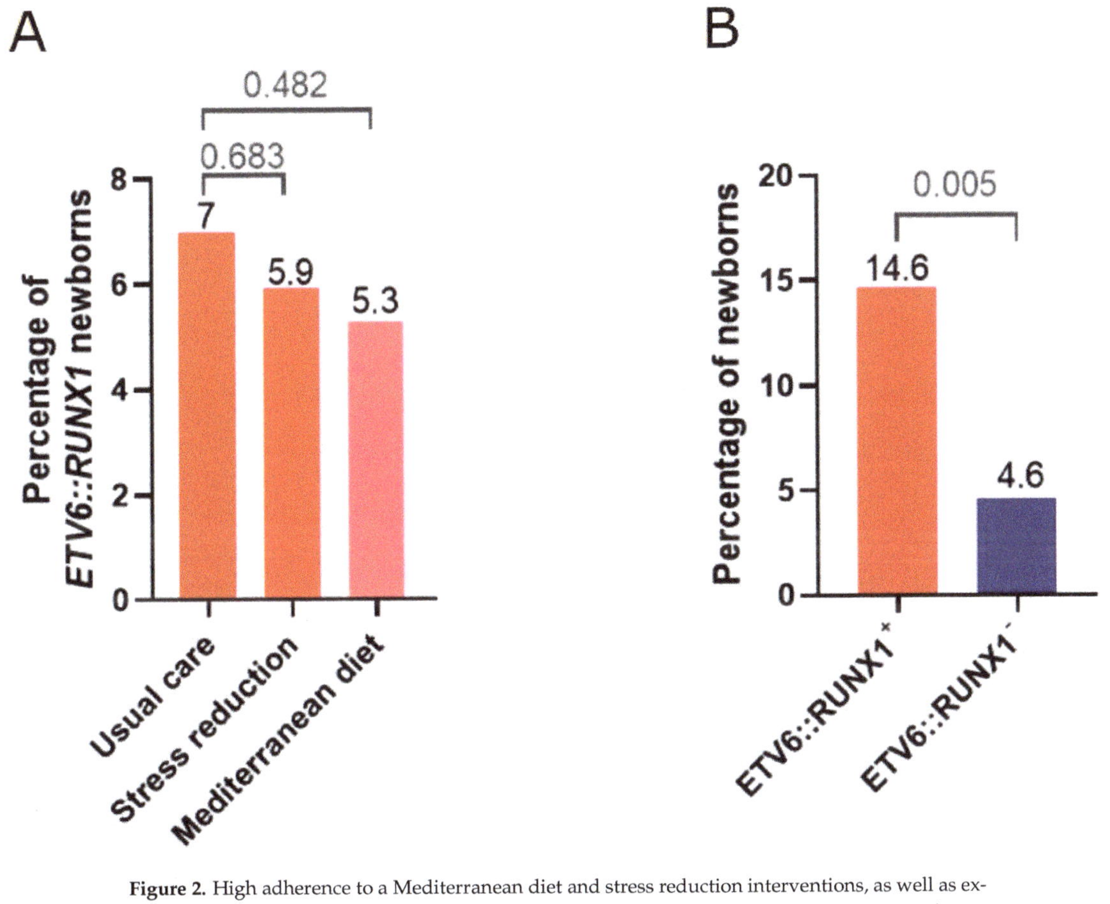

Figure 2. High adherence to a Mediterranean diet and stress reduction interventions, as well as exogenous glucocorticoid application, is associated with a lower prevalence of the *ETV6::RUNX1* fusion as determined by the GIPFEL method. (**A**) The bar graphs present the percentage of *ETV6::RUNX1* fusions considering only participants with high adherence to the interventions showing a non-significant trend toward lower prevalence (stress reduction (OR = 0.8, p = 0.683) and mediterranean diet (OR = 0.7, p = 0.482)). (**B**) The bar graph presents the prevalence of *ETV6::RUNX1* in newborns whose mothers received exogenous corticosteroids during pregnancy (n = 39 in total; *ETV6::RUNX1+* n = 7 of 48; *ETV6::RUNX1−* n = 32 of 693; OR = 3.5, p = 0.005).

Table 1. Prevalence of the neonatal *ETV6::RUNX1* fusion in the intervention groups, including only participants with high adherence to the intervention.

Intervention Group	Neonatal *ETV6::RUNX1+* No. (%)	Neonatal *ETV6::RUNX1−* No. (%)	OR (95% CI)	*p*-Value
Usual care	17 (7.0)	227 (93.0)	reference	NA
Stress reduction	8 (5.9)	128 (94.1)	0.8 (0.3–2.0)	0.683
Mediterranean Diet	9 (5.3)	162 (94.7)	0.7 (0.3–1.7)	0.482

2.2. Univariate Analysis Identifies Corticosteroid Application During Pregnancy as Significantly Associated with the Presence of ETV6::RUNX1 Fusions

Maternal and pregnancy characteristics of the study population were evaluated using univariate analyses based on the presence of the neonatal *ETV6::RUNX1* fusion gene (Table 2 and Table S2). Considering baseline characteristics, univariate analysis identified Maghreb ethnicity (OR 12.8, $p < 0.001$) and study class (OR for secondary/technological studies compared to primary/no studies 0.3, $p = 0.046$; OR for university studies compared to primary/no studies 0.7, $p = 0.525$) as significantly associated with the appearance of preleukemic lesions. Among perinatal characteristics, the administration of exogenous corticosteroids during pregnancy was significantly associated with the preleukemic status (OR 3.5, $p = 0.005$) (Figure 2B), while folate supplementation showed a non-significant preventive tendency (OR 0.6, $p = 0.056$). No specific dietary component or key food was identified as significantly associated with the neonatal *ETV6::RUNX1* fusion gene (Tables S3 and S4).

Table 2. Univariate analysis of the association between maternal and prenatal characteristics with neonatal *ETV6-RUNX1* positivity.

	No. (%)		Crude Odds Ratio for Neonatal *ETV6::RUNX1+*	
	Neonatal *ETV6::RUNX1* Positive (n = 48)	Neonatal *ETV6::RUNX1* Negative (n = 693)	OR (95% CI)	*p*-Value
Maternal Baseline Characteristics				
Maternal age (years)	36.8 (5.0)	36.8 (5.1)	1.00 (0.95–1.06)	0.969
BMI before pregnancy (kg/m^2)	23.6 (4.0)	24.1 (4.8)	0.97 (0.9–1.04)	0.427
Ethnicity				
White	34 (70.8)	544 (78.5)	Reference	Reference
Black	0 (0)	14 (2.0)	NA	NA
Asian	1 (2.1)	15 (2.2)	1.1 (0.1–8.3)	0.951
Indian	0 (0)	5 (0.7)	NA	NA
Latin American	9 (18.8)	110 (15.9)	1.3 (0.6–2.8)	0.489
Maghreb	**4 (8.3)**	**5 (0.7)**	**12.8 (3.3–49.9)**	**<0.001**
Study class				
Primary/no studies	5 (10.4)	43 (6.2)	Reference	Reference
Secondary/technological	**9 (18.7)**	**247 (35.6)**	**0.3 (0.1–0.98)**	**0.046**
University	34 (70.8)	403 (58.2)	0.7 (0.3–1.9)	0.525
Socioeconomic status $^{\psi}$				
Low	5 (10.4)	43 (6.2)	Reference	Reference
Medium	12 (25.0)	270 (39.0)	0.4 (0.1–1.1)	0.084
High	31 (64.6)	380 (54.8)	0.7 (0.3–1.9)	0.485

Table 2. *Cont.*

	No. (%)		Crude Odds Ratio for Neonatal *ETV6::RUNX1+*	
	Neonatal *ETV6::RUNX1* Positive (*n* = 48)	**Neonatal *ETV6::RUNX1* Negative (*n* = 693)**	**OR (95% CI)**	***p*-Value**
Nulliparous	27 (56.3)	399 (57.6)	0.95 (0.5–1.7)	0.857
Pregestational diabetes	5 (10.4)	38 (5.5)	2.0 (0.8–5.4)	0.165
Thyroid disorder	4 (8.3)	90 (13.0)	0.6 (0.2–1.7)	0.353
Autoimmune disease	6 (12.5)	112 (16.2)	0.7 (0.3–1.8)	0.504
Chronic hypertension	1 (2.1)	29 (4.2)	0.5 (0.1–3.7)	0.484
Chronic kidney disease	1 (2.1)	17 (2.5)	0.8 (0.1–6.5)	0.872
Obesity $^{\Phi}$	3 (6.3)	85 (12.3)	0.5 (0.1–1.6)	0.223
Pregnancy and prenatal characteristics				
Intervention group Usual care Stress reduction Mediterranean Diet	 17 (35.4) 16 (33.3) 15 (31.3)	 228 (32.9) 230 (33.3) 235 (33.9)	 Reference 0.9 (0.5–1.9) 0.85 (0.4–1.8)	 Reference 0.847 0.671
Cigarette smoking	2 (4.2)	54 (7.8)	0.5 (0.1–2.2)	0.367
Alcohol intake	0 (0)	17 (2.5)	NA	NA
Recreational drug consumption	0 (0)	3 (0.4)	NA	NA
Folate supplementation	31 (64.6)	532 (76.8)	0.6 (0.3–1.2)	0.059
Exogenous corticosteroids	**7 (14.6)**	**32 (4.6)**	**3.5 (1.5–8.5)**	**0.005**
Gestational diabetes	3 (6.3)	81 (11.7)	0.5 (0.2–1.7)	0.259
Preterm birth	4 (8.3)	37 (5.3)	1.6 (0.5–4.7)	0.386
Preeclampsia	0 (0)	57 (8.2)	NA	NA
Small for gestational age $^{\Upsilon}$	7 (14.6)	27 (3.9)	0.8 (0.3–1.8)	0.591
Gestational age at delivery (weeks)	39.2 (1.9)	39.4 (1.7)	0.9 (0.8–1.1)	0.404
Neonatal sex Female Male	 20 (41.7) 28 (58.3)	 335 (48.4) 357 (51.6)	 Reference 1.3 (0.7–2.4)	 0.367

Results are displayed as n (%) or mean (SD). BMI denotes body mass index. $^{\Psi}$ Socio economic status defined as low if participants reported having never worked or being unemployed for more than 2 years and having a partner with unqualified work or who was unemployed; high if they reported university studies regardless of whether they were working; and medium if any other situations. $^{\Phi}$ Obesity is defined as body mass index above 30. $^{\Upsilon}$ Small for gestational age is defined as birthweight below the 10th centile according to local standards.

2.3. Multivariate Regression Models Report a Higher Probability of ETV6::RUNX1 Occurrence for Corticosteroid Administration Before 26 Weeks of Gestation and Use of Betamethasone

Multivariate regression analysis confirmed the administration of exogenous corticosteroids as significantly associated with the *ETV6::RUNX1* fusion gene (Table 3).

Regarding the use of exogenous corticosteroids, a subanalysis according to gestational age reported a higher estimated probability in those participants receiving corticosteroids before 26 weeks of gestation (estimated probability of neonatal *ETV6::RUNX1* appearance of 0.50 (95% CI 0.1–0.9)) compared to those who received the treatment over 26 weeks of pregnancy (estimated probability 0.11 (95% CI 0.01–0.23); OR for administration before

26 weeks 7.7 (95% CI 1.08–50), $p = 0.04$). Betamethasone ($n = 25$) was the corticosteroid class most commonly used (with most participants receiving 2 doses for fetal lung maturation purposes), followed by methylprednisone ($n = 1$), prednisone ($n = 11$), and topical use ($n = 2$). GR:MR refers to the glucocorticoid receptor (GR) to mineralocorticoid receptor (MR) activity ratio for each exogenous corticosteroid listed (Table 4 and Table S5).

Table 3. Multivariate regression models for maternal and prenatal risk factors for neonatal ETV6::RUNX1 positivity.

	Model 1		Model 2		Model 3	
	Crude OR (95%CI)	*p*-Value	Adjusted OR (95%CI)	*p*-Value	Adjusted OR (95%CI)	*p*-Value
Maghreb ethnicity	12.8 (3.3–9.9)	<0.001	14.8 (3.4–63.9)	<0.001	10.5 (2.4–45.4)	0.002
Exogenous corticosteroids	3.5 (1.5–8.5)	0.005	3.4 (1.4–8.4)	0.007	3.9 (1.6–9.8)	0.003
Study class Primary/no studies Secondary/technological University	Reference 0.3 (0.1–0.98) 0.7 (0.3–1.9)	0.0367	Reference 0.5 (0.2–1.9) 1.4 (0.4–4.6)	0.03	Reference 0.5 (0.1–1.9) 1.4 (0.4–4.4)	0.306
Folate supplementation	0.6 (0.3–1.2)	0.059	0.6 (0.3–1.1)	0.10	0.6 (0.3–1.1)	0.083

Model 1 is unadjusted. Model 2 is adjusted for ethnicity and corticosteroids, study class, and folate supplementation. Model 3 is adjusted for variables in model 2 plus neonatal sex, pregestational diabetes, and smoking habit.

Table 4. Prevalence of neonatal *ETV6::RUNX1* positivity according to the class of exogenous corticosteroid.

Exogenous Corticosteroid Class	Neonatal *ETV6::RUNX1*+ ($n = 7$)	Neonatal *ETV6::RUNX1*− ($n = 32$)	OR (95% CI)	*p*-Value
Betamethasone (GR:MR = 25:1)	5	20	4.0 (1.4–11.3)	0.008
Metilprednisone (GR:MR = 11:1)	1	0	NA	NA
Prednisone (GR:MR = 5:1)	1	10	1.6 (0.2–12.9)	0.653
Topical corticosteroids (GR:MR = 10–100:1, according to steroid class)	0	2	NA	NA

CI denotes confidence interval; NA, not applicable. GR:MR refers to the glucocorticoid receptor (GR) to mineralocorticoid receptor (MR) activity ratio for each exogenous corticosteroid listed.

3. Discussion

This study identified prenatal exposure to corticosteroids as a potential risk factor for prenatal *ETV6::RUNX1* fusions and, thus, for leukemia development later in life. In addition, Maghreb ethnicity was associated with *ETV6::RUNX1* fusions, although absolute numbers were small. There were no other statistically significant associations between maternal lifestyle interventions and the prevalence of preleukemic *ETV6::RUNX1* lesions revealed in our study.

Previous studies have evaluated the association of prenatal risk factors [23–34] that suggest maternal age, ethnicity, some dietary components, folate supplementation, alcohol consumption, X-ray exposure, pesticide exposure, perinatal infections, and fetal growth as probable risk factors. Nevertheless, most available evidence is based on retrospective case-control studies that used self-reported questionnaires, which are susceptible to recall and underreporting bias. These studies were also unable to differentiate prenatal from postnatal exposures for most risk factors.

Our current study adds to previous evidence by reporting the analysis of 741 cord blood samples from a well-phenotyped cohort of pregnant women included in the IMPACT-BCN randomized trial. Maternal and prenatal factors were determined prospectively during the trial by the use of medical records by a trained medical doctor, avoiding recall bias related to self-report [35]. Additionally, assessment of maternal diet was prospectively performed by an expert nutritionist in a prospective manner during pregnancy, with proven validity [31]. Likewise, the use of biomarkers to confirm adherence to interventions enhances the reliability of results. The prevalence of *ETV6::RUNX1* was large enough to allow statistical comparisons.

The neonatal *ETV6::RUNX1* positivity prevalence of 6.5% (48 out of 741) is in line with previously published papers on healthy newborns in the USA, Europe, and Japan, ranging from 0.01 to 7% [14].

There was a significant association between maternal Maghreb ethnicity and neonatal preleukemic fusion genes. Although differences in the incidence of childhood leukemia between ethnicities had been previously described, an association with Maghreb ethnicity has not been reported. Concerning the *ETV6::RUNX1* fusion gene, Western countries and Hispanics have consistently described as presenting higher rates [36] compared with Black and Asian ethnicities. Genomic profiles [37] and folate metabolism polymorphisms [4] have been hypothesized as underlying factors. The difference in Maghreb ethnicity could be explained by genetic polymorphisms or by lifestyle or environmental determinants not recorded during the randomized trial. However, the small number of samples hampered the analysis of dietary components in this specific subpopulation.

Regarding corticosteroids, the main indications for their use in our cohort were fetal lung maturation in cases at risk of preterm birth and maternal autoimmune or dermatological conditions. To address the potential confounding effect of the indication for steroid administration, we conducted additional analyses adjusting for clinical conditions that warranted steroid use. These adjustments confirmed that the association between corticosteroid exposure and *ETV6::RUNX1* positivity remained significant, independent of the underlying indication for treatment. This suggests that the effect is more likely attributable to the drug itself rather than the condition requiring its administration. Prenatally, corticosteroids improve fetal lung maturation but have strong effects on fetal somatic and brain growth. This study identified exogenous corticosteroids as risk factors for preleukemic *ETV6::RUNX1* lesions. To our knowledge, this association has not previously been reported. Most cases had two doses of betamethasone under 26 weeks of gestation. In fact, glucocorticoids represent a key drug in the treatment backbone of B-ALL [38]. However, frequent secondary lesions in *ETV6::RUNX1*-positive ALL may directly affect the glucocorticoid response (e.g., by mutation of the glucocorticoid receptor NR3C1), the cell death pathway induced via its signaling, or components of the mismatch repair pathways [39]. One speculation on the effect of glucocorticoid treatment on the increased incidence of *ETV6::RUNX1* lesions in our study may therefore be an expansion of pre-existing minor *ETV6::RUNX1*-positive clones, due to secondary mutations that arise under the selective pressure of betamethasone. In this case, children would be expected to have a higher likelihood of developing leukemia later in life, which needs to be evaluated in larger studies with long-term follow-up.

Another probable explanation is the generation of the fusion gene itself as a direct consequence of glucocorticoid treatment. Recently, physiological concentrations of glucocorticoids have been shown to induce DNA double-strand breaks via inhibition of topoisomerase II. The effect of different glucocorticoids depended on their affinity for the glucocorticoid receptors (Figure 3) [40]. Betamethasone was significantly associated with an increased incidence of ETV6::RUNX1 fusions in our study and has a higher affinity for the receptor compared to dexamethasone, prednisolone, or endogenous cortisol [40]. Repair of the DNA double-strand breaks by non-homologous end joining could lead to *ETV6::RUNX1* fusion. Interestingly, after glucocorticoid treatment of B-ALL patients, increased numbers of *ETV6::RUNX1*-positive lymphocytes have been detected [41].

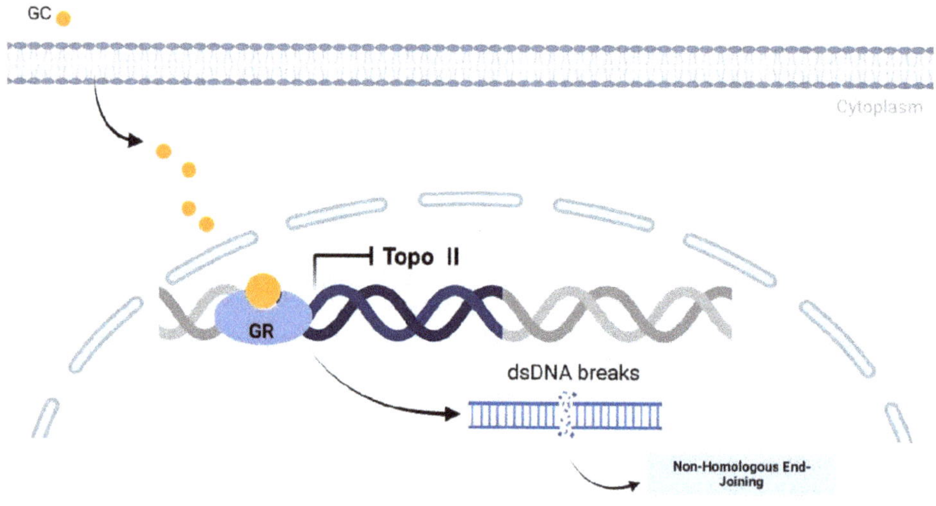

Figure 3. Glucocorticoids can inhibit DNA topoisomerase II (Topo II) activity, potentially by down-regulating its expression or modifying its access to DNA through chromatin remodeling. Reduced Topo II activity compromises its role in resolving DNA supercoiling, untangling chromatids, and repairing double-strand breaks (DSBs). One proposed mechanism for leukemia-causing chromosomal translocations entails chromosomal breakage by DNA topoisomerase II and recombination of DNA free ends from different chromosomes through DNA repair. Topo II inhibition increases the risk of chromosomal translocations, such as those involving the *MLL* gene rearrangements on chromosome 11q23, a hallmark of de novo acute leukemia development. Similarly, dysregulated Topo II activity has been linked to secondary leukemias, particularly therapy-related acute myeloid leukemia (t-AML) [42].

As recently demonstrated by Kim et al., using bisulfite sequencing of DNA extracted from bloodspot "Guthrie" cards, glucocorticoids also induce genome-wide methylation differences following antenatal exposure [43]. Methylation changes have been associated with prenatal environmental exposures and an increased risk of developing childhood leukemia [44], and there is also a correlation of methylation changes and specific genetic subtypes of ALL [45]. Hypomethylated areas are likely transcriptionally active regions and have a more open chromatin that may lend itself to interchromosomal contacts, DNA breaks, and aberrant repair [44]. It is tempting to speculate that the ETV6 and RUNX1 genes that recombine with one another so frequently are spatially localized together in transcriptional or chromatin compartments. The clustering of breakpoints, as seen in ETV6::RUNX1-positive ALL, also likely reflects selection for functional rearrangements during tumorigenesis. Although data from leukemia is lacking, previous studies in prostate cancer [46,47] show a significantly reduced methylation at breakpoints of specific interchromosomal translocations.

Glucocorticoid administration during pregnancy may also lead to permanent changes that impact on the risk of developing leukemia after birth (Figure 4). Glucocorticoid administration during pregnancy leads to higher plasma cortisol levels throughout adult life, indicating persistent reprogramming of the hypothalamus–pituitary–adrenal axis in animal models [48]. In the adrenal hypothesis model proposed by Schmiegelow et al. [49], early childhood infections result in profound changes in the hypothalamus–pituitary–adrenal axis that increase plasma cortisol levels and protect from leukemia by direct elimination of preleukemic cells via immune modulation and promotion of Th1-cytokine response [49]. As a possible explanation, it is conceivable that glucocorticoids may work as a double-edged sword: They may increase the risk of preleukemic fusions arising in utero but may also reduce the risk of leukemic transformation after birth.

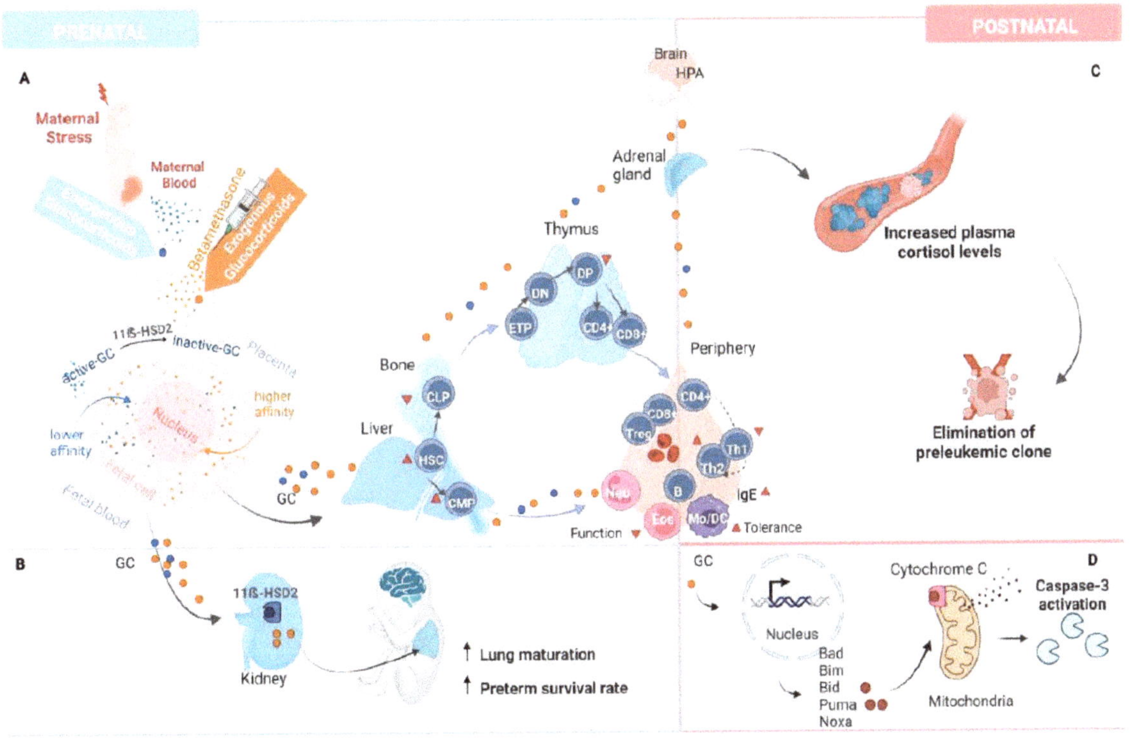

Figure 4. Potential impact of glucocorticoids on fetal prenatal and early postnatal development, as well as ALL therapy. (**A**) Endogenous glucocorticoid (cortisol) levels may increase due to prenatal stress perception during the second trimester. However, only a limited amount of endogenous GC crosses the placental barrier as they serve as a good substrate for the placental 11β-HSD2 enzyme, getting inactivated. Hence, exogenous GCs, such as betamethasone, may be administered to promote fetal lung maturation. These GCs, being steroid compounds, can easily cross the placenta and bind to the glucocorticoid receptor (GR) with a higher affinity compared to endogenous ones. Particularly, betamethasone shows a higher affinity for GR than other GCs. Upon binding, GR translocates to the nucleus, where it acts as either a transcriptional activator or repressor of target genes. Fetal exposure to high levels of GCs, whether from elevated maternal stress or medical treatment, may promote myeloid hematopoiesis and bone marrow or bone erythropoiesis by shifting hematopoietic stem cell (HSC) differentiation toward common myeloid progenitors (CMP) rather than common lymphoid progenitors (CLP). Additionally, GCs can influence bone marrow stromal cells, such as osteoblasts, which release soluble factors that regulate HSC differentiation and proliferation. Such changes in hematopoiesis may correlate with compromised humoral (B cell-derived) immune responses and perinatal neutrophil (Neu) function. GC excess in the thymus acts as a potent inducer of immature double-positive (DP) thymocyte apoptosis, as GCs are, in addition, locally produced here toward the end of the pregnancy, subsequently accelerating the maturation of double-negative (DN) thymocytes

to occupy the available niche. By inhibiting T-cell receptor signaling (TCR) or autoimmune regulator (AIRE)-mediated autoantigen transcription, glucocorticoids may mitigate the apoptosis event, allowing autoreactive CD4 and CD8 single-positive (SP) T cells to circulate. Prenatal GC exposure also programs CD4 T helper (Th) cells toward a Th2 response. Furthermore, prenatal glucocorticoids may increase postnatal HPA axis activity, resulting in elevated levels of corticotropin-releasing hormone (CRH) and arginine vasopressin (AVP). This hyperactivity can enhance both innate and adaptive immune responses, potentially leading to monocyte (Mo), macrophage, and dendritic cell (DC) tolerance to pathogens or excessive mast cell degranulation. These prenatal adaptations in immune function may increase the risk of infections, asthma, and other immune-related disorders in later life. (**B**) Exogenous GC, in contrast to endogenous GC, can bypass the placental and fetal 11SS-HSD2 enzyme, and promote fetal lung maturation and preterm survival rate. (**C**) Increased plasma cortisol levels resulting from excess GC-induced perturbations to the hypothalamus–pituitary–adrenal axis may directly eliminate pre-leukemic cells and suppress leukemia-promoting Th1-cytokine responses. (**D**) Mechanisms of action of glucocorticoid in ALL therapy. Glucocorticoid-mediated apoptosis is thought to be induced via the mitochondrial pathway through caspase activation.

Our data indicate that the effects of glucocorticoids in early pregnancy and childhood need to be carefully studied in larger cohorts, and the pros and cons of their application need to be deliberated by informed clinicians. Additionally, our study indicates promising options for potential strategies to prevent leukemia in childhood.

A previous study has reported an association between maternal diet quality and the risk of childhood leukemia [23]. In the present study, no significant associations were found between the Mediterranean diet or specific dietary components and the rate of *ETV6::RUNX1*. However, there was a trend toward lower rates among pregnancies with a high adherence to the Mediterranean diet, which deserves further investigation in larger sample sizes. Foods containing DNA topoisomerase II inhibitors, including coffee, tea, and other caffeinated beverages, canned food or dried vegetables or legumes, cocoa, red wine, apples, and berries, have been described as potential childhood leukemia triggers [50,51]. Concerning vitamins, deficient maternal folate intake during pregnancy has been extensively described as a risk factor for ALL [4,18,24,27,52,53]. In the present study, there was a non-significant trend toward lower rates of *ETV6::RUNX1* in relation to folate intake, although we cannot separate the intrinsic effect of other vitamins that are commonly taken for supplementation during pregnancy. There was no association between smoking, alcohol, or other toxics and *ETV6::RUNX1* fusion gene, but the prevalence of these risk factors was too low to allow meaningful comparisons. A lower socioeconomic class has been suggested to have a preventive effect on childhood cancer postnatally [54,55]. In this study, we observed a non-significant trend toward the opposite effect on *ETV6::RUNX1* frequency, which could be related to a lower folate intake in this subgroup of women.

Some limitations of our study should be acknowledged. First, the fact that the population of the study was from a single center located in Spain, which, although including several ethnicities, could hamper the external validation of results. Second, given the small sample size of *ETV6::RUNX1* positive individuals, our findings should be considered preliminary and require validation in larger studies to confirm their biological and clinical significance. Third, the variability in reported *ETV6::RUNX1* prevalence in cord blood across studies may stem from differences in detection methods, sample sizes, and population characteristics; however, the rigorous methodology and quality controls in our study support the credibility of our findings within this context. Fourth, we performed a 4-year follow-up in which none of the *ETV6::RUNX1*-positive newborns have been diagnosed with acute lymphoblastic leukemia. However, given the relatively short follow-up period, we acknowledge that longer-term monitoring is needed, and child follow-up should be warranted to confirm or discard the development of overt leukemia.

4. Material and Methods

4.1. Study Design and Participant Selection

This study represents a secondary analysis of the IMPACT-BCN trial, a randomized clinical trial conducted at BCNatal (Hospital Clínic and Hospital Sant Joan de Deu, Barcelona, Spain) from 2017 to 2020 [56]. The trial was approved by the Institutional Review Board (HCB-2016-0830). All participants provided written informed consent. Participants were screened for eligibility during routine second-trimester ultrasound scans (19–23.6 weeks of gestation) for being at high risk of developing small-for-gestational-age newborns [57].

4.2. Interventions

Participants were randomly assigned 1:1:1 to one of the three study groups: a Mediterranean diet intervention, a stress reduction program intervention, or usual care without any additional intervention (control group).

The Mediterranean diet intervention was adapted from the PREDIMED trial [58]. Registered dietitians conducted group meetings, as well as monthly face-to-face and additional telephone interviews to ensure adherence to the diet. All participants received olive oil (2 L every month) and 15 g of walnuts per day (450 g every month) at no cost. The participant received dietary training to encourage an increased intake of whole grain cereals (≥5 servings/d); vegetables and dairy products (≥3 servings/d); fresh fruit (≥2 servings/d); and legumes, nuts, fish, and white meat (≥3 servings/week). A 151-item food frequency questionnaire, a 7-day dietary journal, and a 17-item dietary assessment score (range, 0–17) were used to assess baseline nutrients and vitamin intake of all participants and adherence to the Mediterranean diet for the intervention group. A high adherence was defined as an improvement of at least 3 points in the final score of the 17-item dietary screener compared with the baseline score.

Led by experienced, certified instructors, the stress reduction group received an 8-week-long program of structured intervention (based on mindfulness) tested in clinical trials [59]. It included weekly 2.5-h sessions, 1 full-day session, and daily home practice, formal and informal techniques, with the goal of enhancing awareness and reducing anxiety. The sessions included didactic presentations, formal 45-min meditation practices, yoga, body awareness, and group discussions. Adherence to the stress reduction intervention was considered high if at least 6 of 9 stress reduction sessions were attended.

A detailed description of the interventions was previously published [22,56].

4.3. Cord Blood Sample Collection

Cord blood samples were collected at delivery after cord clamping, usually between 30 s to 1 min following international recommendations for delivery, and immediately stored at 4 °C. Ficoll density-gradient centrifugation (GE Healthcare, headquartered in Chicago, IL, USA) was performed in the following hours by an expert technician who was blinded to the intervention group. Cord blood components (serum, plasma, and mononuclear cells) were stored separately at −80 °C. Mononuclear cells were shipped to the University Hospital Düsseldorf, Department of Pediatric Oncology, Hematology, and Clinical Immunology for detection of *ETV6::RUNX1* fusions by GIPFEL analysis.

4.4. Genomic Inverse PCR for Exploration of Ligated Breakpoints (GIPFEL)

Cryopreserved cord blood samples were screened using the GIPFEL technique to check the neonatal *ETV6::RUNX1* positivity as detailed in the Supplementary Materials and Methods. In brief, genomic DNA was isolated from cord blood samples and digested with the restriction enzyme SacI. After circularization mediated by T4 DNA ligase, ETV6 and RUNX1 joints are ligated if the translocation is present. Circularized/ligated DNA was amplified, and *ETV6::RUNX1* was detected by real-time PCR, agarose gel electrophoresis, and confirmed by Sanger sequencing. Researchers and technicians performing the laboratory analysis were blinded to the intervention group.

4.5. Predictive Variables

The main outcome was the presence of the neonatal *ETV6::RUNX1* fusion gene in CB samples. All maternal and prenatal variables were considered potential predictive variables and were obtained prospectively at the beginning or during the trial, including maternal age, body mass index, ethnicity, level of education, socioeconomic status, parity, medical conditions, cigarette smoking, alcohol or recreational drug intake during pregnancy, folate supplementation, use of exogenous corticosteroids, occurrence of pregnancy complications (gestational diabetes, preterm birth, preeclampsia, small-for-gestational-age), and gestational age at delivery. Ethnicity was self-reported by participants among White, Latin, Afro-American, Maghreb, or others. Maghreb ethnicity was defined as patients from Western and Central North Africa. Study class was divided into primary/no studies, secondary/technology, and university studies and was self-reported by participants. Socioeconomic status was defined as low if participants reported having never worked or being unemployed for more than two years and having a partner with unskilled work or who was unemployed; and high if they reported university studies regardless of whether they were working; and medium if any other situations. Exogenous use of corticosteroids was determined according to medical records, which specified the dose, class, and duration of use. Gestational diabetes was defined as any degree of glucose intolerance with onset during pregnancy, in our setting diagnosed by the presence of two altered values in a 100-g, 3-h oral glucose tolerance test. Preterm birth was defined as delivery before 37 weeks of gestation. Preeclampsia was defined as the concurrence of pregnancy-onset hypertension and proteinuria. Small-for-gestational-age was defined as birthweight below the 10th centile according to local standards. Steroid use included betamethasone for preterm labor and oral or topical prednisone for maternal autoimmune and skin conditions that were administered during the entire pregnancy. Maternal dietary key food and nutrient intake were obtained by a specialized nutritionist at the beginning and end of the intervention with several questionnaires validated for the present study population [60].

4.6. Statistical Analysis

Maternal and prenatal data are presented as mean (standard deviation, SD) or number (percentage), as appropriate. Statistical analysis for comparison of clinical and prenatal characteristics included the use of the Student's *t*-test for continuous variables, the Chi-square test for categorical variables, and logistic univariate and multivariate regression analyses. Data were expressed using odds ratios and their corresponding 95% confidence intervals. Multivariate regression models were adjusted for significant variables in our study and described confounders. Estimated probability (or estimated marginal means) was computed with emmeans library version 1.8.6. Differences were considered significant when the *p*-value was lower than 0.05. Statistical analyses were conducted using Stata 15.1 (StataCorp LP, Houston, TX, USA) and R version 4.2.1 (Vienna, Austria).

5. Conclusions

Prenatal exposure to corticosteroids within a critical time window may therefore increase the risk of developing *ETV6::RUNX1*+ preleukemic clones and potentially leukemia after birth. Taken together, this study indicates that the prevalence of *ETV6::RUNX1* preleukemia may be modulated and potentially prevented.

Supplementary Materials: The following supporting information can be downloaded at: https://www.mdpi.com/article/10.3390/ijms26072971/s1.

Author Contributions: Conceptualization, L.B., A.B., E.G. and F.C. (Fàtima Crispi); methodology, E.K., F.C. (Francesca Crovetto) and F.C. (Fàtima Crispi); software, E.K. and L.B.; validation, L.B., A.B. and E.G.; formal analysis, E.K. and L.B.; investigation, L.B.; resources, L.B.; data curation, L.B. and R.B.; writing—original draft preparation, L.B., S.C.-B., E.K. and F.C. (Fàtima Crispi); writing—review and editing, A.B., E.G., U.F., E.K., F.C. (Fàtima Crispi), F.C. (Francesca Crovetto), S.C.-B., M.L., L.Y., H.C., C.B., R.C., R.B., E.V., R.E. and P.M.; visualization, L.B.; supervision, E.V., R.E., P.M., A.B. and E.G.; project administration, L.B.; funding acquisition, A.B. All authors have read and agreed to the published version of the manuscript.

Funding: The project was partially funded by a grant from "la Caixa" Foundation (LCF/PR/GN18/10310003); Cerebra Foundation for the Brain Injured Child (Carmarthen, Wales, UK); AGAUR under grant 2017 SGR No. 1531. Dr. Benítez was supported by a research grant from the Instituto de Salud Carlos III (CM21/00058). Dr. Castro-Barquero has received support from the Margarita Salas fellowship, University of Barcelona. Dr. Crovetto has received support from Centro de Investigaciones Biomédicas en Red sobre Enfermedades Raras (CIBERER) and Primary care interventions to prevent maternal and child chronic diseases of perinatal and developmental origin RD21/0012/0003, Instituto de Salud Carlos III, Spain. Dr. Youssef has received support from the Juan de la Cierva grant FJC2021-048123-I, funded by MCIN/AEI/10.13039/501100011033 and by the European Union "NextGenerationEU"/PRTR. Dr. Crispi has received support from the Instituto Carlos III (INT21/00027 and PI20/00246) and Fundació Jesus Serra (Spain). Dr. Fischer was supported by the Deutsche José-Carreras Leukämie-Stiftung (DJCLS 18R/2021), the Deutsche Forschungsgemeinschaft (DFG, German Research Foundation)—grant no. 495318549, the Deutsche Krebshilfe (DKH, German Cancer Aid) within the "Cancer Prevention—Graduate School" (CPGS)—grant no. 70114736, and the Deutsche Kinderkrebsstiftung (DKKS)—grant no. A2023/31. Dr. Fischer and Prof. Borkhardt were supported by the Bundesministerium für Bildung und Forschung (BMBF, the German Ministry for Education and Research)—grant no. 01KD2410A (EDI-4-ALL) and the Bundesamt für Strahlenschutz (BfS)—grant no. 3622S32231. Dr. Casas and Dr. Estruch received a grant from the INSA-Ma María de Maeztu Unit of Excellence (grant CEX2021-001234-M funded by MICIN/AEI/FEDER, UE).

Institutional Review Board Statement: The study was approved by the Hospital Clínic institutional review board (HCB-2016-0830) and the Ethical Commission of the Medical Faculty of Heinrich-Heine University.

Informed Consent Statement: All individuals who agreed to participate provided written informed consent before randomization.

Data Availability Statement: The data presented in this study are available on request from the corresponding author due to local IRB requirements.

Conflicts of Interest: The authors declare no conflict of interest.

Abbreviations

ALL	Acute Lymphoblastic Leukemia
B-ALL	B-cell Acute Lymphoblastic Leukemia
CB	Cord Blood
CBGs	Cord Blood Genomes
CMP	Common Myeloid Progenitor
CRH	Corticotropin-Releasing Hormone
DNA	Deoxyribonucleic Acid
DP	Double-Positive (thymocytes)
ETV6::RUNX1	Gene fusion associated with preleukemic clones

GCs	Glucocorticoids
GIPFEL	Genomic Inverse PCR for Exploration of Ligated Breakpoints
GR	Glucocorticoid Receptor
HPA	Hypothalamic–pituitary–adrenal (axis)
HSC	Hematopoietic Stem Cells
IMPACT-BCN	Improving Mothers for a Better Prenatal Care Trial Barcelona
MLL	Mixed-Lineage Leukemia (gene)
Mo	Monocyte
Neu	Neutrophils
NR3C1	Nuclear Receptor Subfamily 3 Group C Member 1 (Glucocorticoid Receptor Gene)
OR	Odds Ratio
PBX1	Pre-B-Cell Leukemia Homeobox 1
PCR	Polymerase Chain Reaction
PREDIMED	Prevención con Dieta Mediterránea (Prevention with Mediterranean Diet)
RCT	Randomized Controlled Trial
RNA	Ribonucleic Acid
SGA	Small for Gestational Age
SP	Single-positive (thymocytes)
TCF3	Transcription Factor 3
Th	T-helper Cell
Topo II	DNA Topoisomerase II
t-AML	Therapy-Related Acute Myeloid Leukemia

References

1. Hunger, S.P.; Mullighan, C.G. Acute Lymphoblastic Leukemia in Children. *N. Engl. J. Med.* **2015**, *373*, 1541–1552. [PubMed]
2. Steliarova-Foucher, E.; Colombet, M.; Ries, L.A.G.; Moreno, F.; Dolya, A.; Bray, F.; Hesseling, P.; Shin, H.Y.; Stiller, C.A.; Bouzbid, S.; et al. International Incidence of Childhood Cancer, 2001–10: A Population-Based Registry Study. *Lancet Oncol.* **2017**, *18*, 719–731. [CrossRef]
3. Siegel, R.L.; Miller, K.D.; Fuchs, H.E.; Jemal, A. Cancer statistics, 2022. *CA Cancer J. Clin.* **2022**, *72*, 7–33. [PubMed]
4. Whitehead, T.P.; Metayer, C.; Wiemels, J.L.; Singer, A.W.; Miller, M.D. Childhood Leukemia and Primary Prevention. *Curr. Probl. Pediatr. Adolesc. Health Care* **2016**, *46*, 317–352. [PubMed]
5. Greaves, M.; Cazzaniga, V.; Ford, A. Can we prevent childhood Leukaemia? *Leukemia* **2021**, *35*, 1258–1264. [PubMed]
6. Hauer, J.; Fischer, U.; Borkhardt, A. Towards prevention of childhood ALL by early-life immune training. *Blood* **2021**, *138*, 1412–1428.
7. Knudson, A.G. The genetics of childhood cancer. *Bull. Cancer* **1988**, *75*, 135–138.
8. Greaves, M. A causal mechanism for childhood acute lymphoblastic leukaemia. *Nat. Rev. Cancer* **2018**, *18*, 471–484.
9. Bueno, C.; Tejedor, J.R.; Bashford-Rogers, R.; Gonzalez-Silva, L.; Valdes-Mas, R.; Agraz-Doblas, A.; Diaz de la Guardia, R.; Ribera, J.; Zamora, L.; Bilhou-Nabera, C.; et al. Natural history and cell of origin of TC F3-ZN F384 and PTPN11 mutations in monozygotic twins with concordant BCP-ALL. *Blood* **2019**, *134*, 900–905.
10. Wiemels, J.L.; Cazzaniga, G.; Daniotti, M.; Eden, O.B.; Addison, G.M.; Masera, G.; Saha, V.; Biondi, A.; Greaves, M.F. Prenatal origin of acute lymphoblastic leukaemia in children. *Lancet* **1999**, *354*, 1499–1503.
11. Taub, J.W.; Konrad, M.A.; Ge, Y.; Naber, J.M.; Scott, J.S.; Matherly, L.H.; Ravindranath, Y. High frequency of leukemic clones in newborn screening blood samples of children with B-precursor acute lymphoblastic leukemia. *Blood* **2002**, *99*, 2992–2996. [PubMed]
12. Eguchi-Ishimae, M.; Eguchi, M.; Ishii, E.; Miyazaki, S.; Ueda, K.; Kamada, N.; Mizutani, S. Breakage and fusion of the TEL (ETV6) gene in immature B lymphocytes induced by apoptogenic signals. *Blood* **2001**, *97*, 737–743.
13. Mori, H.; Colman, S.M.; Xiao, Z.; Ford, A.M.; Healy, L.E.; Donaldson, C.; Hows, J.M.; Navarrete, C.; Greaves, M. Chromosome translocations and covert leukemic clones are generated during normal fetal development. *Proc. Natl. Acad. Sci. USA* **2002**, *99*, 8242–8247. [PubMed]
14. Hein, D.; Borkhardt, A.; Fischer, U. Insights into the prenatal origin of childhood acute lymphoblastic leukemia. *Cancer Metastasis Rev.* **2020**, *39*, 161–171.
15. Wiemels, J.L.; Ford, A.M.; Van Wering, E.R.; Postma, A.; Greaves, M. Protracted and variable latency of acute lymphoblastic leukemia after TEL-AML1 gene fusion in utero. *Blood* **1999**, *94*, 1057–1062. [PubMed]

16. Papaemmanuil, E.; Rapado, I.; Li, Y.; Potter, N.E.; Wedge, D.C.; Tubio, J.; Alexandrov, L.B.; Van Loo, P.; Cooke, S.L.; Marshall, J.; et al. RAG-mediated recombination is the predominant driver of oncogenic rearrangement in ETV6-RUNX1 acute lymphoblastic leukemia. *Nat. Genet.* **2014**, *46*, 116–125. [PubMed]
17. Benitez, L.; Castro-Barquero, S.; Crispi, F.; Youssef, L.; Crovetto, F.; Fischer, U.; Kameri, E.; Bueno, C.; Camos, M.; Menendez, P.; et al. Maternal Lifestyle and Prenatal Risk Factors for Childhood Leukemia: A Review of the Existing Evidence. *Fetal Diagn. Ther.* **2024**, *51*, 395–410.
18. Marcotte, E.L.; Spector, L.G.; Mendes-de-Almeida, D.P.; Nelson, H.H. The Prenatal Origin of Childhood Leukemia: Potential Applications for Epidemiology and Newborn Screening. *Front. Pediatr.* **2021**, *9*, 639479.
19. Fueller, E.; Schaefer, D.; Fischer, U.; Krell, P.F.; Stanulla, M.; Borkhardt, A.; Slany, R.K. Genomic inverse PCR for exploration of ligated breakpoints (GIPFEL), a new method to detect translocations in leukemia. *PLoS ONE* **2014**, *9*, e104419.
20. Schafer, D.; Olsen, M.; Lahnemann, D.; Stanulla, M.; Slany, R.; Schmiegelow, K.; Borkhardt, A.; Fischer, U. Five percent of healthy newborns have an ETV6-RUNX1 fusion as revealed by DNA-based GIPFEL screening. *Blood* **2018**, *131*, 821–826.
21. Hein, D.; Dreisig, K.; Metzler, M.; Izraeli, S.; Schmiegelow, K.; Borkhardt, A.; Fischer, U. The preleukemic TCF3-PBX1 gene fusion can be generated in utero and is present in approximately 0.6% of healthy newborns. *Blood* **2019**, *134*, 1355–1358. [PubMed]
22. Crovetto, F.; Crispi, F.; Casas, R.; Martin-Asuero, A.; Borras, R.; Vieta, E.; Estruch, R.; Gratacos, E.; Investigators, I.B.T. Effects of Mediterranean Diet or Mindfulness-Based Stress Reduction on Prevention of Small-for-Gestational Age Birth Weights in Newborns Born to At-Risk Pregnant Individuals: The IMPACT BCN Randomized Clinical Trial. *JAMA* **2021**, *326*, 2150–2160.
23. Jensen, C.D.; Block, G.; Buffler, P.; Ma, X.; Selvin, S.; Month, S. Maternal Dietary Risk Factors in Childhood Acute Lymphoblastic Leukemia (United States). *Cancer Causes Control* **2004**, *15*, 559–570. [CrossRef]
24. Thompson, J.R.; Gerald, P.F.; Willoughby, M.L.; Armstrong, B.K. Maternal folate supplementation in pregnancy and protection against acute lymphoblastic leukaemia in childhood: A case-control study. *Lancet* **2001**, *358*, 1935–1940.
25. Orsi, L.; Rudant, J.; Ajrouche, R.; Leverger, G.; Baruchel, A.; Nelken, B.; Pasquet, M.; Michel, G.; Bertrand, Y.; Ducassou, S.; et al. Parental smoking, maternal alcohol, coffee and tea consumption during pregnancy, and childhood acute leukemia: The ESTELLE study. *Cancer Causes Control* **2015**, *26*, 1003–1017.
26. MacArthur, A.C.; McBride, M.L.; Spinelli, J.J.; Tamaro, S.; Gallagher, R.P.; Theriault, G. Risk of childhood leukemia associated with parental smoking and alcohol consumption prior to conception and during pregnancy: The cross-Canada childhood leukemia study. *Cancer Causes Control* **2008**, *19*, 283–295. [PubMed]
27. Metayer, C.; Milne, E.; Dockerty, J.D.; Clavel, J.; Pombo-de-Oliveira, M.S.; Wesseling, C.; Spector, L.G.; Schuz, J.; Petridou, E.; Ezzat, S.; et al. Maternal supplementation with folic acid and other vitamins and risk of leukemia in offspring: A Childhood Leukemia International Consortium study. *Epidemiology* **2014**, *25*, 811–822. [PubMed]
28. Karalexi, M.A.; Dessypris, N.; Thomopoulos, T.P.; Ntouvelis, E.; Kantzanou, M.; Diamantaras, A.A.; Moschovi, M.; Baka, M.; Hatzipantelis, E.; Kourti, M.; et al. Parental alcohol consumption and risk of leukemia in the offspring: A systematic review and meta-analysis. *Eur. J. Cancer Prev.* **2017**, *26*, 433–441.
29. O'Neill, K.A.; Murphy, M.F.; Bunch, K.J.; Puumala, S.E.; Carozza, S.E.; Chow, E.J.; Mueller, B.A.; McLaughlin, C.C.; Reynolds, P.; Vincent, T.J.; et al. Infant birthweight and risk of childhood cancer: International population-based case control studies of 40,000 cases. *Int. J. Epidemiol.* **2015**, *44*, 153–168. [PubMed]
30. Klimentopoulou, A.; Antonopoulos, C.N.; Papadopoulou, C.; Kanavidis, P.; Tourvas, A.D.; Polychronopoulou, S.; Baka, M.; Athanasiadou-Piperopoulou, F.; Kalmanti, M.; Sidi, V.; et al. Maternal smoking during pregnancy and risk for childhood leukemia: A nationwide case-control study in Greece and meta-analysis. *Pediatr. Blood Cancer* **2012**, *58*, 344–351.
31. Milne, E.; Royle, J.A.; Bennett, L.C.; de Klerk, N.H.; Bailey, H.D.; Bower, C.; Miller, M.; Attia, J.; Scott, R.J.; Kirby, M.; et al. Maternal consumption of coffee and tea during pregnancy and risk of childhood ALL: Results from an Australian case-control study. *Cancer Causes Control* **2011**, *22*, 207–218. [CrossRef] [PubMed]
32. Menegaux, F.; Ripert, M.; Hemon, D.; Clavel, J. Maternal alcohol and coffee drinking, parental smoking and childhood leukaemia: A French population-based case-control study. *Paediatr. Perinat. Epidemiol.* **2007**, *21*, 293–299. [CrossRef]
33. Petridou, E.; Trichopoulos, D.; Kalapothaki, V.; Pourtsidis, A.; Kogevinas, M.; Kalmanti, M.; Koliouskas, D.; Kosmidis, H.; Panagiotou, J.P.; Piperopoulou, F.; et al. The risk profile of childhood leukaemia in Greece: A nationwide case-control study. *Br. J. Cancer* **1997**, *76*, 1241–1247. [CrossRef] [PubMed]
34. McKinney, P.A.; Juszczak, E.; Findlay, E.; Smith, K.; Thomson, C.S. Pre- and perinatal risk factors for childhood leukaemia and other malignancies: A Scottish case control study. *Br. J. Cancer* **1999**, *80*, 1844–1851. [CrossRef]
35. Jurek, A.M.; Greenland, S.; Spector, L.G.; Roesler, M.A.; Robison, L.L.; Ross, J.A. Self-report versus medical record—Perinatal factors in a study of infant leukaemia: A study from the Children's Oncology Group. *Paediatr. Perinat. Epidemiol.* **2011**, *25*, 540–548. [CrossRef]
36. Liang, D.-C.; Shih, L.-Y.; Yang, C.-P.; Hung, I.-J.; Liu, H.-C.; Jaing, T.-H.; Yeh, T.-C.; Liang, S.-T.; Chang, C.-L.; Lee, E.-H.; et al. Frequencies of ETV6-RUNX1 Fusion and Hyperdiploidy in Pediatric Acute Lymphoblastic Leukemia Are Lower in Far East than West. *Pediatr. Blood Cancer* **2010**, *55*, 430–433. [CrossRef] [PubMed]

37. Hsu, L.-I.; Briggs, F.; Shao, X.; Metayer, C.; Wiemels, J.L.; Chokkalingam, A.P.; Barcellos, L.F. Pathway Analysis of Genome-Wide Association Study in Childhood Leukemia among Hispanics. *Cancer Epidemiol. Biomark. Prev.* **2016**, *25*, 815–822. [CrossRef] [PubMed]
38. Gaynon, P.S.; Lustig, R.H. The use of glucocorticoids in acute lymphoblastic leukemia of childhood. Molecular, cellular, and clinical considerations. *J. Pediatr. Hematol. Oncol.* **1995**, *17*, 1–12. [CrossRef]
39. Kuster, L.; Grausenburger, R.; Fuka, G.; Kaindl, U.; Krapf, G.; Inthal, A.; Mann, G.; Kauer, M.; Rainer, J.; Kofler, R.; et al. ETV6/RUNX1-positive relapses evolve from an ancestral clone and frequently acquire deletions of genes implicated in glucocorticoid signaling. *Blood* **2011**, *117*, 2658–2667. [CrossRef]
40. Akter, S.; Shimba, A.; Ikuta, K.; Mahmud, M.R.A.; Yamada, S.; Sasanuma, H.; Tsuda, M.; Sone, M.; Ago, Y.; Murai, K.; et al. Physiological concentrations of glucocorticoids induce pathological DNA double-strand breaks. *Genes Cells* **2023**, *28*, 53–67. [CrossRef]
41. Brassesco, M.S.; Camparoto, M.L.; Tone, L.G.; Sakamoto-Hojo, E.T. Analysis of ETV6/RUNX1 fusions for evaluating the late effects of cancer therapy in ALL (acute lymphoblastic leukemia) cured patients. *Cytogenet. Genome Res.* **2004**, *104*, 346–351. [PubMed]
42. Greaves, M.F. Aetiology of acute leukaemia. *Lancet* **1997**, *349*, 344–349. [PubMed]
43. Kim, B.; Sasaki, A.; Murphy, K.; Matthews, S.G. DNA methylation signatures in human neonatal blood following maternal antenatal corticosteroid treatment. *Transl. Psychiatry* **2022**, *12*, 132. [PubMed]
44. Timms, J.A.; Relton, C.L.; Sharp, G.C.; Rankin, J.; Strathdee, G.; McKay, J.A. Exploring a potential mechanistic role of DNA methylation in the relationship between in utero and post-natal environmental exposures and risk of childhood acute lymphoblastic leukaemia. *Int. J. Cancer* **2019**, *145*, 2933–2943.
45. Nordlund, J.; Backlin, C.L.; Zachariadis, V.; Cavelier, L.; Dahlberg, J.; Ofverholm, I.; Barbany, G.; Nordgren, A.; Overnas, E.; Abrahamsson, J.; et al. DNA methylation-based subtype prediction for pediatric acute lymphoblastic leukemia. *Clin. Epigenetics* **2015**, *7*, 11.
46. Berger, M.F.; Lawrence, M.S.; Demichelis, F.; Drier, Y.; Cibulskis, K.; Sivachenko, A.Y.; Sboner, A.; Esgueva, R.; Pflueger, D.; Sougnez, C.; et al. The genomic complexity of primary human prostate cancer. *Nature* **2011**, *470*, 214–220. [CrossRef]
47. Lin, P.C.; Giannopoulou, E.G.; Park, K.; Mosquera, J.M.; Sboner, A.; Tewari, A.K.; Garraway, L.A.; Beltran, H.; Rubin, M.A.; Elemento, O. Epigenomic alterations in localized and advanced prostate cancer. *Neoplasia* **2013**, *15*, 373–383.
48. Seckl Prenatal Glucocorticoids and Long-Term Programming. *Eur. J. Endocrinol.* **2004**, *151*, U49–U62. [CrossRef]
49. Schmiegelow, K.; Vestergaard, T.; Nielsen, S.M.; Hjalgrim, H. Etiology of common childhood acute lymphoblastic leukemia: The adrenal hypothesis. *Leukemia* **2008**, *22*, 2137–2141.
50. Ross, J.A.; Potter, J.D.; Reaman, G.H.; Pendergrass, T.W.; Robison, L.L. Maternal exposure to potential inhibitors of DNA topoisomerase II and infant leukemia (United States): A report from the Children's Cancer Group. *Cancer Causes Control* **1996**, *7*, 581–590.
51. Spector, L.G.; Xie, Y.; Robison, L.L.; Heerema, N.A.; Hilden, J.M.; Lange, B.; Felix, C.A.; Davies, S.M.; Slavin, J.; Potter, J.D.; et al. Maternal diet and infant leukemia: The DNA topoisomerase II inhibitor hypothesis: A report from the children's oncology group. *Cancer Epidemiol. Biomark. Prev.* **2005**, *14*, 651–655.
52. Bailey, H.D.; Miller, M.; Langridge, A.; de Klerk, N.H.; van Bockxmeer, F.M.; Attia, J.; Scott, R.J.; Armstrong, B.K.; Milne, E. Maternal dietary intake of folate and vitamins B6 and B12 during pregnancy and the risk of childhood acute lymphoblastic leukemia. *Nutr. Cancer* **2012**, *64*, 1122–1130.
53. Chokkalingam, A.P.; Chun, D.S.; Noonan, E.J.; Pfeiffer, C.M.; Zhang, M.; Month, S.R.; Taggart, D.R.; Wiemels, J.L.; Metayer, C.; Buffler, P.A. Blood levels of folate at birth and risk of childhood leukemia. *Cancer Epidemiol. Biomark. Prev.* **2013**, *22*, 1088–1094.
54. Poole, C.; Greenland, S.; Luetters, C.; Kelsey, J.L.; Mezei, G. Socioeconomic status and childhood leukaemia: A review. *Int. J. Epidemiol.* **2006**, *35*, 370–384. [PubMed]
55. Adam, M.; Rebholz, C.E.; Egger, M.; Zwahlen, M.; Kuehni, C.E. Childhood Leukaemia and Socioeconomic Status: What Is the Evidence? *Radiat. Prot. Dosim.* **2008**, *132*, 246–254. [CrossRef]
56. Crovetto, F.; Crispi, F.; Borras, R.; Paules, C.; Casas, R.; Martin-Asuero, A.; Arranz, A.; Vieta, E.; Estruch, R.; Gratacos, E. Mediterranean diet, Mindfulness-Based Stress Reduction and usual care during pregnancy for reducing fetal growth restriction and adverse perinatal outcomes: IMPACT BCN (Improving Mothers for a better PrenAtal Care Trial BarCeloNa): A study protocol for a randomized controlled trial. *Trials* **2021**, *22*, 362.
57. Morris, R.K.; Johnstone, E.; Lees, C.; Morton, V.; Smith, G.; Royal College of Obstetricians and Gynaecologists. Investigation and Care of a Small-for-Gestational-Age Fetus and a Growth Restricted Fetus (Green-top Guideline No. 31). *BJOG Int. J. Obstet. Gynaecol.* **2024**, *131*, e31–e80.
58. Toledo, E.; Salas-Salvado, J.; Donat-Vargas, C.; Buil-Cosiales, P.; Estruch, R.; Ros, E.; Corella, D.; Fito, M.; Hu, F.B.; Aros, F.; et al. Mediterranean Diet and Invasive Breast Cancer Risk Among Women at High Cardiovascular Risk in the PREDIMED Trial: A Randomized Clinical Trial. *JAMA Intern. Med.* **2015**, *175*, 1752–1760.

59. Ludwig, D.S.; Kabat-Zinn, J. Mindfulness in medicine. *JAMA* **2008**, *300*, 1350–1352.
60. Juton, C.; Castro-Barquero, S.; Casas, R.; Freitas, T.; Ruiz-Leon, A.M.; Crovetto, F.; Domenech, M.; Crispi, F.; Vieta, E.; Gratacos, E.; et al. Reliability and Concurrent and Construct Validity of a Food Frequency Questionnaire for Pregnant Women at High Risk to Develop Fetal Growth Restriction. *Nutrients* **2021**, *13*, 1629. [CrossRef]

International Journal of *Molecular Sciences*

MDPI

Article

Regulation of Neuronal Chloride Homeostasis by Pro- and Mature Brain-Derived Neurotrophic Factor (BDNF) via KCC2 Cation–Chloride Cotransporters in Rat Cortical Neurons

Mira Hamze [1,2,3], Cathy Brier [1,2,3], Emmanuelle Buhler [1,2,3], Jinwei Zhang [4], Igor Medina [1,2,3] and Christophe Porcher [1,2,3,*]

1 INMED, INSERM, Aix-Marseille University, 13273 Marseille, France; mira.hamze@inserm.fr (M.H.); cahty.brier@inserm.fr (C.B.); emmanuelle.buhler@inserm.fr (E.B.); igor.medyna@inserm.fr (I.M.)
2 INSERM (Institut National de la Santé et de la Recherche Médicale), Unité 1249, Parc Scientifique de Luminy, 13273 Marseille, France
3 INMED (Institut de Neurobiologie de la Méditerranée), Parc Scientifique de Luminy, 13273 Marseille, France
4 Shanghai Institute of Organic Chemistry, Chinese Academy of Sciences, 345 Ling Ling Road, Shanghai 200032, China; jinweizhang@sioc.ac.cn
* Correspondence: christophe.porcher@inserm.fr; Tel.: +33-4-91-82-81-29

Abstract: The strength of inhibitory neurotransmission depends on intracellular neuronal chloride concentration, primarily regulated by the activity of cation–chloride cotransporters NKCC1 (Sodium–Potassium–Chloride Cotransporter 1) and KCC2 (Potassium–Chloride Cotransporter 2). Brain-derived neurotrophic factor (BDNF) influences the functioning of these co-transporters. BDNF is synthesized from precursor proteins (proBDNF), which undergo proteolytic cleavage to yield mature BDNF (mBDNF). While previous studies have indicated the involvement of BDNF signaling in the activity of KCC2, its specific mechanisms are unclear. We investigated the interplay between both forms of BDNF and chloride homeostasis in rat hippocampal neurons and in utero electroporated cortices of rat pups, spanning the behavioral, cellular, and molecular levels. We found that both pro- and mBDNF play a comparable role in immature neurons by inhibiting the capacity of neurons to extrude chloride. Additionally, proBDNF increases the endocytosis of KCC2 while maintaining a depolarizing shift of E_{GABA} in maturing neurons. Behaviorally, proBDNF-electroporated rat pups in the somatosensory cortex exhibit sensory deficits, delayed huddling, and cliff avoidance. These findings emphasize the role of BDNF signaling in regulating chloride transport through the modulation of KCC2. In summary, this study provides valuable insights into the intricate interplay between BDNF, chloride homeostasis, and inhibitory synaptic transmission, shedding light on the underlying cellular mechanisms involved.

Keywords: GABA; KCC2 activity; chloride homeostasis; BDNF

1. Introduction

KCC2 (Potassium–Chloride Cotransporter 2) and NKCC1 (Sodium–Potassium–Chloride Cotransporter 1) are crucial for maintaining chloride ion balance inside and outside neurons, playing complementary roles in the regulation of GABAergic (gamma-aminobutyric acid) inhibition and chloride homeostasis in the nervous system. NKCC1 intrudes chloride ions (Cl^-), leading to a higher intracellular chloride concentration ($[Cl^-]_i$), while KCC2 actively extrudes Cl^-, establishing a negative Cl^- gradient across the cell membrane [1,2]. The Cl^- gradient created by KCC2 is essential for GABAergic inhibition as it determines the direction of chloride flow through ionotropic $GABA_A$ receptors ($GABA_AR$). In mature neurons with functional KCC2, the low $[Cl^-]_i$ maintained by the transporter ensures that the influx of Cl^- ions through $GABA_A$Rs hyperpolarizes the neuron, resulting in inhibition [1,3]. This mechanism dampens neuronal excitability. During early brain development, NKCC1 expression is predominant, resulting in higher $[Cl^-]_i$ in immature neurons. Consequently, GABAergic receptor activation leads to depolarizing responses and contributes to the excitatory actions of GABA during this critical period [1,4]. These depolarizing responses are essential for synaptic plasticity and the refinement of neural circuits in response to environmental stimuli. As the nervous system matures, there is a developmental shift in the relative functional expression of NKCC1 and KCC2. This shift is vital for the transition from depolarizing to hyperpolarizing responses to GABA. Increased KCC2 expression promotes Cl^- extrusion from neurons, reducing $[Cl^-]_i$ and strengthening inhibitory signaling [5–7]. This transition establishes an appropriate balance between excitation and inhibition, thereby contributing to the closure of the critical period and normal brain functioning [8]. Changes in the expression or functionality of NKCC1 and KCC2 are associated with a range of neurological disorders. These conditions include neurodevelopmental disorders (NDDs) like epilepsy, autism spectrum disorder (ASD), schizophrenia, Down syndrome, and Rett syndrome [9,10] and, more recently, neurodegenerative diseases [11,12]. All these conditions have been linked to disruptions in the expression or function of these cotransporters. Dysregulated NKCC1 or impaired KCC2 maturation can disrupt chloride homeostasis, leading to neuronal hyperexcitability and altered synaptic plasticity. The expression patterns of NKCC1 and KCC2 are tightly regulated by various factors, including neurohormones, hormones, trophic polypeptides, and neurotrophins such as BDNF (brain-derived neurotrophic factor) [13–15]. BDNF exists in two forms: proBDNF (the precursor of mature BDNF) and mature BDNF (mBDNF). These two forms of BDNF interact with different receptors and elicit distinct cellular responses [16]. In the neurons of rodents, the expression of proBDNF and p75 neurotrophin receptor ($p75^{NTR}$) undergoes developmental regulation, with the highest levels observed in the early postnatal weeks, followed by a reduction at 6 weeks after birth [17]. This developmental pattern correlates with increased $[Cl^-]_i$ and the depolarizing action of GABA. In contrast, mBDNF is minimally detectable during the first postnatal week but reaches full expression around the third postnatal week [18,19]. These two forms of BDNF play distinct roles and have the potential to interact with KCC2. However, the intracellular pathways involving BDNF receptors and KCC2 remain unclear [15]. Several studies have suggested that proBDNF can reduce KCC2 protein expression and impair its function in neurons. This downregulation of KCC2 disrupts chloride ion homeostasis, leading to impaired inhibitory neurotransmission. Consequently, proBDNF promotes neuronal hyperexcitability and can contribute to conditions associated with reduced inhibitory signaling, such as epilepsy [20,21]. On the other hand, mature BDNF has been shown to either upregulate KCC2 expression and enhance its activity in immature neurons through the BDNF-TrkB-ERK1/2 (extracellular signal-regulated kinase 1/2) signaling pathway [14,22] or reduce KCC2 function in mature neurons via the BDNF-TrkB-PlC pathway [13]. This dual effect of mature BDNF can promote seizure susceptibility and contribute to neuropathic pain [13,23,24]. These findings collectively support the notion that mBDNF, proBDNF, and KCC2 interact at different stages of brain development to regulate neuronal survival, synaptic development and plasticity, and the balance between excitation and inhibition. The dysregulation of this interplay, such as an imbalance between proBDNF and mature BDNF or alterations in KCC2 expression or function, has been implicated in various neurological disorders and conditions. Therefore, understanding the intricate mechanisms underlying their interactions is crucial for unraveling the role of these factors in normal brain function and their implications for neurological disorders. In this study, we aimed to investigate the respective roles of both forms of BDNF in regulating chloride homeostasis and GABAergic inhibitory strength during early developmental stages. We achieved this by examining the functional expression of KCC2 in cultured rat hippocampal neurons. Additionally, we explored the behavioral consequences of proBDNF expression using electroporated rat pups.

2. Results

2.1. Functional Activity of KCC2 and BDNF Signaling Pathways in Maturing Cultured Hippocampal Neurons

The developmental excitatory–inhibitory GABA sequence mediated by the upregulation of KCC2 is closely paralleled by the downregulation of proBDNF and $p75^{NTR}$ expression in cortical regions [4,18]. Furthermore, an interplay between depolarizing GABA, the KCC2/NKCC1 imbalance, and $p75^{NTR}$ has previously been shown after neuronal injury [25]. These observations led us to explore whether the developmental expression of KCC2 was present in our primary cultured hippocampal neurons. We initially investigated the developmental expression of KCC2 by measuring KCC2 fluorescence intensity from 1 day in vitro (DIV) to 9 DIV. Our findings revealed a very low expression of KCC2 from 1 DIV to 3 DIV, succeeded by a swift and sustained escalation in KCC2 fluorescence intensity from 5 DIV to 9 DIV (Figure 1A,B). Next, our objective was to confirm the functionality of the TrkB signaling pathway at 6–7 DIV, induced by the exogenous application of mature BDNF (mBDNF), in comparison to cleavage-resistant proBDNF (CR-proBDNF) application and control conditions. We quantified the ratio of the phosphorylated form of the TrkB receptor (pTrkB) to MAP2 intensity in primary cultured hippocampal neurons (Figure 1C,D). This was achieved by utilizing primary antibodies targeting the phosphorylated form relative to the total expression of MAP2 fluorescence intensity in cultures treated with CR-proBDNF (25 ng/mL, 2 h), mBDNF (25 nM, 2 h), TAT-Pep5 (2 μM, 2 h), or TrkB-IgG (1 μg/mL, 2 h). We observed a significant increase in the pTrkB/MAP2 ratio intensity in neurons treated with mBDNF compared to other conditions. (Mean values were 1.87 ± 0.99 a.u. for mBDNF vs. 1.01 ± 0.66 a.u. for CR-proBDNF, 0.82 ± 0.63 a.u. for TAT-Pep5, and 0.81 ± 0.44 a.u. for TrkB-IgG when compared to the control condition; see Figure 1C,D. Here, and throughout the results, all numbers indicate mean ± S.D. values.) These results further suggest the specificity of the signaling pathway activated by mBDNF compared to CR-proBDNF, as proBDNF failed to activate the TrkB pathway.

To assess the developmental activity of KCC2 in cultured hippocampal neurons, we performed the $NH4^+$ flux assay. This assay offers an estimation of the ion transport efficacy of KCC2, providing valuable insights into its functionality across different developmental stages [26,27]. Changes in $NH4^+$-dependent intracellular pH (pH_i) were monitored using the ratiometric fluorescent probe [28] composed of pH-sensitive pHluorine [29] and pH-insensitive mCherry (Figure 2A,B). The advantage of this genetically encoded pH-sensitive probe (pH-sensor) is that it allows for visualization exclusively in transfected cells. When the $NH4^+$-containing media were applied to hippocampal cultured neurons, there was a progressive acidification of the pH_i following a slight and brief (10–30 s) alkalinization event (Figure 2C,D). We observed higher rates of $NH4^+$-dependent acidification in older (more mature) neurons (12–13 DIV) compared to their younger counterparts (6–7 DIV), indicating heightened KCC2 transport activity operating in reverse mode, facilitating the influx of $NH4^+$ (mean values were -173 ± 107 a.u. at 6–7 DIV vs. -649 ± 207 a.u. at 12–13 DIV for the control condition; see Figure 2E). This effect was diminished in the presence of 70 μM of bumetanide, a diuretic that inhibits KCC2 at this concentration (mean values were -176 ± 114 a.u. at 6–7 DIV vs. -328 ± 164 a.u. at 12–13 DIV for the bumetanide condition; see Figure 2E). Overall, these results align with the progressive increase in KCC2 intensity and confirm the heightened activity of KCC2 during the early development of neuronal cell cultures.

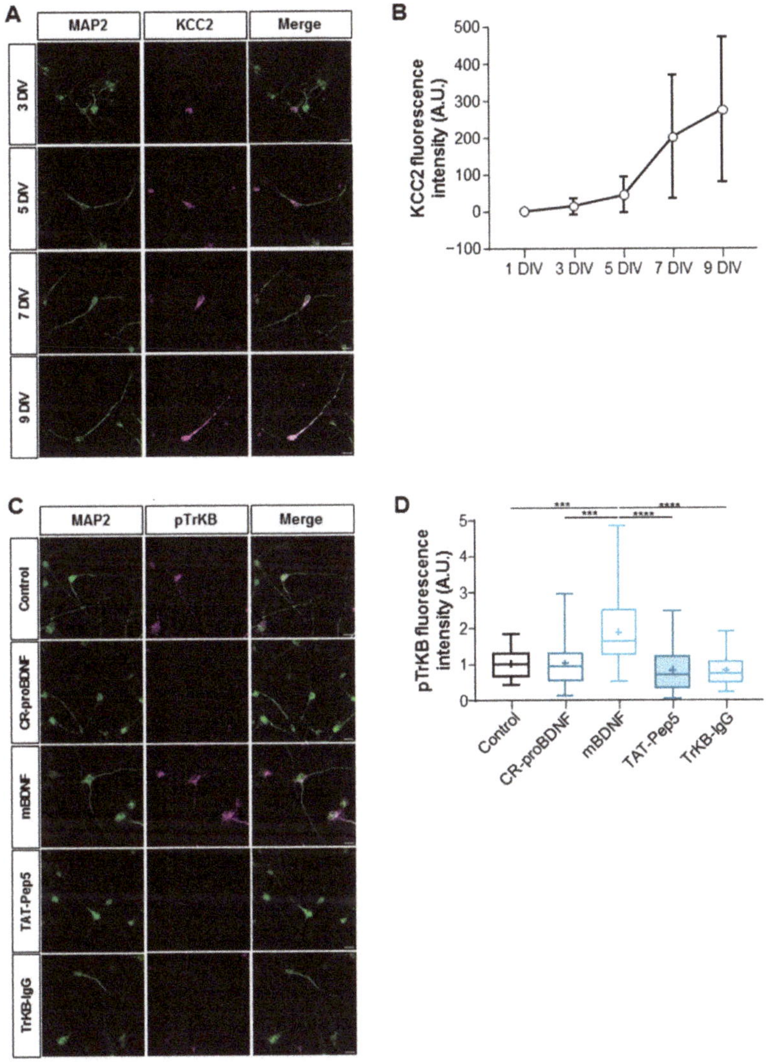

Figure 1. Developmental profile of KCC2 expression and BDNF signaling pathway activation in maturing cultured hippocampal neurons. (**A**) Representative images of immunofluorescence showing MAP2 (green) and KCC2 (purple) expression in hippocampal neurons from 3 DIV to 9 DIV. Scale bar: 10 µm. (**B**) The graph shows that the fluorescence intensity of KCC2 expression increased over time from 1 to 9 DIV. N represents the number of cultures = 3; n represents the number of cells = 35 for 1 DIV; n = 45 for 3 DIV; n = 32 for 5 DIV; n = 45 for 7 DIV; n = 35 for 9 DIV. For each condition, the mean ± SD was represented. (**C**) Representative images of immunofluorescence showing MAP2 (green) and pTrkB (purple) expression in neurons from 6–7 DIV in control and Cr-proBDNF-, mBDNF-, TAT-Pep5-, and TrKB-IgG-treated neurons. Scale bar: 10 µm. (**D**) Box plot showing the quantification of pTrKB fluorescence intensity in the indicated conditions. N = 3; n = 36 for control; n = 38 for CR-proBDNF; n = 41 for mBDNF; n = 37 for TAT-Pep5; n = 45 for TrKB-IgG. Box plots show 25th, 50th, and 75th percentiles, minimum and maximum as whiskers, with the mean indicated as "+". Compared using the Kruskal–Wallis test followed by Dunn's post hoc test, *** $p = 0.0008$ for mBDNF vs. control, *** $p = 0.0001$ for mBDNF vs. CR-proBDNF, **** $p < 0.0001$ for mBDNF vs. TrKB-IgG, **** $p < 0.0001$ for mBDNF vs. TAT-Pep5.

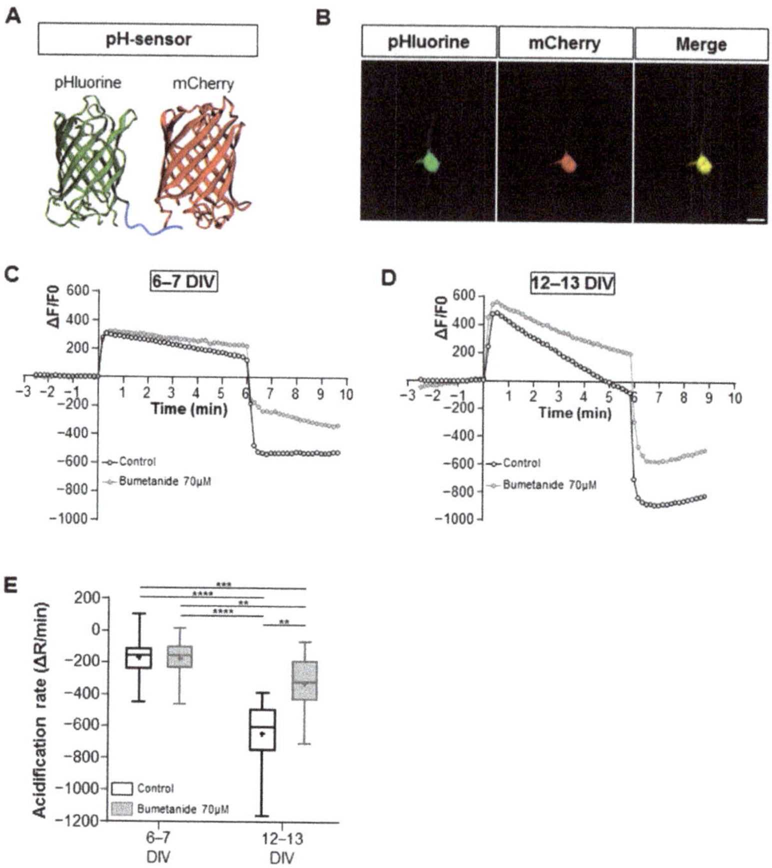

Figure 2. Functional activity of KCC2 in maturing cultured hippocampal neurons. (**A**) A 3D structure of the pH sensor, composed of pH-sensitive fluorine in green and pH-insensitive mCherry in red. (**B**) Representative images of immunofluorescence illustrating the pH sensor (green) in a transfected neuron (red). Scale bar: 150 μm. (**C**,**D**) Graph showing the ΔF/F of recorded neurons in control and bumetanide 70 μM conditions at 6–7 DIV (**C**) and 12–13 DIV (**D**). Neurons were perfused with a 10 mM NH_4Cl-containing solution at t = 0. Swiftly after NH_4Cl application, the intracellular space alkalized as NH3 diffused through the plasma membrane. KCC2 began to transport $NH4^+$ into the cell, acidifying the intracellular space. The rate of pHi changes during $NH4^+$ transportation, visualized by the fluorescence intensity of pHluorin, was dependent on the ion transportation activity of KCC2. (**E**) The acidification rate of neurons in the indicated conditions is visualized as box plots; N = 9 and n = 63 for control 6–7 DIV; N = 5 and n = 31 for bumetanide 70 μM 6–7 DIV; N = 5 and n = 26 for control 12–13 DIV; N = 5 and n = 30 for bumetanide 70 μM 12–13 DIV. Box plots show 25th, 50th, and 75th percentiles, minimum and maximum as whiskers, with the mean indicated as "+". Compared using Kruskal–Wallis test followed by Dunn's post hoc test, **** $p < 0.0001$ for control 6–7 DIV vs. control 12–13 DIV, *** $p = 0.001$ for control 6–7 DIV vs. bumetanide 70 μM 12–13 DIV, **** $p < 0.0001$ bumetanide 70 μM 6–7 DIV vs. control 12–13 DIV, ** $p = 0.0061$ bumetanide 70 μM 6–7 DIV vs. bumetanide 70 μM 12–13 DIV, ** $p = 0.0012$ for control 12–13 DIV vs. bumetanide 70 μM 12–13 DIV.

2.2. proBDNF Maintains a Depolarized GABA Response in Hippocampal Neurons

To investigate the effects of BDNF on GABA polarity shift and neuronal chloride homeostasis, we performed the gramicidin-perforated patch-clamp technique to measure the

$GABA_A$ reversal potential (E_{GABA}) and calculated the intracellular chloride concentration ($[Cl^-]_i$) in dissociated hippocampal cultures in immature neurons at 6–7 DIV and more mature neurons at 12–13 DIV. We measured E_{GABA} in neurons transfected with GFP (control), GFP-BDNF-Cherry (BDNF), or GFP-CR-proBDNF-Cherry (CR-proBDNF). Because BDNF has been shown to influence the activity of NKCC1 [30], expressed in neurons and astrocytes [31], all measurements of E_{GABA} were performed in the presence of bumetanide, an inhibitor of NKCC1 at low concentrations (10 μM). As previously described, in our preparations of cultured hippocampal neurons, 10 μM bumetanide produced a 5 mV negative shift of E_{GABA} in immature neurons (6–7 DIV) and an 8 mV negative shift in more mature cells (13–15 DIV) [32]. Our results revealed significant differences in E_{GABA} and intracellular chloride concentration ($[Cl^-]_i$) in both immature and mature neurons between the control condition and CR-proBDNF-transfected cells (mean values were -80.3 ± 8.54 mV and 6.9 ± 2.11 mM for the control vs. -72.1 ± 6.43 mV and 9.3 ± 2.4 mM for CR-proBDNF at 6–7 DIV and -85.3 ± 8.64 mV and 5.8 ± 1.85 mM for the control vs. -79.1 ± 3.5 mV and 6.9 ± 0.95 mM for CR-proBDNF at 12–13 DIV; see Figure 3C–F) and between CR-proBDNF- and BDNF-transfected cells (-81.3 ± 5.72 mV and 6.5 ± 1.42 mM at 6–7 DIV vs. -86.3 ± 7.2 mV and 5.4 ± 1.4 mM at 12–13 DIV for the BDNF condition; see Figure 3C–F). Conversely, E_{GABA} and $[Cl^-]_i$ in BDNF-transfected neurons showed no significant differences in both immature and mature neurons when compared to control conditions. The absence of significant differences between the control and BDNF conditions could be attributed to various factors. Firstly, BDNF-transfected neurons are likely to express both forms of BDNF (pro and mature), potentially obscuring the specific effects of mature BDNF alone. Additionally, it is important to consider the presence of trophic factors in the culture media, which could also influence the observed outcomes. Furthermore, these findings suggest that pro-BDNF may play a role in maintaining E_{GABA} and intracellular chloride concentration in an immature state.

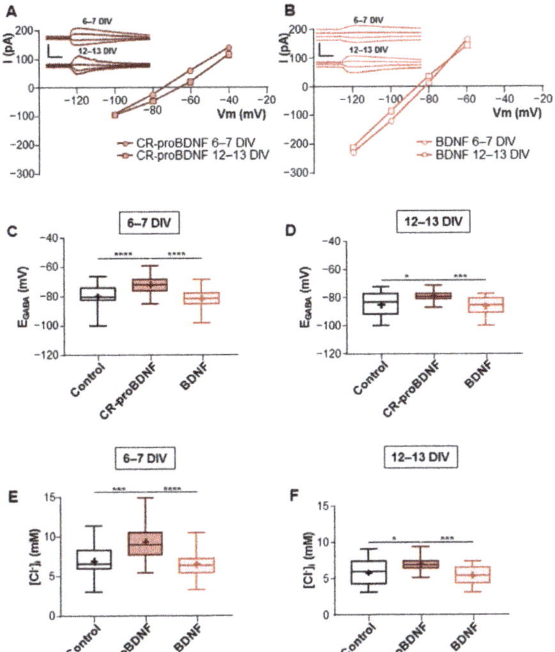

Figure 3. CR-proBDNF promotes GABA depolarization polarity in hippocampal neurons. **(A)** Gramicidin-perforated patch-clamp recording current–voltage (I–V) relationships for isoguvacine-

response currents in hippocampal primary cultures at 6–7 and 12–13 DIV transfected with CR-proBDNF-mCherry. Inserts depict the isoguvacine currents at both ages. Scale bars: 200 ms, 200 pA. (**B**) I-V and inserts depict the BDNF-mCherry transfected neurons. Scale bars: 200 ms, 200 pA. (**C**,**D**) Box plots of E_{GABA} for the indicated conditions at 6–7 and 12–13 DIV. (**E**,**F**) Box plots of $[Cl^-]_i$ for the indicated conditions at 6–7 and 12–13 DIV. For 6–7 DIV: N = 10 and n = 29 for control; N = 8 and n = 40 for CR-proBDNF; N = 10 and n = 34 for BDNF. For 12–13 DIV: N = 7 and n = 22 for control; N = 5 and n = 26 for CR-proBDNF; N = 8 and n = 25 for BDNF. Box plots show 25th, 50th, and 75th percentiles, minimum and maximum as whiskers, with the mean indicated as "+". Compared using the Kruskal–Wallis test followed by Dunn's post hoc test. At 6–7 DIV: *** $p = 0.0002$ for E_{GABA} and $[Cl^-]_i$ of CR-proBDNF vs. control, **** $p < 0.0001$ for E_{GABA} and $[Cl^-]_i$ of CR-proBDNF vs. BDNF. At 12–13 DIV: * $p = 0.0295$ for E_{GABA} of CR-proBDNF vs. control, *** $p = 0.0008$ for E_{GABA} of CR-proBDNF vs. BDNF. Compared using the one-way ANOVA test followed by Holm–Sidak's post hoc test at 12–13 DIV for $[Cl^-]_i$, * $p = 0.0104$ for CR-proBDNF vs. control, *** $p = 0.0007$ for CR-proBDNF vs. BDNF.

2.3. Effects of Pro- and Mature-BDNF on KCC2 Cell Trafficking

Given that low $[Cl^-]_i$ is KCC2-dependent, we conducted a live-staining analysis to assess the surface expression and internalization of KCC2-pH$_{ext}$. The stability of KCC2 at the cell surface is dependent on the process of phosphorylation or dephosphorylation of amino acid residues in the C-terminal part of the protein. For instance, the activation of Serine^{940} (Ser^{940}) residue increases the stability of KCC2 in the plasma membrane, whereas the dephosphorylation of Ser^{940} or the phosphorylation of Threonine 906/1007 (Thr$^{906/1007}$) residues alter surface expression abilities and promote KCC2 endocytosis [33,34]. We, therefore, assessed whether CR-proBDNF and BDNF could regulate KCC2 stability at the cell surface of hippocampal neurons by measuring KCC2 expression in different cell compartments at 9 DIV. As a tool, we used a KCC2 construct tagged in an external loop with a fluorescent protein pHluorine (KCC2-pH$_{ext}$) [34]. As detailed in the Methods section, this construct allowed for the visualization and quantification of KCC2-pH$_{ext}$ molecules that were decorated with a specific antibody on the surface of living neurons over two hours (F_{all}), the number of molecules labeled and internalized during this period (F_i), the amount of surface-expressed molecules at a given instant (F_m), and the total amount of KCC2-pH$_{ext}$ overexpressed by a neuron (F_t) (Figure 4A–D). Before proceeding with the live staining of KCC2-pH$_{ext}$, we co-transfected neuronal cultures at 7 DIV with KCC2-pH$_{ext}$ and BDNF isoform constructs (CR-proBDNF and BDNF). As positive controls, we employed KCC2 mutant constructs to either retain KCC2 at the neuronal membrane (A/A KCC2 mutant) or prevent its targeting to the neuronal membrane (ΔNTD-KCC2 mutant) [35]. Our findings indicated a significant increase in all the clusters of KCC2 expression (F_{all}) in neurons transfected with A/A or BDNF isoform constructs compared to the control group (mean values were 1 ± 0.5 a.u. for the control vs. 2.01 ± 1.24 a.u. for A/A, 2.14 ± 1.11 a.u. for CR-proBDNF, and 1.32 ± 1.04 a.u. for BDNF; see Figure 4A,B). As expected, the A/A KCC2 mutant condition led to an increase in the KCC2 membrane pool (mean values were 2.07 ± 1.61 a.u. for A/A vs. 1 ± 0.42 a.u. for the control; see Figure 4A,C). Regarding the ΔNTD-KCC2 mutant, the multi-step immunolabeling protocol did not detect any membrane expression (0.013 ± 0.014 a.u. for ΔNTD; Figure 4A,C) or internalization (0.01 ± 0.58 a.u. for ΔNTD vs. 0.99 ± 0.7 a.u. for the control; Figure 4A,D), which confirmed the integrity of the cell membrane during the experiment. In comparison to the control group, neurons transfected with the CR-proBDNF construct exhibited a significant increase in the amount of endocytosed KCC2-pH$_{ext}$ (F_i) (0.99 ± 0.7 a.u. for control vs. 2.7 ± 1.93 a.u. for CR-proBDNF; Figure 4A,D), while the amount of KCC2 internalization (F_i) remained similar between the BDNF and control groups (1.73 ± 0.93 a.u. for BDNF; Figure 4A,D). In summary, our results revealed an elevated rate of KCC2 endocytosis in the presence of

CR-proBDNF, underscoring the potential of this immature form of BDNF to trigger the increased internalization of KCC2.

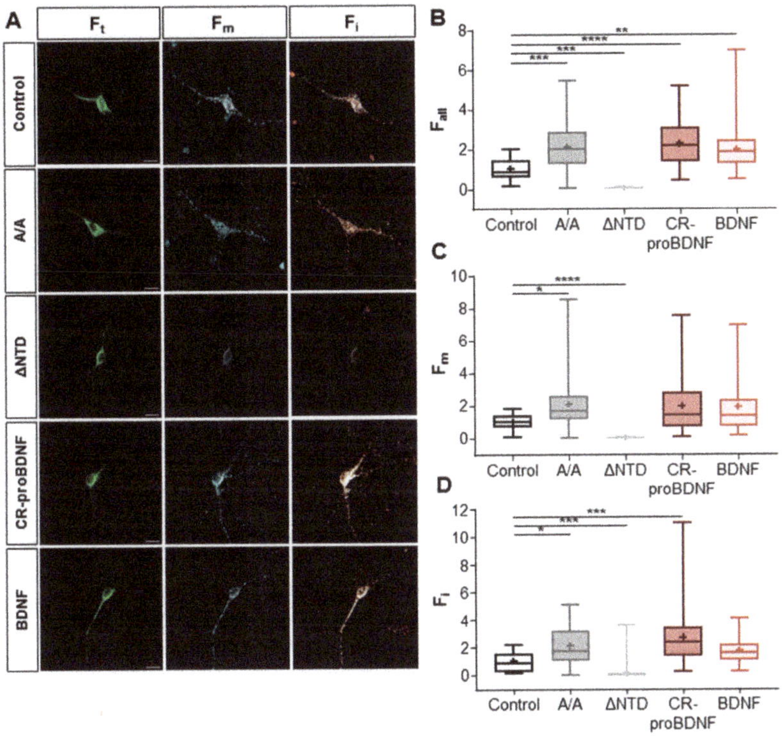

Figure 4. CR-proBDNF increases the internalization of KCC2. (**A**) Representative images of immunofluorescence illustrating total (F_t; green) membrane (F_m; blue) and internalized (F_i; red) pools of KCC2 with an external tag (KCC2-pH_{ext}) for the control, CR-ProBDNF-mCherry, BDNF-mCherry, A/A-KCC2-pH_{ext}, and ΔNTD-KCC2-pH_{ext} conditions in hippocampal primary culture neurons. Scale bar: 10 μm. Box plots of (**B**) all clusters (F_{all}) and (**C**) membranes (F_m) and (**D**) internalized pool (F_i) fluorescence normalized per cell in cultured neurons of the indicated conditions. N = 4; n = 27 for control; n = 56 for A/A; n = 39 for ΔNTD; n = 57 for CR-proBDNF; n = 55 for BDNF. Box plots show 25th, 50th, and 75th percentiles, minimum and maximum as whiskers, with the mean indicated as "+". Compared using the Kruskal–Wallis test followed by Dunn's post hoc test. F_{all}: *** $p = 0.0005$ for control vs. A/A, *** $p = 0.0009$ for control vs. ΔNTD, **** $p < 0.0001$ for CR-proBDNF vs. control, ** $p = 0.0026$ for BDNF vs. control; F_m: * $p = 0.019$ for control vs. A/A, **** $p < 0.0001$ for control vs. ΔNTD; F_i: * $p = 0.015$ for control vs. A/A, *** $p = 0.0003$ for control vs. ΔNTD, *** $p = 0.0001$ for CR-proBDNF vs. control, $p = 0.074$ for BDNF vs. control.

2.4. KCC2-Dependent Ammonium Transport in BDNF-Treated Hippocampal Neurons

To further investigate whether treatment with both forms of BDNF affected the Cl^- extrusion capability of the KCC2 transporter, we conducted the $NH4^+$ flux assay. As described before, this assay allows us to assess changes in ion transport efficacy, providing insights into the modulation of KCC2 function by BDNF. Given the potential masking effects of trophic factors present in the culture media or endogenously secreted BDNF by the cell cultures, in this set of experiments, we chose to block the $p75^{NTR}$ and TrkB receptor signaling pathways to uncover the specific action of pro- and mBDNF. Hippocampal neuronal cultures were treated at 6–7 DIV with either the $p75^{NTR}$ signaling inhibitor TAT-Pep5 or the mBDNF scavenger TrkB-IgG. These treatments were administered in the presence or absence of 70 μM of bumetanide to block the co-transporters NKCC1 and KCC2. Neurons treated with TAT-Pep5 showed a significantly faster $NH4^+$-induced acidification rate compared to the control group (−484 ± 216.2 a.u. for TAT-Pep5 vs. −173 ± 107 a.u. for the control condition; Figure 5A,C). Notably, the presence of 70 μM of bumetanide effectively abolished this effect, underscoring the specificity of the $p75^{NTR}$ signaling pathway's impact on KCC2 (−201.5 ± 161.5 a.u. for + bumetanide 70 μM vs. −484 ± 216.2 a.u. for the TAT-Pep5 condition; Figure 5C). Interestingly, neurons treated with TrKB-IgG displayed a similar significant increase in acidification rate compared to the control (−291 ± 117 a.u. for TrkB-IgG vs. −173 ± 107 for the control; Figure 5B,D). This effect was blocked in the presence of 70 μM of bumetanide (−191 ± 77 a.u. for TrkB-IgG + bumetanide 70 μM vs. −291 ± 117 a.u. for TrkB-IgG; Figure 5D). Taken together, these findings suggest that during this specific developmental stage (6–7 DIV), pro- and mBDNF serve a comparable function through their respective receptors, $p75^{NTR}$ and TrkB, in diminishing the neuronal cell's capacity to extrude chloride, thereby resulting in the maintenance of a depolarizing E_{GABA}.

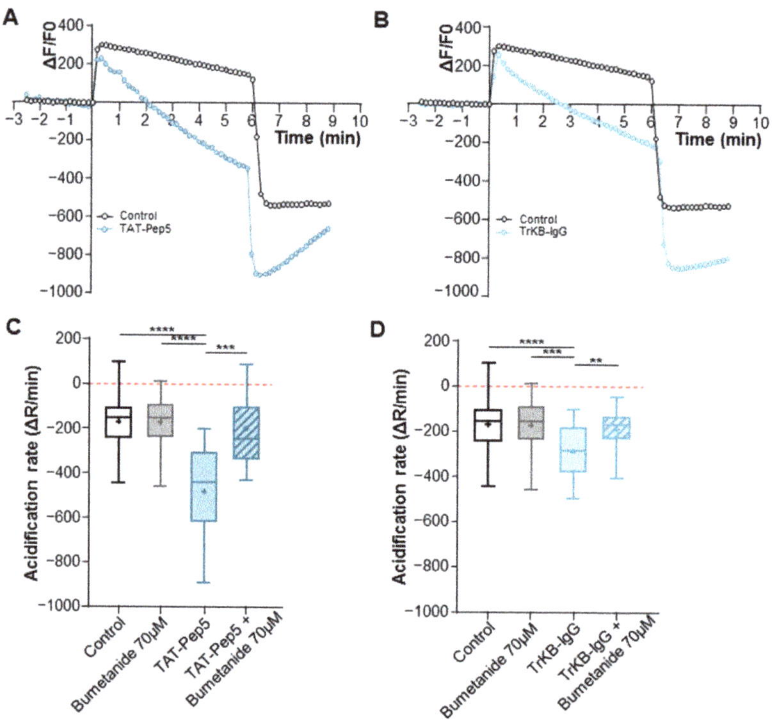

Figure 5. The activity of KCC2 is increased with the inhibitors of both forms of BDNF. (**A**) Graph showing the ΔF/F of recorded neurons at 6–7 DIV for the control and TAT-Pep5 conditions. (**B**) Graph showing the ΔF/F of recorded neurons at 6–7 DIV for the control and TrKB-IgG conditions. (**C**,**D**) The acidification rate of neurons in the indicated conditions at 6–7 DIV is visualized as box plots. N = 9 and n = 63 for control; N = 5 and n = 31 for bumetanide 70 μM; N = 6 and n = 20 for TAT-Pep5; N = 4 and n = 26 for TAT-Pep5 + bumetanide 70 μM; N = 6 and n = 24 for TrKB-IgG; N = 4 and n = 26 for TrKB-IgG + bumetanide 70 μM. Box plots show 25th, 50th, and 75th percentiles, minimum and maximum as whiskers, with the mean indicated as "+". Compared using the Kruskal–Wallis test followed by Dunn's post hoc test, **** $p < 0.0001$ for TAT-Pep5 vs. control, **** $p < 0.0001$ for TAT-Pep5 vs. bumetanide 70 μM, *** $p = 0.0008$ for TAT-Pep5 vs. TAT-Pep5 + bumetanide 70 μM. Compared using the one-way ANOVA test followed by Holm–Sidak's post hoc test, **** $p < 0.0001$ for TrKB-IgG

vs. control, *** $p = 0.0006$ for TrKB-IgG vs. bumetanide 70 μM, ** $p = 0.0065$ for TrKB-IgG vs. TrKB-IgG + bumetanide 70 μM.

2.5. CR-proBDNF-Electroporated Rat Pups Exhibited Behavioral Deficits

Building upon the data obtained from our in vitro experiments, our investigation was expanded to in vivo analysis, specifically examining the developmental milestones of in utero electroporated (IUE) rats with either GFP or CR-proBDNF or BDNF constructs. Given the known role of BDNF in regulating energy balance and promoting anorectic signals [36], our initial focus was to analyze the weight of electroporated rats during the first two postnatal weeks. No weight loss was observed in rats electroporated with either CR-proBDNF or BDNF (Figure 6A). Furthermore, IUE with BDNF constructs did not show a significant impact on other physical landmarks, such as eye opening, incisor eruption, and fur development (Figure 6B). In addition, we aimed to explore meaningful behaviors indicative of cortical circuit maturation. We observed huddling behavior (as depicted in Figure 6C,D), a valuable model for studying early social interaction with peers [37,38]. We fitted our results in a non-linear regression fit model ($R^2 = 0.9184$ for GFP, $R^2 = 0.9047$ for CR-proBDNF, $R^2 = 0.9066$ for BDNF; Figure 6D). Despite no observed difference between the control group (GFP) and BDNF-electroporated rodents, we noted a significant decrease in huddling behavior in rats electroporated with CR-proBDNF when compared to both the control group and the BDNF groups (Figure 6C,D). In summary, these findings demonstrate that the overexpression of CR-proBDNF in the somatosensory cortex leads to sensory deficits, as evidenced by the results of the huddling test.

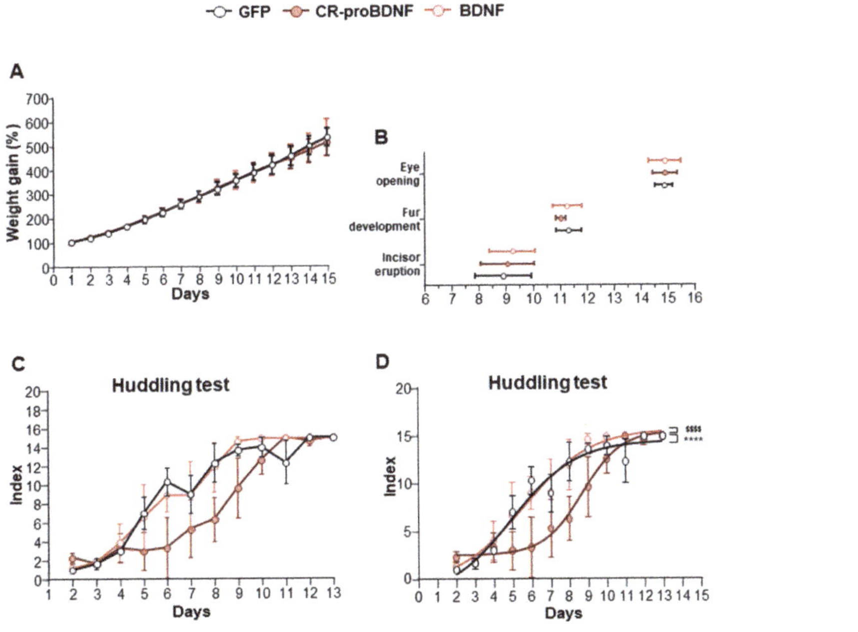

Figure 6. CR-proBDNF-electroporated rats exhibited normal physical landmarks and behavioral deficits. (**A**) Percentage of weight gain from PN1 to PN15 in GFP and CR-proBDNF- and BDNF-electroporated rats. (**B**) Physical landmarks: eye pening, fur development, and incisor eruption in the indicated conditions. n represents the number of pups for each condition = 32 for GFP; n = 30 for CR-proBDNF; n = 28 for BDNF. (**C**) Index of aggregon patterns from PN2 to PN13 in the indicated conditions. (**D**) Nonlinear regression fit model of data from (**C**). N = 3 dams of rats for each condition. For each group, the mean ± SD was represented. **** $p < 0.0001$ and F = 16.6 for CR-proBDNF vs. GFP; $$$$ $p < 0.0001$ and F = 14.7 for CR-proBDNF vs. BDNF.

2.6. CR-proBDNF-Electroporated Rats Exhibited Deficits in Sensory–Motor Maturation

To evaluate the influence of CR-proBDNF or BDNF overexpression in the somatosensory cortices of IUE rats on the emergence of developmental reflexes, we conducted the righting reflex test. Overall, the results revealed that rat pups electroporated with BDNF plasmid displayed a reduced time to correct positioning compared to the control group between postnatal ages 8 (PN 8) and 10 (PN 10) (mean values 1.99 ± 0.55, 1.97 ± 0.76, and 1.56 ± 0.41 s for the control vs. 1.57 ± 0.6, 1.37 ± 0.48, and 1.34 ± 0.53 s for BDNF between PN 8–10; Figure 7A) and to CR-proBDNF rodents at PN 9 (mean values 1.37 ± 0.48 s for BDNF vs. 1.73 ± 0.57 s for CR-proBDNF; Figure 7A). Additionally, we found that CR-proBDNF exhibited a reduced time at PN 13 when compared to the control group (mean values 1.02 ± 0.076 s for CR-proBDNF vs. 1.14 ± 0.21 s for the control; Figure 7A). To further investigate potential sensory deficits in electroporated rats, we employed the cliff avoidance test. The results revealed that rat pups electroporated with the CR-proBDNF construct exhibited lower success rates in completing the test and spent more time turning away or retracting from the edge compared to both the GFP control group and rodents electroporated with BDNF (refer to Tables 1 and 2 for cliff avoidance time and score; Figure 7B,C). These findings suggest an alteration in sensory–motor maturation in the CR-proBDNF condition.

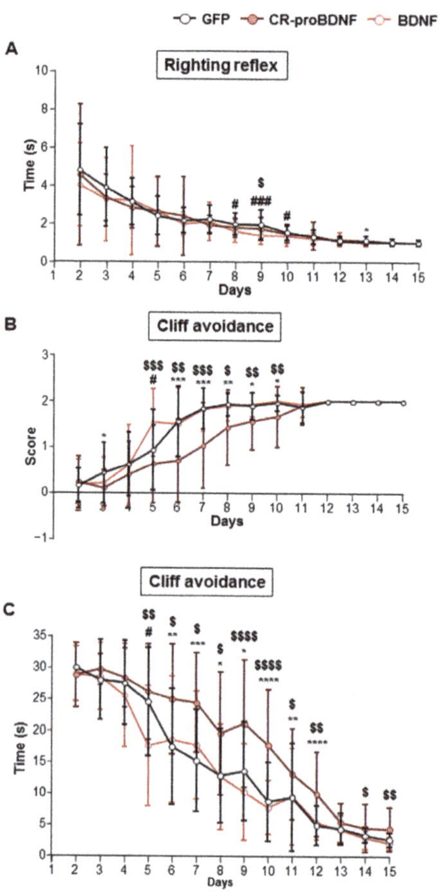

Figure 7. CR-proBDNF-electroporated rats exhibited deficits in sensory maturation. (**A**) Time spent to complete the righting reflex from PN2 to PN15 for GFP and CR-proBDNF- and BDNF-electroporated

rats. Compared using the Kruskal–Wallis test followed by Dunn's post hoc test. PN8: # $p = 0.02$ for BDNF vs. GFP; PN9: ### $p = 0.0006$ for BDNF vs. GFP, \$ $p = 0.03$ for CR-proBDNF vs. BDNF; PN10: # $p = 0.017$ for BDNF vs. GFP; PN13: * $p = 0.016$ for CR-proBDNF vs. GFP. (**B**) Score of cliff avoidance success from PN2 to PN15 in the indicated conditions. (**C**) Time spent to avoid the cliff from PN2 to PN15 in the indicated conditions. n = 32 for GFP; n = 30 for CR-proBDNF; n = 28 for BDNF. For each group, the mean ± SD is represented. Compared using the Kruskal–Wallis test followed by Dunn's post hoc test. * $p < 0.05$, ** $p < 0.01$, *** $p < 0.001$, and **** $p < 0.0001$ for CR-proBDNF vs. GFP. # $p < 0.05$ for BDNF vs. GFP. \$ $p < 0.05$, \$\$ $p < 0.01$, \$\$\$ $p < 0.001$, and \$\$\$\$ $p < 0.0001$ for CR-proBDNF vs. BDNF.

Table 1. Score of cliff avoidance test.

Days/Condition	GFP	CR-proBDNF	BDNF
2	0.16 ± 0.37	0.23 ± 0.57	0.18 ± 0.55
3	0.44 ± 0.67	0.1 ± 0.4 *	0.21 ± 0.57
4	0.63 ± 0.7	0.4 ± 0.72	0.61 ± 0.88
5	0.94 ± 0.88	0.63 ± 0.85	1.54 ± 0.74 #
6	1.56 ± 0.76	0.7 ± 0.92 ***	1.5 ± 0.79 $$
7	1.84 ± 0.45	1.03 ± 0.93 ***	1.82 ± 0.48 $$$
8	1.94 ± 0.25	1.43 ± 0.82 **	1.9 ± 0.31 $
9	1.91 ± 0.3	1.6 ± 0.62 *	1.93 ± 0.26 $$
10	1.97 ± 0.18	1.7 ± 0.66 *	2 ± 0 $$
11	1.88 ± 0.34	1.9 ± 0.4	1.93 ± 0.26
12	2 ± 0	2 ± 0	2 ± 0
13	2 ± 0	2 ± 0	2 ± 0
14	2 ± 0	2 ± 0	2 ± 0
15	2 ± 0	2 ± 0	2 ± 0

Data are presented as mean ± SD, compared using the Kruskal–Wallis test followed by Dunn's post hoc test. * $p < 0.05$, ** $p < 0.01$, and *** $p < 0.001$ for CR-proBDNF vs. GFP. # $p < 0.05$ for BDNF vs. GFP. \$ $p < 0.05$, \$\$ $p < 0.01$, and \$\$\$ $p < 0.001$ for CR-proBDNF vs. BDNF.

Table 2. Time spent to accomplish the cliff avoidance test.

Days/Condition	GFP	CR-proBDNF	BDNF
2	30 ± 0	28.7 ± 5.08	28.96 ± 4.31
3	28 ± 6.37	29.6 ± 2.37	28.54 ± 5.39
4	27.53 ± 6.7	28.23 ± 4.75	25.54 ± 8.24
5	24.5 ± 8.67	26.03 ± 7.62	17.5 ± 9.68 $$ #
6	17.4 ± 9.23	24.9 ± 8.79 **	18.54 ± 10.11
7	15.16 ± 8.15	24.3 ± 7.95 ***	17.64 ± 8.63 $
8	12.78 ± 7.51	19.4 ± 9.77 *	12.61 ± 8.45 $
9	13.5 ± 7.93	21.03 ± 10.16 *	10.18 ± 7.65 $$$$
10	8.7 ± 6.3	17.5 ± 8.95 ****	7.71 ± 4.31 $$$$
11	9.31 ± 8.59	13 ± 7.27 **	9.5 ± 7.71 $
12	4.9 ± 3.07	9.83 ± 6.74 ****	5.32 ± 2.68 $$
13	4.4 ± 2.47	5.3 ± 3.01	4.14 ± 2.37
14	3.31 ± 1.18	4.4 ± 3.78	2.9 ± 1.75 $
15	2.63 ± 1.16	4.23 ± 3.45	2.11 ± 0.95 $$

Data are presented as mean ± SD, compared using the Kruskal–Wallis test followed by Dunn's post hoc test. * $p < 0.05$, ** $p < 0.01$, *** $p < 0.001$, and **** $p < 0.0001$ for CR-proBDNF vs. GFP. # $p < 0.05$ for BDNF vs. GFP. \$ $p < 0.05$, \$\$ $p < 0.01$, and \$\$\$\$ $p < 0.0001$ for CR-proBDNF vs. BDNF.

3. Discussion

In altricial rodents, neonates' brains display a depolarizing GABA activity that is particularly evident during the initial postnatal days in vivo or the first two postnatal weeks in vitro. This efflux of chloride (Cl^-) ions is crucial for spontaneous network activity, such as giant depolarizing potentials, which play a role in the maturation of functional networks [4]. Consequently, during this early developmental period, chloride homeostasis shifts from an initially elevated intracellular chloride concentration ($[Cl^-]_i$) to lower $[Cl^-]_i$.

This shift subsequently changes GABA polarity from depolarizing to inhibitory. Chloride homeostasis is primarily regulated by electroneutral secondary active co-transporters, NKCC1 and KCC2, functioning in opposing directions. Additionally, it is influenced by both sodium (Na^+)-dependent and Na^+-independent exchangers [39]. Thus, in healthy mature neurons, KCC2 activity dynamically regulates the constant cytoplasmic low levels of chloride ions. In this study, our primary objective was to elucidate how the principal brain neurotrophic factor (BDNF) regulates the expression and function of KCC2 during the early developmental period. Our primary findings, as observed in hippocampal neuronal cultures, demonstrate that proBDNF maintains a positive E_{GABA} in both mature and immature neurons. Additionally, the analysis of KCC2 trafficking correlated with electrophysiological data, revealing an increased endocytosis rate in CR-proBDNF-expressing neurons. Interestingly, the developmental period of GABAergic maturation during the first few postnatal weeks, transitioning from depolarizing to inhibitory responses, coincides with a surge in proBDNF [17,18] and low KCC2 expression [20]. In contrast, BDNF, acting through the TrkB signaling pathway, promotes the functional maturation of KCC2 [14,22]. Consequently, it is conceivable that both forms of BDNF play opposing roles in KCC2 expression and function in maturing neurons. The function of KCC2 is primarily dependent on the trafficking and targeting processes, which are intricately governed by post-translational regulations involving the (de)phosphorylation of specific amino acid residues. As mentioned earlier, Ser^{940} [33] and Thr^{1007} [32,40] have undergone extensive study, emphasizing their crucial roles in the regulation of surface stability and internalization, respectively. Our in vitro results did not demonstrate the capacity of both forms of BDNF to increase the cell surface expression of KCC2. However, proBDNF enhanced KCC2 turnover, implying a mechanism that could potentially constrain the functional efficacy of KCC2 while favoring NKCC1 activity. This finding highlights a regulatory dynamic wherein proBDNF may exert a modulatory influence on chloride homeostasis, potentially impacting neuronal excitability and synaptic transmission. Further investigations into the specific molecular mechanisms underlying this effect will be crucial for a comprehensive understanding of how proBDNF contributes to the intricate balance of chloride ion regulation in the brain. Moreover, a recent study revealed distinct membrane dynamics for NKCC1 compared to KCC2. NKCC1 tends to be more localized in endocytic zones rather than the membrane and exhibits a rapid transition from endocytic zones to the membrane [41]. The interplay of these various effects may contribute to a reversal of chloride flux in pathological conditions. Both forms of BDNF activate distinct signaling pathways. ProBDNF activates the RhoA-Rock pathway through $p75^{NTR}$ [5,20,42], while mature BDNF activates the ERK1/2 or PLC signaling cascade through TrkB [22,43]. To gain a comprehensive understanding of the roles played by pro- and mBDNF in regulating downstream targets affecting the cell surface expression or stability of KCC2, further experimentation is warranted. It is established that the cell surface expression of KCC2 relies on the phosphorylation status of the KCC2 Thr^{1007} residue, a target of phosphorylation by WNK1. Several studies, particularly in hippocampal neuronal cultures in vitro, have demonstrated that the phosphorylation of these amino acids by the WNK1 pathway leads to the inhibition of KCC2 transport activity and increased internalization of the protein [32,44,45]. However, the precise contribution of mBDNF in facilitating the maturation of Cl^- homeostasis and GABA polarity through the potential downregulation of WNK1 remains to be elucidated. WNK1 is a pivotal player involved in KCC2 endocytosis and NKCC1 activation [46]. Complementary investigations are essential to unravel the intricate interplay between mBDNF signaling and the regulation of WNK1 activity, providing insights into the mechanisms governing chloride homeostasis and GABAergic signaling maturation. Additionally, the complexity of this regulatory mechanism is compounded by the potential involvement of proBDNF/$p75^{NTR}$ in facilitating the mBDNF/TrkB signaling pathway [47]. This also suggests a scenario where $p75^{NTR}$ and TrkB may collaboratively work to regulate the transcriptional and post-translational amplification of the mBDNF/TrkB pathway. This added layer of intricacy significantly contributes to regulating the expression and function of KCC co-transporters. At the cognitive level, both proBDNF and mBDNF are implicated in various neurological disorders and play a role in modulating learning and memory processes. For example, mBDNF has been shown to enhance synaptic transmission and promote synaptic plasticity, which is essential for optimal learning and memory [48–50]. Conversely, increased levels of proBDNF could potentially impede synaptic plasticity, thereby hindering memory consolidation and impacting cognitive functions, ultimately leading to deficits in behavioral tasks [51]. Importantly, these effects are age-dependent. The disruption of proBDNF signaling during the postnatal period has been shown to lead to disrupted spatial memory consolidation and a reduction in spine density [17]. The role of KCC2 and NKCC1 in maintaining chloride homeostasis extends beyond brain development, as dysfunctions of these co-transporters are also observed in various neurodegenerative disorders, including Alzheimer's disease (AD), amyotrophic lateral sclerosis, Parkinson's disease, and Huntington's disease [10]. Notably, in Alzheimer's disease, alterations in KCC2 activity occur during the presymptomatic phase and are linked to cognitive deficits [11]. Overall, while the specific mechanisms linking KCC2 to neurodegenerative diseases are still being elucidated, the dysregulation of KCC2-mediated chloride homeostasis appears to be a common feature in various neurodegenerative conditions. In this study, our focus was on the developmental period, and, to understand the impact of both forms of BDNF in a more physiological context, we utilized in utero electroporation (IUE) as a method to introduce either proBDNF or BDNF into the somatosensory cortex of rat pups. This approach enabled us to examine the potential behavioral effects resulting from the targeted introduction of this neurotrophic factor. During the neonatal period, typically encompassing the first few weeks after birth, the somatosensory cortex undergoes rapid and intricate development, establishing the neural foundation for sensory perception and motor control. Studies involving mice lacking BDNF have revealed significant deficiencies in coordination and balance (ataxia) and sensory deficits, underlining the critical role of BDNF in these processes [52,53]. Additionally, the expression of BDNF in the cortex has been identified as crucial for motor learning [54] and plasticity [55]. In our study, the observed results aligned with cellular-level data, clearly illustrating the contrasting effects of both forms of BDNF. Specifically, within the CR-proBDNF condition, we noted a deficit in sensory–motor responses in huddling and cliff avoidance tasks. These results strongly suggest that an imbalance between pro- and mature BDNF can potentially alter the functional maturation of cortical circuits during this critical developmental period.

4. Materials and Methods

4.1. Reagents and Treatments

Isoguvacine was purchased from Tocris Cookson (Bristol, UK). Bumetanide was purchased from Sigma (St-Louis, MO, USA). Human CR-ProBDNF and human mBDNF were purchased from alomone labs (Jerusalem, Israel). TAT-Pep5 was purchased from Millipore (Molsheim, France). TrkB-IgG was purchased from the R & D system (Minneapolis, MN, USA).

4.2. Primary Cultures of Rat Hippocampal Neurons

Neurons from 18-day-old rat embryos were dissociated with trypsin and plated at 300,000 cells/mL in a minimal essential medium (MEM) with supplements, as previously described [56]. On days 7, 10, and 13 of culture (DIV), half of the medium was replaced with MEM containing 2% B27 (Invitrogen, Carlsbad, CA, USA) and 2 mM L-Glutamine. For electrophysiology and surface labeling, neurons were plated on 12 mm coverslips in 35 mm dishes coated with polyethyleneimine (10 mg/mL).

4.3. Transfections

Neurons were transfected using 300 µL Opti-MEM, 7 µL Lipofectamine 2000, 1.5 µL CombiMag (OZ Biosciences, Marseille, France) per µg DNA, and 1.5 µg DNAs of interest, following the previously described method [35]. After a 15 min incubation at room temperature (RT), the mix was added to the culture. Dishes were then placed on a magnetic plate for 2 h at 37 °C with 5% CO_2. Transfection was ended by replacing 50% of the solution with a fresh medium. Cells were used 24–48 h post-transfection for immature neurons and 96 h post-transfection for mature neurons. Generally, one plasmid was used per well, except for co-transfection with KCC2-pH_{ext} (0.6 µg) and BDNF constructs (0.9 µg).

4.4. $NH^+{}_4$ Flux Assay

Primary hippocampal neurons were transiently transfected using Lipofectamine 2000 and CombiMag with the ratiometric fluorescent probe composed of pH-independent mCherry and pH-sensitive pHluorine (called a pH sensor) [28,29]. Coverslips were treated for 2 h with either TAT-Pep5 (2 µM) or TrKB-IgG (1 µg/mL) in the presence or absence of bumetanide 70 µM and for 1 h with bumetanide 70 µM alone. For recordings, coverslips were submerged in a recording chamber and perfused with an extracellular Hepes-buffered solution (HBS) containing (in mM) 150 NaCl, 2.5 KCl, 5 HEPES, 2.0 $CaCl_2$, and 2.0 $MgCl_2$, at pH 7.4 and supplemented with 0.45% glucose and bumetanide 10 µM. The ratiometric fluorescence of pHluorine/mCherry was measured using an epifluorescence imaging setup on an inverted Olympus IX71 microscope (Tokyo, Japan) with a FITC/CY3 Dualband ET Filterset and additional single-band filters. pH-sensitive pHluorine fluorescence (F480) was obtained with a 480/20 excitation filter and 520/40 emission filter, while pH-insensitive mCherry-KCC2 fluorescence (F577) was obtained with a 577/25 excitation filter and 645/75 emission filter. Fluorescence was sampled at 0.1 Hz using a CoolSNAPHQ CCD camera and MetaMorph software (Version 7.7.5.0, Molecular Devices, San Jose, CA, USA). Excitation lasted 100 ms for F480 and 50 ms for F577. Recordings were performed with a 40× objective (NA 0.6). Baseline fluorescence was acquired for 5 min, followed by a 6 min perfusion with 10 mM NH_4Cl solution and 2–5 min washout imaging. The ΔF/F ratio was calculated from the images, and the acidification rate (ΔR/min) was determined by the change in ΔF/F values between 0.5 and 5 min during $NH4^+$ presence. For further details, please refer to the article [28].

4.5. Gramicidin-Perforated Patch-Clamp Recordings

Patch-clamp recordings using gramicidin were conducted on primary hippocampal neurons transfected with constructs encoding GFP and BDNF-piTracer (GFP-BDNF-mCherry) or CR-proBDNF-piTracer (GFP-CR-ProBDNF-mCherry), 1 to 4 days post-transfection (6–7 or 12–13 DIV). Neurons were perfused with an external Hepes-buffered solution (HBS) containing 150 mM NaCl, 2.5 mM KCl, 5 mM HEPES, 2.0 mM $CaCl_2$, and 2.0 mM $MgCl_2$, with 0.45% glucose and osmolarity of 300 mOsm. For recording, the external HBS contained 10 µM bumetanide. Recording micropipettes (5 MΩ) were filled with a solution containing 140 mM KCl, 5 mM HEPES, and 20 µg/mL gramicidin. Isoguvacine (30 µM) was applied to recorded cells via a micropipette connected to a Picospritzer. Recordings were performed in voltage-clamp mode using an Axopatch-200A amplifier (Molecular Devices, San Jose, CA, USA) and pCLAMP acquisition software (Version 8, Molecular Devices, San Jose, CA, USA). Data were low-pass filtered at 2 kHz and acquired at 10 kHz. Isoguvacine responses were recorded at different voltages depending on neuron GABA reversal potential. Linear regression was used to calculate the voltage dependence of the isoguvacine responses, and the Nernst equation was used to calculate $[Cl^-]_i$.

4.6. Immunocytochemistry and Confocal Microscopy

Hippocampal cultures were treated as described above and fixed in 4% PFA–sucrose, followed by PBS or HBS washes and incubation in 0.3 M glycine. Blocking was performed with 1% BSA, 5% goat serum, and 0.3% Triton X-100 to reduce nonspecific binding. Next, cultures were incubated overnight at 4 °C with rabbit anti-KCC2 (1:3000, US-K0120–07, US Biological, Salem, MA, USA) or rabbit anti-phosphoTrkB (1:500, ABN1381, Millipore, Molsheim, France) along with chicken anti-MAP2 (1:500, ab5392, Abcam, Cambridge, UK). Primary antibodies were visualized with goat anti-rabbit Alexa-488 (1:500, 4412, Cell

Signaling, Danvers, MA, USA)) and donkey anti-chicken Alexa-647 (1:500, AP194SA6, Millipore, Molsheim, France). After washing, coverslips were mounted with Vectashield. Immunoreactivity was sequentially acquired using a laser-scanning confocal microscope (Olympus Fluorview-500, Tokyo, Japan) with ×63 oil-immersion objective and adjustments to avoid saturation. Images were processed with MetaMorph software (version 7.7.5.0, Molecular Devices, San Jose, CA, USA).

4.7. Surface Immunolabeling on Living Neurons and Analysis of KCC2-pH$_{ext}$ Proteins

For the immunolabeling of KCC2-pH$_{ext}$ proteins on living neurons, rabbit anti-GFP (1:250, Invitrogen, A-6455) antibodies were applied to neurons in culture media for 2 h at 37 °C, with 5% CO_2. Neurons were then rinsed three times with HBS at room temperature (RT) and labeled with goat anti-rabbit Alexa-647 conjugated antibody (1:250, Invitrogen, A-21244) for 30 min at 13 °C at the surface where KCC2-pH$_{ext}$ was located at the moment of the cooling down of the cells (F_m, membrane fluorescence pool). After fixing in 4% PFA–sucrose at 4 °C, cells were permeabilized and blocked with a mixture of 0.3 M glycine, 0.3% Triton X-100, and 5% GS for 1 h at RT. To reveal all labeled clusters (F_{all}, surface + internalized fluorescence pool), cells were incubated with goat anti-rabbit Alexa-555 conjugated antibody (1:400, Invitrogen, A-21428) for 1 h at RT. For visualization of the total pool of overexpressed KCC2-pH$_{ext}$ (F_t, total expression pool), cells were labeled with chicken anti-GFP antibody (1:500, 1020, Avès lab Davis, CA, USA) and goat anti-chicken Alexa-488 conjugated antibody (1:1000, A-11039, Invitrogen, Carlsbad, CA, USA) for 1 h each at RT. Internalized KCC2-pH$_{ext}$ (F_i, internalized fluorescence pool) was quantified post-hoc using MetaMorph software, considering only Alexa 555-positive molecules. Images were acquired with an Olympus Fluorview-500 confocal microscope (Tokyo, Japan) using a 60× oil immersion objective (NA1.4, zoom 2.5). Each Z-stack included 9 planes taken with a 0.7 µm distance between planes. Absolute fluorescence values were used for statistical analysis, normalized by their respective control-KCC2-pH$_{ext}$ mean values.

4.8. In Utero Electroporation

In utero injections and electroporations (IUE) were performed on embryos from timed pregnant rats at embryonic day 15, as previously described [20]. Briefly, pregnant rats received buprenorphine (0.03 mg/kg, Buprecare, Axience, Pantin, France) and Rimadyl (5 mg/Kg, Zoetis, Malakoff, France) followed by anesthesia with Isoflurane (4% for induction, then 2.5%) after 30 min. Embryos' lateral ventricles were injected with Fast Green (2 mg/mL; Sigma, St. Louis, MO, USA) combined with DNA constructs encoding GFP or BDNF-piTracer or CR-proBDNF-piTracer [20]. Electroporation was carried out using 40 V voltage pulses (BTX ECM 830 electroporator Harvard Apparatus, Holliston, MA, USA) delivered via tweezer-type electrodes (Nepa Gene Co., Chiba, Japan) across the uterine wall. IUE was performed at E15, an active period for the radial and tangential migration of newborn neurons in the neocortex [57]. Successfully electroporated pups were selected postnatally based on GFP reporter fluorescence. Morphological analysis of electroporated tissues was conducted using fluorescent stereomicroscopy. No alterations in cortical layer position or morphology were observed in GFP, BDNF, and CR-proBDNF conditions. However, approximately 30% of electroporations failed due to the absence of transfected cells, resulting in exclusion from the study.

4.9. Developmental Landmarks and Behavioral Tests

For all behavioral tests, animals were first acclimated to the behavioral room for 30 min. Control (GFP), CR-pro-, and BNDF rats were assessed by physical landmarks of rat development, the cliff avoidance test, the huddling behavior test, and the righting reflex test.

Physical landmarks of rat development. The developmental milestones measured in this study included weight, eye opening, incisor eruption, and fur development [58].

Righting reflex evaluation (RR). We assessed the RR by initially placing unrestrained rodents on their backs (supine position). Then, we examined whether animals restored the RR (flipped to the prone position). We measured the time it took for the animals to right themselves. This process was repeated 3 times consecutively and we recorded the average time for each attempt. We also scored the behavior: 0 if the animal stayed on its back, 1 if laid on its side, and 2 if the animal flipped onto its belly. A maximum time of 30 s was given for the animals to perform the reflex test.

Cliff avoidance test. Cliff avoidance tests were conducted by placing rat pups on the edge of a platform (30 cm × 30 cm × 30 cm), with their forepaws and chest extending over the edge. The latency of the rats to turn away or retreat from the edge was recorded. We scored the test: 0 for no movement or falling off the edge, 1 for attempts to move away from the cliff but with hanging limbs, and 2 for successful movement away from the cliff. A maximum time of 30 s was given for the animals to perform the test.

Huddling behavior test (HBT). Rat pups were carefully removed from their dam and uniformly spaced in the arena in a radial position at room temperature. They were free to move and huddle for 10 min. At the end, we hand-scored the aggregon pattern. Each possible combination of numbers of bodies in contact was an aggregon. We had 15 aggregons that could be formed by seven individuals, and each aggregon pattern corresponded to an index. The chamber was cleaned and wiped with H_2O and 70% ethanol between each session.

4.10. Statistical Analysis

No statistical methods predetermined sample sizes. We ensured consistency by repeating trials in different cell cultures from at least three animals per condition. Using GraphPad Prism 8, we assessed normal distribution with the Shapiro–Wilk test. For normally distributed data, we used one-way ANOVA with the Holm–Sidak post-hoc test; for non-normal data, we used the Kruskal–Wallis test with Dunn's post-hoc test. The results are represented as the mean ± standard deviation (SD). In figure legends, N is the number of cultures or dams and n is the number of cells or pups per condition. Statistical analyses were conducted on the data from all cells/pups (n).

5. Conclusions

In conclusion, our findings underscore the critical role of timing and developmental stages in elucidating how both forms of BDNF influence the regulation of KCC2 cell membrane trafficking. Particularly, our data strongly suggest that proBDNF delays the GABA shift polarity, thereby maintaining neurons in an immature state. This consequence could be linked to the behavioral deficits observed in electroporated rats. These actions carry significant implications for cognitive processes and neural circuitry, providing insights into the intricate interplay between neurotrophic factors and neuronal functions. Furthermore, a deeper investigation into the specific mechanisms through which BDNF modulates GABA function and chloride homeostasis during development is imperative. This understanding holds potential implications for our comprehension of neurodevelopmental processes and could pave the way for the development of targeted therapies benefiting individuals with brain disorders.

Author Contributions: M.H., I.M. and C.P. designed the research. M.H., C.B., E.B., I.M., J.Z. and C.P. performed the research. M.H., C.B., C.P. and I.M. analyzed the data. M.H., I.M., C.B. and C.P. wrote the paper. All authors have read and agreed to the published version of the manuscript.

Funding: This research was funded by the National Institute of Health and Medical Research (INSERM), the National Center for Scientific Research (CNRS), and the French National Agency for Research (ANR, grant number R07066AS 2008-2011 to M.H., C.B., C.P. and I.M.). This work was supported by the INMED core facilities Animalerie, InMAGIC, and In Utero injection.

Institutional Review Board Statement: All animal procedures were carried out according to the guidelines set by the European Communities Council Directives (2010/63/UE) and approved by the Ethical Committee for Animal Experimentation (APAFIS number 02331.02). Male and female Wistar rats were purchased from Janvier Labs (https://www.janvier-labs.com/fiche_produit/rat_wistar, Accessed on 10 March 2024). Animals were raised and mated at the INMED A2 animal facility and housed under a 12 h light/dark cycle at 22–24 °C and had access to food and water ad libitum.

Data Availability Statement: Data supporting the findings of this study are available in this manuscript. All other data that support the findings of this study are available from the corresponding authors upon reasonable request.

Acknowledgments: We thank F. Michel at InMAGIC (INMED Imaging Centre) for technical assistance.

Conflicts of Interest: The authors declare no conflicts of interest.

References

1. Ben-Ari, Y. Excitatory Actions of Gaba during Development: The Nature of the Nurture. *Nat. Rev. Neurosci.* **2002**, *3*, 728–739. [CrossRef] [PubMed]
2. Blaesse, P.; Airaksinen, M.S.; Rivera, C.; Kaila, K. Cation-Chloride Cotransporters and Neuronal Function. *Neuron* **2009**, *61*, 820–838. [CrossRef] [PubMed]
3. Rivera, C.; Voipio, J.; Kaila, K. Two Developmental Switches in GABAergic Signalling: The K+–Cl− Cotransporter KCC2 and Carbonic Anhydrase CAVII. *J. Physiol.* **2005**, *562*, 27–36. [CrossRef]
4. Ben-Ari, Y.; Gaiarsa, J.-L.; Tyzio, R.; Khazipov, R. GABA: A Pioneer Transmitter That Excites Immature Neurons and Generates Primitive Oscillations. *Physiol. Rev.* **2007**, *87*, 1215–1284. [CrossRef]
5. Medina, I.; Friedel, P.; Rivera, C.; Kahle, K.T.; Kourdougli, N.; Uvarov, P.; Pellegrino, C. Current View on the Functional Regulation of the Neuronal K^+-Cl^- Cotransporter KCC2. *Front. Cell. Neurosci.* **2014**, *8*, 27. [CrossRef] [PubMed]
6. Khalilov, I.; Minlebaev, M.; Mukhtarov, M.; Khazipov, R. Dynamic Changes from Depolarizing to Hyperpolarizing GABAergic Actions during Giant Depolarizing Potentials in the Neonatal Rat Hippocampus. *J. Neurosci.* **2015**, *35*, 12635–12642. [CrossRef] [PubMed]
7. Kirmse, K.; Kummer, M.; Kovalchuk, Y.; Witte, O.W.; Garaschuk, O.; Holthoff, K. GABA Depolarizes Immature Neurons and Inhibits Network Activity in the Neonatal Neocortex in Vivo. *Nat. Commun.* **2015**, *6*, 7750. [CrossRef] [PubMed]
8. Virtanen, M.A.; Uvarov, P.; Hübner, C.A.; Kaila, K. NKCC1, an Elusive Molecular Target in Brain Development: Making Sense of the Existing Data. *Cells* **2020**, *9*, E2607. [CrossRef] [PubMed]
9. Ben-Ari, Y. NKCC1 Chloride Importer Antagonists Attenuate Many Neurological and Psychiatric Disorders. *Trends Neurosci.* **2017**, *40*, 536–554. [CrossRef]
10. Lam, P.; Newland, J.; Faull, R.L.M.; Kwakowsky, A. Cation-Chloride Cotransporters KCC2 and NKCC1 as Therapeutic Targets in Neurological and Neuropsychiatric Disorders. *Molecules* **2023**, *28*, 1344. [CrossRef]
11. Keramidis, I.; McAllister, B.B.; Bourbonnais, J.; Wang, F.; Isabel, D.; Rezaei, E.; Sansonetti, R.; Degagne, P.; Hamel, J.P.; Nazari, M.; et al. Restoring Neuronal Chloride Extrusion Reverses Cognitive Decline Linked to Alzheimer's Disease Mutations. *Brain* **2023**, *146*, 4903–4915. [CrossRef] [PubMed]
12. Chen, M.; Wang, J.; Jiang, J.; Zheng, X.; Justice, N.J.; Wang, K.; Ran, X.; Li, Y.; Huo, Q.; Zhang, J.; et al. APP Modulates KCC2 Expression and Function in Hippocampal GABAergic Inhibition. *eLife* **2017**, *6*, e20142. [CrossRef] [PubMed]
13. Rivera, C.; Li, H.; Thomas-Crusells, J.; Lahtinen, H.; Viitanen, T.; Nanobashvili, A.; Kokaia, Z.; Airaksinen, M.S.; Voipio, J.; Kaila, K.; et al. BDNF-Induced TrkB Activation down-Regulates the K+-Cl- Cotransporter KCC2 and Impairs Neuronal Cl- Extrusion. *J. Cell Biol.* **2002**, *159*, 747–752. [CrossRef] [PubMed]
14. Aguado, F.; Carmona, M.A.; Pozas, E.; Aguiló, A.; Martínez-Guijarro, F.J.; Alcantara, S.; Borrell, V.; Yuste, R.; Ibañez, C.F.; Soriano, E. BDNF Regulates Spontaneous Correlated Activity at Early Developmental Stages by Increasing Synaptogenesis and Expression of the K+/Cl- Co-Transporter KCC2. *Dev. Camb. Engl.* **2003**, *130*, 1267–1280. [CrossRef]
15. Porcher, C.; Medina, I.; Gaiarsa, J.-L. Mechanism of BDNF Modulation in GABAergic Synaptic Transmission in Healthy and Disease Brains. *Front. Cell. Neurosci.* **2018**, *12*, 273. [CrossRef] [PubMed]
16. Greenberg, M.E.; Xu, B.; Lu, B.; Hempstead, B.L. Symposium: New Insights in the Biology of BDNF Synthesis and Release: Implications in CNS Function. *J. Neurosci.* **2009**, *29*, 12764. [CrossRef] [PubMed]
17. Sun, W.; Cheng, H.; Yang, Y.; Tang, D.; Li, X.; An, L. Requirements of Postnatal proBDNF in the Hippocampus for Spatial Memory Consolidation and Neural Function. *Front. Cell Dev. Biol.* **2021**, *9*, 678182. [CrossRef] [PubMed]
18. Menshanov, P.N.; Lanshakov, D.A.; Dygalo, N.N. proBDNF Is a Major Product of Bdnf Gene Expressed in the Perinatal Rat Cortex. *Physiol. Res. Acad. Sci. Bohemoslov.* **2015**, *64*, 923–934. [CrossRef] [PubMed]
19. Dincheva, I.; Lynch, N.B.; Lee, F.S. The Role of Bdnf in The Development of Fear Learning. *Depress. Anxiety* **2016**, *33*, 907–916. [CrossRef]

20. Riffault, B.; Kourdougli, N.; Dumon, C.; Ferrand, N.; Buhler, E.; Schaller, F.; Chambon, C.; Rivera, C.; Gaiarsa, J.-L.; Porcher, C. Pro-Brain-Derived Neurotrophic Factor (proBDNF)-Mediated p75NTR Activation Promotes Depolarizing Actions of GABA and Increases Susceptibility to Epileptic Seizures. *Cereb. Cortex* **2016**, *28*, 510–527. [CrossRef]
21. Kourdougli, N.; Pellegrino, C.; Renko, J.-M.; Khirug, S.; Chazal, G.; Kukko-Lukjanov, T.-K.; Lauri, S.E.; Gaiarsa, J.-L.; Zhou, L.; Peret, A.; et al. Depolarizing γ-Aminobutyric Acid Contributes to Glutamatergic Network Rewiring in Epilepsy. *Ann. Neurol.* **2017**, *81*, 251–265. [CrossRef]
22. Ludwig, A.; Uvarov, P.; Soni, S.; Thomas-Crusells, J.; Airaksinen, M.S.; Rivera, C. Early Growth Response 4 Mediates BDNF Induction of Potassium Chloride Cotransporter 2 Transcription. *J. Neurosci. Off. J. Soc. Neurosci.* **2011**, *31*, 644–649. [CrossRef]
23. Mapplebeck, J.C.S.; Lorenzo, L.-E.; Lee, K.Y.; Gauthier, C.; Muley, M.M.; De Koninck, Y.; Prescott, S.A.; Salter, M.W. Chloride Dysregulation through Downregulation of KCC2 Mediates Neuropathic Pain in Both Sexes. *Cell Rep.* **2019**, *28*, 590–596.e4. [CrossRef]
24. Patel, D.C.; Thompson, E.G.; Sontheimer, H. Brain-Derived Neurotrophic Factor Inhibits the Function of Cation-Chloride Cotransporter in a Mouse Model of Viral Infection-Induced Epilepsy. *Front. Cell Dev. Biol.* **2022**, *10*, 961292. [CrossRef]
25. Shulga, A.; Magalhães, A.C.; Autio, H.; Plantman, S.; di Lieto, A.; Nykjær, A.; Carlstedt, T.; Risling, M.; Arumäe, U.; Castrén, E.; et al. The Loop Diuretic Bumetanide Blocks Posttraumatic p75NTR Upregulation and Rescues Injured Neurons. *J. Neurosci. Off. J. Soc. Neurosci.* **2012**, *32*, 1757–1770. [CrossRef]
26. Medina, I.; Pisella, L.I. Methods for Investigating the Activities of Neuronal Chloride Transporters. In *Neuronal Chloride Transporters in Health and Disease*; Elsevier: Amsterdam, The Netherlands, 2020; pp. 21–41. ISBN 978-0-12-815318-5.
27. Hershfinkel, M.; Kandler, K.; Knoch, M.E.; Dagan-Rabin, M.; Aras, M.A.; Abramovitch-Dahan, C.; Sekler, I.; Aizenman, E. Intracellular Zinc Inhibits KCC2 Transporter Activity. *Nat. Neurosci.* **2009**, *12*, 725–727. [CrossRef]
28. Järvelä, V.; Hamze, M.; Komulainen-Ebrahim, J.; Rahikkala, E.; Piispala, J.; Kallio, M.; Kangas, S.M.; Nickl, T.; Huttula, M.; Hinttala, R.; et al. A Novel Pathogenic SLC12A5 Missense Variant in Epilepsy of Infancy with Migrating Focal Seizures Causes Impaired KCC2 Chloride Extrusion. *Front. Mol. Neurosci.* **2024**, *17*, 1372662. [CrossRef]
29. Miesenböck, G.; De Angelis, D.A.; Rothman, J.E. Visualizing Secretion and Synaptic Transmission with pH-Sensitive Green Fluorescent Proteins. *Nature* **1998**, *394*, 192–195. [CrossRef]
30. Eftekhari, S.; Mehrabi, S.; Soleimani, M.; Hassanzadeh, G.; Shahrokhi, A.; Mostafavi, H.; Hayat, P.; Barati, M.; Mehdizadeh, H.; Rahmanzadeh, R.; et al. BDNF Modifies Hippocampal KCC2 and NKCC1 Expression in a Temporal Lobe Epilepsy Model. *Acta Neurobiol. Exp.* **2014**, *74*, 276–287. [CrossRef]
31. Ishibashi, M.; Egawa, K.; Fukuda, A. Diverse Actions of Astrocytes in GABAergic Signaling. *Int. J. Mol. Sci.* **2019**, *20*, 2964. [CrossRef]
32. Friedel, P.; Kahle, K.T.; Zhang, J.; Hertz, N.; Pisella, L.I.; Buhler, E.; Schaller, F.; Duan, J.; Khanna, A.R.; Bishop, P.N.; et al. WNK1-Regulated Inhibitory Phosphorylation of the KCC2 Cotransporter Maintains the Depolarizing Action of GABA in Immature Neurons. *Sci. Signal.* **2015**, *8*, ra65. [CrossRef]
33. Lee, H.H.C.; Walker, J.A.; Williams, J.R.; Goodier, R.J.; Payne, J.A.; Moss, S.J. Direct Protein Kinase C-Dependent Phosphorylation Regulates the Cell Surface Stability and Activity of the Potassium Chloride Cotransporter KCC2. *J. Biol. Chem.* **2007**, *282*, 29777–29784. [CrossRef]
34. Kahle, K.T.; Merner, N.D.; Friedel, P.; Silayeva, L.; Liang, B.; Khanna, A.; Shang, Y.; Lachance-Touchette, P.; Bourassa, C.; Levert, A.; et al. Genetically Encoded Impairment of Neuronal KCC2 Cotransporter Function in Human Idiopathic Generalized Epilepsy. *EMBO Rep.* **2014**, *15*, 766–774. [CrossRef]
35. Friedel, P.; Ludwig, A.; Pellegrino, C.; Agez, M.; Jawhari, A.; Rivera, C.; Medina, I. A Novel View on the Role of Intracellular Tails in Surface Delivery of the Potassium-Chloride Cotransporter KCC2. *eNeuro* **2017**, *4*, 1–19. [CrossRef]
36. Baeza-Raja, B.; Sachs, B.D.; Li, P.; Christian, F.; Vagena, E.; Davalos, D.; Le Moan, N.; Ryu, J.K.; Sikorski, S.L.; Chan, J.P.; et al. P75 Neurotrophin Receptor Regulates Energy Balance in Obesity. *Cell Rep.* **2016**, *14*, 255–268. [CrossRef]
37. Wilson, S.P. Modelling the Emergence of Rodent Filial Huddling from Physiological Huddling. *R. Soc. Open Sci.* **2017**, *4*, 170885. [CrossRef]
38. Naskar, S.; Narducci, R.; Balzani, E.; Cwetsch, A.W.; Tucci, V.; Cancedda, L. The Development of Synaptic Transmission Is Time-Locked to Early Social Behaviors in Rats. *Nat. Commun.* **2019**, *10*, 1195. [CrossRef]
39. Hübner, C.A.; Holthoff, K. Anion Transport and GABA Signaling. *Front. Cell. Neurosci.* **2013**, *7*, 177. [CrossRef]
40. Inoue, K.; Furukawa, T.; Kumada, T.; Yamada, J.; Wang, T.; Inoue, R.; Fukuda, A. Taurine Inhibits K+-Cl− Cotransporter KCC2 to Regulate Embryonic Cl- Homeostasis via with-No-Lysine (WNK) Protein Kinase Signaling Pathway. *J. Biol. Chem.* **2012**, *287*, 20839–20850. [CrossRef]
41. Pol, E.; Côme, E.; Merlaud, Z.; Gouhier, J.; Russeau, M.; Scotto-Lomassese, S.; Moutkine, I.; Marques, X.; Lévi, S. NKCC1 and KCC2 Chloride Transporters Have Different Membrane Dynamics on the Surface of Hippocampal Neurons. *Cells* **2023**, *12*, 2363. [CrossRef]
42. KOURDOUGLI, N.; Varpula, S.; Chazal, G.; Rivera, C. Detrimental Effect of Post Status Epilepticus Treatment with ROCK Inhibitor Y-27632 in a Pilocarpine Model of Temporal Lobe Epilepsy. *Front. Cell. Neurosci.* **2015**, *9*, 413. [CrossRef]
43. Porcher, C.; Hatchett, C.; Longbottom, R.E.; McAinch, K.; Sihra, T.S.; Moss, S.J.; Thomson, A.M.; Jovanovic, J.N. Positive Feedback Regulation between Gamma-Aminobutyric Acid Type A (GABA(A)) Receptor Signaling and Brain-Derived Neurotrophic Factor (BDNF) Release in Developing Neurons. *J. Biol. Chem.* **2011**, *286*, 21667–21677. [CrossRef]

44. de Los Heros, P.; Alessi, D.R.; Gourlay, R.; Campbell, D.G.; Deak, M.; Macartney, T.J.; Kahle, K.T.; Zhang, J. The WNK-Regulated SPAK/OSR1 Kinases Directly Phosphorylate and Inhibit the K+-Cl− Co-Transporters. *Biochem. J.* **2014**, *458*, 559–573. [CrossRef]
45. McMoneagle, E.; Zhou, J.; Zhang, S.; Huang, W.; Josiah, S.S.; Ding, K.; Wang, Y.; Zhang, J. Neuronal K+-Cl− Cotransporter KCC2 as a Promising Drug Target for Epilepsy Treatment. *Acta Pharmacol. Sin.* **2023**, *45*, 1–22. [CrossRef]
46. Gagnon, K.B.; Delpire, E. On the Substrate Recognition and Negative Regulation of SPAK, a Kinase Modulating Na+-K+-2Cl− Cotransport Activity. *Am. J. Physiol. Cell Physiol.* **2010**, *299*, C614–C620. [CrossRef]
47. Zanin, J.P.; Montroull, L.E.; Volosin, M.; Friedman, W.J. The P75 Neurotrophin Receptor Facilitates TrkB Signaling and Function in Rat Hippocampal Neurons. *Front. Cell. Neurosci.* **2019**, *13*, 485. [CrossRef]
48. Lu, B. BDNF and Activity-Dependent Synaptic Modulation. *Learn. Mem.* **2003**, *10*, 86–98. [CrossRef]
49. Kowiański, P.; Lietzau, G.; Czuba, E.; Waśkow, M.; Steliga, A.; Moryś, J. BDNF: A Key Factor with Multipotent Impact on Brain Signaling and Synaptic Plasticity. *Cell. Mol. Neurobiol.* **2018**, *38*, 579–593. [CrossRef]
50. Lee, F.S.; Hempstead, B.L. New Roles for an Ancient Factor. *Trends Neurosci.* **2018**, *41*, 765–767. [CrossRef] [PubMed]
51. Yang, J.; Harte-Hargrove, L.C.; Siao, C.-J.; Marinic, T.; Clarke, R.; Ma, Q.; Jing, D.; Lafrancois, J.J.; Bath, K.G.; Mark, W.; et al. proBDNF Negatively Regulates Neuronal Remodeling, Synaptic Transmission, and Synaptic Plasticity in Hippocampus. *Cell Rep.* **2014**, *7*, 796–806. [CrossRef] [PubMed]
52. Ernfors, P.; Lee, K.-F.; Jaenisch, R. Mice Lacking Brain-Derived Neurotrophic Factor Develop with Sensory Deficits. *Nature* **1994**, *368*, 147–150. [CrossRef]
53. Jones, D.L. A Potential Role for Nerve Growth Factor in Regulating the Maturation of Inhibitory Neurotransmission. *J. Neurosci.* **2010**, *30*, 6813–6814. [CrossRef]
54. Andreska, T.; Lüningschrör, P.; Sendtner, M. Regulation of TrkB Cell Surface Expression—A Mechanism for Modulation of Neuronal Responsiveness to Brain-Derived Neurotrophic Factor. *Cell Tissue Res.* **2020**, *382*, 5–14. [CrossRef]
55. Genoud, C.; Knott, G.; Sakata, K.; Lu, B.; Welker, E. Altered Synapse Formation in the Adult Somatosensory Cortex of Brain-Derived Neurotrophic Factor Heterozygote Mice. *J. Neurosci. Off. J. Soc. Neurosci.* **2004**, *24*, 2394–2400. [CrossRef]
56. Buerli, T.; Pellegrino, C.; Baer, K.; Lardi-Studler, B.; Chudotvorova, I.; Fritschy, J.-M.; Medina, I.; Fuhrer, C. Efficient Transfection of DNA or shRNA Vectors into Neurons Using Magnetofection. *Nat. Protoc.* **2007**, *2*, 3090–3101. [CrossRef]
57. Kriegstein, A.R.; Noctor, S.C. Patterns of Neuronal Migration in the Embryonic Cortex. *Trends Neurosci.* **2004**, *27*, 392–399. [CrossRef]
58. Heyser, C.J. Assessment of Developmental Milestones in Rodents. *Curr. Protoc. Neurosci.* **2004**, *25*, 8–18. [CrossRef]

International Journal of
Molecular Sciences

MDPI

Article

Retinoic Acid-Mediated Control of Energy Metabolism Is Essential for Lung Branching Morphogenesis

Hugo Fernandes-Silva [1,2,3], Marco G. Alves [4], Marcia R. Garcez [5,6], Jorge Correia-Pinto [1,2,7], Pedro F. Oliveira [8], Catarina C. F. Homem [5] and Rute S. Moura [1,2,*]

1 Life and Health Sciences Research Institute (ICVS), School of Medicine, University of Minho, 4710-057 Braga, Portugal; hugomiguelfsilva@gmail.com (H.F.-S.); jcp@med.uminho.pt (J.C.-P.)
2 ICVS/3B's–PT Government Associate Laboratory, 4710-057 Braga/Guimarães, Portugal
3 PhDOC PhD Program, ICVS/3B's, School of Medicine, University of Minho, 4710-057 Braga, Portugal
4 Institute of Biomedicine (iBiMED), Department of Medical Sciences, University of Aveiro, 3810-193 Aveiro, Portugal; alvesmarc@gmail.com
5 iNOVA4Health, NOVA Medical School/Faculdade de Ciências Médicas (NMS/FCM), Universidade Nova de Lisboa, 1449-011 Lisbon, Portugal; marcia.garcez@nms.unl.pt (M.R.G.); catarina.homem@nms.unl.pt (C.C.F.H.)
6 Graduate Program in Areas of Basic and Applied Biology (GABBA), Universidade do Porto, 4050-313 Porto, Portugal
7 Department of Pediatric Surgery, Hospital of Braga, 4710-243 Braga, Portugal
8 LAQV-REQUIMTE & Department of Chemistry, University of Aveiro, 3810-193 Aveiro, Portugal; pfobox@gmail.com
* Correspondence: rutemoura@med.uminho.pt

Abstract: Lung branching morphogenesis relies on intricate epithelial–mesenchymal interactions and signaling networks. Still, the interplay between signaling and energy metabolism in shaping embryonic lung development remains unexplored. Retinoic acid (RA) signaling influences lung proximal–distal patterning and branching morphogenesis, but its role as a metabolic modulator is unknown. Hence, this study investigates how RA signaling affects the metabolic profile of lung branching. We performed ex vivo lung explant culture of embryonic chicken lungs treated with DMSO, 1 μM RA, or 10 μM BMS493. Extracellular metabolite consumption/production was evaluated by using ^{1}H-NMR spectroscopy. Mitochondrial respiration and biogenesis were also analyzed. Proliferation was assessed using an EdU-based assay. The expression of crucial metabolic/signaling components was examined through Western blot, qPCR, and in situ hybridization. RA signaling stimulation redirects glucose towards pyruvate and succinate production rather than to alanine or lactate. Inhibition of RA signaling reduces lung branching, resulting in a cystic-like phenotype while promoting mitochondrial function. Here, RA signaling emerges as a regulator of tissue proliferation and lactate dehydrogenase expression. Furthermore, RA governs fatty acid metabolism through an AMPK-dependent mechanism. These findings underscore RA's pivotal role in shaping lung metabolism during branching morphogenesis, contributing to our understanding of lung development and cystic-related lung disorders.

Keywords: respiratory system; glycolysis; pyruvate; mitochondria; chicken embryo; cystic lung disease

1. Introduction

Pulmonary branches are formed during the early stages of embryonic lung development through an intricate process known as lung branching morphogenesis. Lung branching shapes the pulmonary airway conducting system and is characterized by epithelial–mesenchymal interactions mediated by complex signaling [1,2]. Lung branching is a highly regulated process and, when disrupted, leads to congenital malformations such as congenital pulmonary airway malformation (CPAM) [3,4].

In the chicken embryo (*Gallus gallus*), the lung arises from the primitive foregut around day 3 of embryogenesis [5]. The primary bronchus (mesobronchus) grows distally, and the secondary bronchi (buds) sprout laterally into the surrounding mesenchyme [6]. This lateral/monopodial branching is similar to the domain branching subroutine observed during mammalian lung development [7,8]. In addition to their structural similarities, the avian respiratory system shares highly conserved molecular mechanisms with mammals, indicating comparable functions [9]. For example, FGF (Fibroblast Growth Factor), SHH (Sonic Hedgehog), WNT (Wingless-related Integration Site), and RA (retinoic acid) play crucial roles in both chicken and mammalian pulmonary branching morphogenesis [10–13]. These similarities highlight the embryonic chicken lung as an exceptional model for studying lung organogenesis and particularly the early branching processes.

RA signaling is fundamental for vertebrate embryonic development and is a major player in lung organogenesis [14–16]. Intracellularly, RA binds to specific nuclear retinoic acid receptors (RARs), of which there are three subtypes (RARα, RARβ, and RARγ). Then, RARs form a heterodimer with retinoic X receptors (RXRs) that mediate downstream cellular signaling and the transcription of RA target genes [17]. RA modulates multiple aspects of embryonic lung development, specifically proximal–distal tissue patterning and branching morphogenesis [13,18,19].

In addition to intercellular signaling, tissue growth involves energy consumption processes and the synthesis of biomolecules. In the embryonic lung, it has been recently described by our group that chicken lung branching morphogenesis gradually adapts to a glycolytic lactate-based metabolism to sustain the lung's energetic demands, revealing the importance of metabolic regulation in this phase [20]. In the adult lung, metabolic requirements are achieved mainly through the uptake and catabolism of glucose [21–23]. However, how signaling networks and energy metabolism cooperate to shape embryonic lung development remains largely unexplored.

The interaction between cell signaling and energy metabolism is starting to emerge as a key concept for understanding developmental processes and developmental abnormalities. Hence, it is crucial to investigate how metabolism influences cellular and developmental decisions, and how metabolism is dynamically regulated during development [24–27]. Recent studies have demonstrated that metabolism interacts with conserved signaling pathways during development. For instance, in *Drosophila* wing discs, Notch signaling activates glycolysis and suppresses the Krebs cycle, resembling the Warburg effect. Conversely, absence of Notch signaling leads to a decrease in glycolytic gene expression [28]. The Hedgehog signaling pathway can enhance glycolytic ATP production in *Drosophila* wing discs [29]. During neural tissue development in *Zebrafish*, Notch signaling regulates the expression of glycolysis-related genes in a stage-specific manner [30]. Also, an FGF/Wnt signaling coordinated glycolytic gradient regulates cell motility and controls specification, contributing to presomitic mesoderm development during vertebrate embryo somite formation [31].

Considering the well-known role of RA as a modulator of lung organogenesis, we aimed to investigate how the RA signaling pathway modulates the energy metabolism of early stages of chicken lung development. Our study reveals that RA signaling inhibition decreases lung branching and induces a cystic-like phenotype. At the metabolic level, RA signaling is involved in the regulation of glycolysis and pyruvate metabolism, mitochondrial function, and fatty acid metabolism. Here, we show that RA signaling modulates metabolism during early lung organogenesis and that proper RA levels contribute to an adequate metabolic profile during early lung branching. Overall, this report reveals new insights on the RA signaling–metabolism interaction during embryonic lung development that can contribute to understanding the etiology of cystic-related congenital lung disorders.

2. Results

2.1. Retinoic Acid Signaling Downregulation Decreases Lung Branching and Induces a Cystic-like Phenotype

RA signaling stimulation triggers an increase in lung branching in mouse, rat, and chicken models [13,32,33]. Conversely, the information regarding RA signaling inhibition on lung branching is scarce, even though it is known that its impairment leads to catastrophic effects on the respiratory system development [18]. BMS493 (BMS) is a pan-RAR inverse agonist used to inhibit RA signaling; it specifically suppresses the RA signaling pathway by establishing associations between the RARs and transcriptional co-repressors, thus blocking RA signaling-dependent transcription [14,34].

To determine the impact of BMS on lung branching, we tested several doses that were selected according to the literature [34–37]. The dose-dependent inhibition effect of BMS on the RA signaling pathway and lung branching morphogenesis led to the choice of the 10 μM of BMS as the best dose to proceed with the studies (please refer to Figure S1 for further details).

To compare the effect of RA signaling stimulation versus inhibition on lung branching morphogenesis, a 48 h ex vivo lung explant culture was performed using stage b2 lungs (2 secondary buds per bronchus). In vitro lung explants were treated with 1 μM of RA (stimulation) or 10 μM of BMS (inhibition); DMSO, the solvent, was used as the control. Afterward, D0 (0 h) and D2 (48 h) explants were morphometrically analyzed, and the results were represented as a D2/D0 ratio. The RA signaling pathway activation status was assessed by performing in situ hybridization for *rarβ*, a recognized target of this cascade [13,32,38].

RA treatment increased *rarβ* expression, suggesting an activation of the RA signaling cascade (Figure 1A). On the other hand, BMS treatment decreased *rarβ* expression (Figure 1A), implying a downregulation of the RA pathway. To evaluate branching morphogenesis, the lung's epithelial compartment was outlined and analyzed, as shown in Figure 1A. There was an increase in the epithelial perimeter of RA-treated lungs compared to DMSO, whereas BMS treatment induced a significant decrease compared to the DMSO (Figure 1B). The changes in the epithelial perimeter were due to alterations in branching morphogenesis since no variations were detected regarding the epithelial area (Figure 1C).

RA is a known stimulator of lung branching morphogenesis, and these results are in agreement with previous RA supplementation studies that have used the embryonic lung [13,32,33]. Furthermore, we show here that BMS treatment downregulates the RA signaling pathway and decreases branching morphogenesis of the embryonic chicken lung. While RA-treated lungs maintained proper branching morphology, BMS-treated lungs displayed a wider primary bronchus (Figure 1A; arrow) and larger epithelial pouches (Figure 1A; asterisk), resembling cystic-like structures. This phenotype is comparable to rat and mouse cystic malformations [3,4].

2.2. High Proliferation Is Associated with Active Branching Regions

Considering the effect of RA signaling stimulation vs. inhibition on lung branching morphogenesis, we wondered whether RA signaling modulation could influence the proliferation status of lung branching morphogenesis. For that purpose, we assessed EdU incorporation into new DNA strands using Alexa Fluor 488 (Green) in the three experimental conditions. The nuclei were counterstained with Hoechst 33342 (Red) (Figure 2).

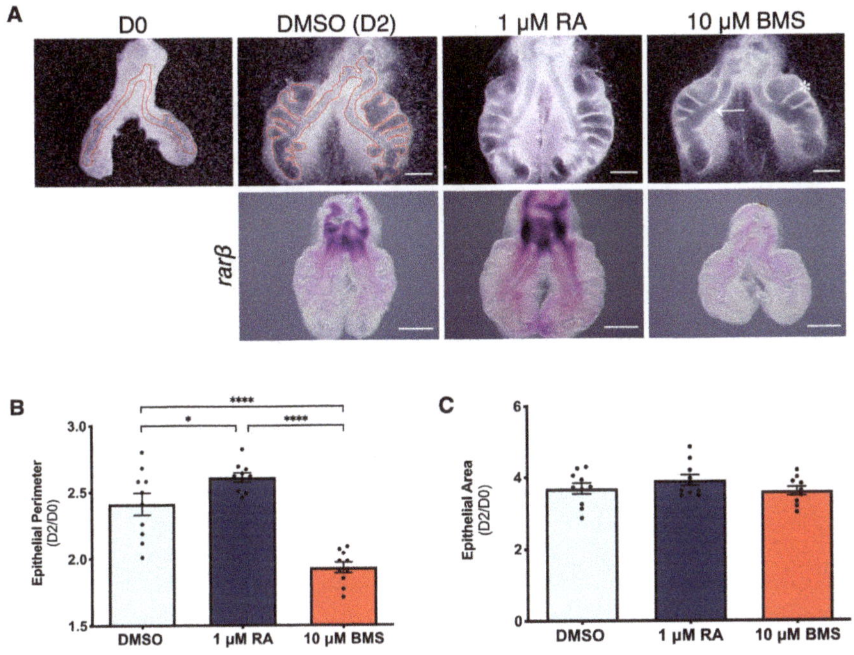

Figure 1. Effect of RA signaling stimulation versus inhibition on lung branching morphogenesis. (**A**) Representative examples of b2 lungs at 0 h (D0) and after 48 h (D2) of explant culture (DMSO, 1 μM of RA or 10 μM of BMS) (upper panel); probed for *rarβ* (lower panel) ($n \geq 4$/condition). Scale bar: 500 μm. Red line: epithelial compartment outline. Arrow: wider primary bronchus. Asterisk: larger epithelial pouches. The epithelial compartment was outlined (0 h and 48 h DMSO examples) and the epithelial perimeter and area determined. Results are expressed as D2/D0 and represented as mean ± SEM (n = 10/condition) for (**B**) epithelial perimeter and (**C**) epithelial area. One-Way ANOVA and Fisher's LSD tests were performed. Significantly different results are indicated as * $p < 0.05$; **** $p < 0.0001$.

DMSO-treated explants revealed high proliferative areas in the trachea region (arrowhead), in the distal area of the lung (arrow), and at active branching sites (asterisk) (Figure 2B). These results match the previous data described in [20]. Moreover, the proliferation pattern remained unaltered for the 1 μM of RA and 10 μM of BMS conditions (Figure 2B,F,J). Since RA stimulation increases lung branching, and high proliferation is characteristic of active branching sites (asterisk), the regions of high proliferation expand substantially (Figure 2F). In contrast, 10 μM of BMS decreases lung branching morphogenesis and induces a cystic-like phenotype. Thus, in this condition, high proliferation levels are restricted to the branching structures (asterisk) and generally decrease compared to other conditions (Figure 2J).

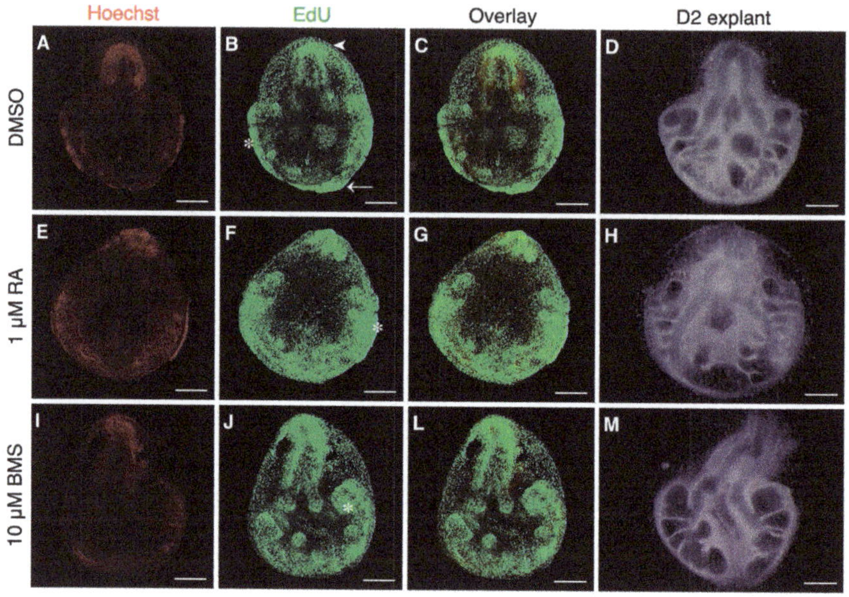

Figure 2. Proliferation analysis of lung branching morphogenesis upon RA signaling stimulation/inhibition. Representative confocal microscopy fluorescence images of b2 lung explants after 48 h in culture (D2), supplemented with (**A–D**) DMSO, (**E–H**) 1 μM of RA, and (**I–M**) 10 μM of BMS. (**A**,**E**,**I**) Nuclei were stained with Hoechst 33342 (Red). (**B**,**F**,**J**) Proliferation assessed through EdU incorporation in DNA, followed by detection using Alexa Fluor 488 (Green). (**C**,**G**,**L**) Merged images. (**D**,**H**,**M**) Lung explants after 48 h of culture and before Hoechst 33,342 staining or EdU incorporation protocols; it is worth mentioning that tissue size reduces after undergoing the proliferation procedure protocol ($n \geq 4$/condition). Scale bar: 500 μm. Arrowhead: trachea region. Asterisk: active branching sites. Arrow: distal tip.

2.3. Retinoic Acid Signaling Stimulation Requires Less Glucose Consumption

Recent data associate intercellular signaling with energy metabolism processes during the tissue growth of embryonic systems [28–31]. In this sense, we explored potential metabolic alterations induced by RA signaling stimulation versus inhibition. Since both embryonic and adult lungs preferentially use glucose as the primary energy source [20–23], we started by focusing on glucose metabolism (Figure 3A). For that purpose, lung explants were exposed to one of the following conditions: DMSO, 1 μM of RA, or 10 μM of BMS. The explant culture was performed for 48 h and refreshed at D1 (24 h). The medium was collected at D0 (0 h; reference/control), D1, and D2 and analyzed by using ^{1}H-NMR spectroscopy. Extracellular metabolite production/consumption was calculated following the mathematical formula |(D1-D0) + (D2-D0)|, as described in [20], and expressed in fold variation to DMSO. D2 lung tissues were collected to perform qPCR for phosphofructokinase 1 (*pfk1*), glucose-6-phosphate dehydrogenase (*g6pd*), and 6-phosphogluconate dehydrogenase (*pgd*).

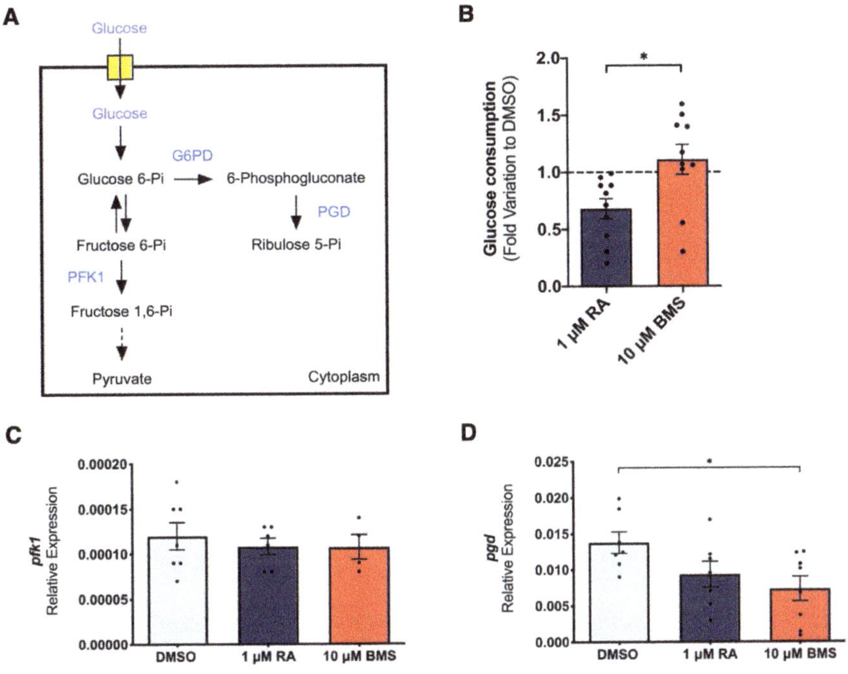

Figure 3. Impact of RA signaling modulation on glucose metabolism during lung branching morphogenesis. (**A**) Schematic representation of glucose metabolism through the glycolytic and pentose phosphate pathways. Blue labeling indicates the molecular targets analyzed; ^{1}H-NMR analysis of (**B**) glucose consumption ($n \geq 8$/condition) during 48 h of lung explant culture (DMSO, 1 μM of RA, or 10 μM of BMS). Metabolite consumption/production was calculated following the mathematical formula |(D1-D0) + (D2-D0)| and represented in fold variation to DMSO. D2 lung mRNA relative expression levels of (**C**) *pfk1* ($n \geq 4$/condition) and (**D**) *pgd* ($n \geq 7$/condition). All results are expressed as mean ± SEM. One-Way ANOVA and Fisher's LSD tests were performed. Significantly different results are indicated as * $p < 0.05$.

The ^{1}H-NMR analysis revealed that both RA signaling stimulation and inhibition do not alter lung glucose consumption compared to the control (Figure 3B). Nevertheless, glucose consumption is significantly lower under RA stimulation compared to RA inhibition (Figure 3B). We have previously shown that the embryonic chicken lung has the molecular machinery for the transport and uptake of glucose, and that the glucose consumption profile adjusts to cope with specific energy and nutrient requirements to sustain proper branching morphogenesis [20]. To assess the main enzyme controlling glycolysis [39], we measured *pfk1* expression. *pfk1* expression levels remained unaltered in the three experimental conditions (Figure 3C), meaning that RA signaling modulation does not impact the *pfk1* transcript levels. However, *pfk1* expression displays minimal variations throughout the early stages of lung branching [20].

The main catabolic fate of glucose 6-phosphate is the glycolytic breakdown into pyruvate through glycolysis. However, part of glucose 6-phosphate can be oxidized into pentose phosphates by the pentose phosphate pathway (PPP) (according to the cell needs or NADP+/NADPH concentrations) [39]. To study how RA signaling modulation affects PPP, the expression levels of *g6pd* and *pgd* were evaluated. Both transcripts encode enzymes responsible for the oxidative reactions of PPP that produce NADPH. In the embryonic lung, *g6pd* expression levels are very low, regardless of RA stimulation or inhibition conditions. As for the *pgd*, the expression levels decrease from DMSO to BMS-treated lungs, and RA stimulation displays the same tendency but without statistically significant differences (Figure 3D). According to our results, PPP is not favored under RA signaling stimulation or inhibition, which indicates that glucose 6-phosphate is available to be used by glycolysis. To conclude, under RA signaling stimulation, lung explants seem to adapt to make better use of glucose or to utilize other metabolic substrates, since RA signaling stimulation increases lung branching morphogenesis (Figure 1A,B) with less glucose consumption (Figure 3B).

2.4. Retinoic Acid Signaling Controls Pyruvate Metabolism

Since glucose consumption between RA signaling stimulation and inhibition was different, we searched for additional glycolysis-related products and further investigated other metabolite alterations promoted by RA signaling modulation. In the ^{1}H-NMR spectra, it was possible to detect the final product of glycolysis, pyruvate, and pyruvate-derived metabolites, namely alanine, lactate, acetate, and succinate (Figure 4A).

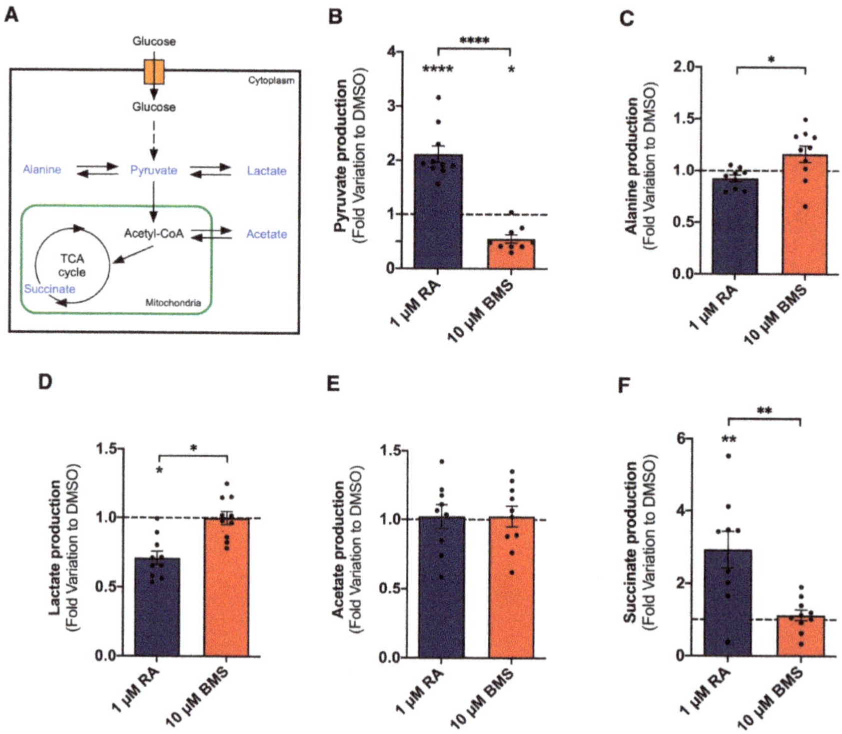

Figure 4. Extracellular metabolite changes induced by RA signaling modulation during lung branching morphogenesis. (**A**) Schematic representation of pyruvate-derived metabolites. Blue labeling indicates the metabolites detected and quantified in the ^{1}H-NMR spectra; ^{1}H-NMR analysis of (**B**) pyruvate production; (**C**) alanine production, (**D**) lactate production; (**E**) acetate production; and (**F**) succinate production during 48 h of lung explant culture (DMSO, 1 µM of RA, or 10 µM of BMS). Metabolite consumption/production was calculated following the mathematical formula $|(D1-D0) + (D2-D0)|$ and represented in fold variation to DMSO. Results are expressed as mean ± SEM ($n \geq 8$/condition). One-Way ANOVA and Fisher's LSD tests were performed. Significantly different results are indicated as * $p < 0.05$; ** $p < 0.01$; **** $p < 0.0001$.

RA signaling stimulation promotes a sharp increase in pyruvate production compared to the DMSO (≃111% increase) and BMS-treated groups (Figure 4B). In addition, under BMS treatment, pyruvate production is lower than in the DMSO group (≃45% decrease)

(Figure 4B). These data suggest that RA signaling stimulation promotes pyruvate production and its accumulation in the extracellular space because it may not be used by mitochondria and may have other metabolic or signaling functions.

No differences were detected in alanine production compared to the DMSO condition; yet, 1 μM of RA produces less alanine than in the 10 μM of BMS-treated group (Figure 4C). This decrease in alanine production is similar to the decrease observed in glucose consumption (Figure 3B).

Furthermore, pyruvate can be inter-converted into lactate through an enzymatic reaction catalyzed by lactate dehydrogenase (LDH), which regenerates NAD^+ [39]. The conversion of pyruvate to lactate is crucial for lung branching morphogenesis [20]. Moreover, it facilitates the uptake and incorporation of nutrients to form new biomass and sustain active tissue growth [20,40]. Here, we showed that RA signaling stimulation leads to a decrease in lactate production compared to the control and the BMS groups (Figure 4D). This lactate decrease seems to follow the decrease in glucose consumption promoted by RA signaling stimulation (Figure 3B).

After entering the mitochondria, pyruvate is converted into acetyl-CoA, which can be shuttled into acetate or directly fuel the Krebs cycle. Acetate can be used as a substrate for lipid synthesis and might incorporate new cellular membranes to sustain embryonic lung growth [20]. However, no alterations were observed in acetate production among the experimental groups (Figure 4E). Citrate was not detected in the ^{1}H-NMR spectra, but succinate production greatly increased upon RA signaling stimulation compared to the control ($\simeq$193% increase) and the BMS-treated group (Figure 4F). Succinate is the only direct link between the Krebs cycle and the mitochondrial respiratory chain. During oxidative phosphorylation, the electrons obtained from the succinate/fumarate oxidation through mitochondrial complex II/Succinate dehydrogenase (SDH) are used to reduce ubiquinone to ubiquinol [39]. In addition, the export of succinate from the mitochondrial matrix to the cytosol can act as a signal of mitochondrial status and might contribute to regulating overall metabolic homeostasis [41]. Nevertheless, succinate accumulation can trigger numerous cellular events and act as a metabolic signaling molecule [41,42]. We conclude that upon RA signaling stimulation, there is an increase in the amount of glucose that is directed into pyruvate production rather than into the alanine or lactate branches. In addition, mitochondrial succinate seems to increase at the expense of pyruvate availability.

2.5. Retinoic Acid Signaling Modulates Lactate Dehydrogenase Expression

Since RA signaling stimulation promotes a decrease in lactate production during lung branching (Figure 4D), we hypothesized that RA signaling controls lactate dehydrogenase (LDH) expression during branching morphogenesis. To further explore the molecular mechanism underlying lactate production, we analyzed LDHA and LDHB isoforms under RA signaling modulation. The treated explants were processed for in situ hybridization to analyze *ldha* and *ldhb* spatial localizations and relative expression levels. D2 whole lungs were also collected to assess LDHA and LDHB protein expression levels using Western blot.

The *ldha* transcript is limited to the ventral region of the lung (Figure 5A; arrow), and its expression levels are dependent on RA signaling. RA pathway stimulation considerably increases the expression levels of *ldha* compared to DMSO (Figure 5A). Conversely, the BMS treatment decreases *ldha* expression compared to both the DMSO and RA-treated lungs (Figure 5A). Regarding *ldhb*, its expression pattern is restricted to active branching sites (Figure 5B; asterisk). Upon exposure to 1 μM of RA, the expression pattern of *ldhb* is maintained, but its expression levels increase significantly (Figure 5B). On the other hand, lung explants treated with 10 μM of BMS displayed a decrease in *ldhb* expression levels compared to DMSO and RA-exposed lungs (Figure 5B). These results match previous data from our team, wherein both *ldha* and *ldhb* isoforms were present at earlier stages of lung branching and were expressed in a region-specific manner [20]. Moreover, *ldhb* transcript localization was previously associated with active branching regions [20].

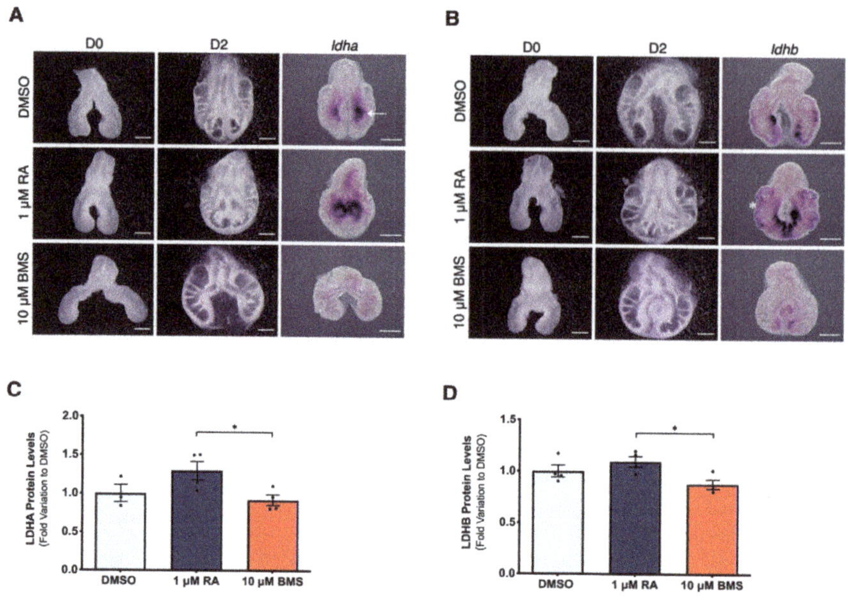

Figure 5. LDHA and LDHB expression alterations induced by RA signaling modulation during lung branching morphogenesis. Representative examples of b2 lung explant culture at D0 and D2, treated with DMSO, 1 μM of RA, and 10 μM of BMS. D2 lungs were probed for (**A**) *ldha* and (**B**) *ldhb* ($n \geq 4$/condition). Scale bar: 500 μm. Arrow: ventral region. Asterisk: active branching sites; D2 lungs were analyzed for (**C**) LDHA and (**D**) LDHB protein expression levels. LDHA and LDHB immunoblots and total protein are presented in Figure S2. Results are represented in fold variation to DMSO. Results are expressed as mean ± SEM ($n \geq 3$/condition). One-Way ANOVA and Fisher's LSD tests were performed. Significantly different results are indicated as * $p < 0.05$.

LDHA and LDHB protein expression changes are not as pronounced as in the transcripts and this can be explained since Western blot samples comprise whole lungs rather than specific tissue regions (Figure 5C and 5D, respectively). LDHA and LDHB protein levels increase between DMSO and RA-treated lungs, but without statistically significant differences; protein expression levels decrease between RA signaling stimulation and RA signaling inhibition conditions in both isoforms (Figure 5C,D). Here, we describe how LDH expression is modulated by RA signaling and that both *ldha* and *ldhb* isoforms are downstream targets of the RA signaling pathway. Furthermore, high proliferation sites match *ldhb* expression localization when the RA signaling pathway is stimulated (Figure 2B). In this sense, we conclude that high proliferation is associated with active branching regions and that, by modulating lung branching, RA signaling influences the overall lung proliferation rates.

2.6. Retinoic Acid Signaling Downregulation Increases Mitochondrial Function

Pyruvate and succinate metabolite data (Figure 4) suggest a potential role of RA signaling in regulating mitochondria function during lung branching morphogenesis. To test this hypothesis, we assessed mitochondrial respiration upon RA signaling stimulation and inhibition. In detail, 48 h lung explant tissues were processed for the real-time measurement of oxygen consumption rate (OCR), and the results were expressed in pmol/min/mg protein (Figure S3).

Regarding basal respiration, the OCR from BMS-treated lungs increased compared to the DMSO and RA-treated lungs (Figure 6A). Similarly, for ATP production, the 10 μM of BMS condition displayed higher OCR levels than for DMSO and 1 μM of RA (Figure 6B).

Likewise, the maximal respiration component revealed an increase in the OCR levels of the RA inhibition group compared to DMSO and 1 μM of RA (Figure 6C). These results point towards RA signaling stimulation presenting an overall mitochondrial respiration profile similar to DMSO, whereas the BMS-treated lungs display increased mitochondrial function. Overall, these results point to a more oxidative metabolism when RA signaling is inhibited compared to control and RA-treated lungs.

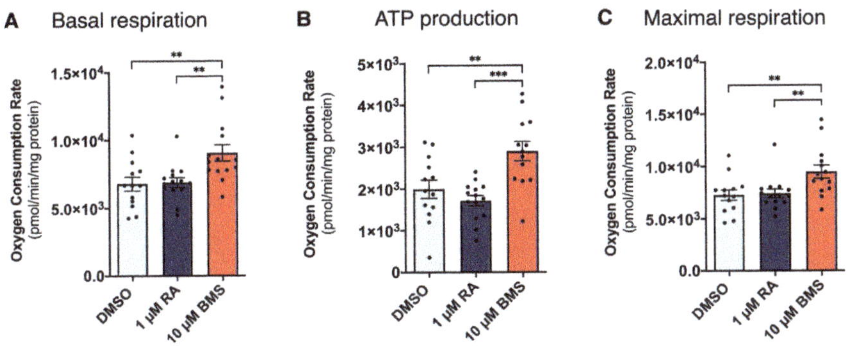

Figure 6. Mitochondrial oxygen consumption of lung explants upon RA signaling stimulation/inhibition. Real-time measurement of oxygen consumption rate (OCR) corresponding to (**A**) basal respiration; (**B**) component of OCR corresponding to ATP production; (**C**) component of OCR corresponding to maximal respiration after 48 h of lung explant culture (DMSO, 1 μM of RA, or 10 μM of BMS). Results are represented in pmol/min and normalized to the total amount of protein. Results are expressed as mean ± SEM ($n \geq 13$/condition). One-Way ANOVA and Fisher's LSD tests were performed. Significantly different results are indicated as ** $p < 0.01$; *** $p < 0.001$.

To assess if these differences were due to changes in mitochondrial biogenesis, mtDNA copy number and transcription factor A mitochondrial (*tfam*) expression levels were evaluated by qPCR [43,44]. No differences were observed in mtDNA copy numbers among the three experimental conditions (Figure S4A). Similarly, the expression levels of *tfam* remained unaltered between conditions (Figure S4B). These results suggest the same number of mitochondria among experimental groups, meaning that RA signaling does not affect mitochondrial biogenesis during lung branching morphogenesis.

2.7. Retinoic Acid Signaling Controls Fatty Acid Metabolism through AMPK

Considering that RA signaling inhibition promotes mitochondria function and mitochondria plays a pivotal role in lipid metabolism, we decided to investigate how RA signaling modulation impacts lipid metabolism during lung branching. In fact, RA is a known modulator of lipid metabolism [45–47]. Moreover, AMP-activated protein kinase (AMPK) is a master regulator of metabolism and a sensor of cellular energy status. AMPK can restore energy balance, modulate glucose and lipid metabolism, and aid mitochondria homeostasis [48,49]. In this sense, we explored whether AMPK and lipid metabolism could be affected by RA signaling modulation during lung branching morphogenesis.

Our results revealed that RA stimulation upregulated the AMPK pathway through increased pAMPK/AMPK protein expression levels (Figure 7A). Still, under RA signaling inhibition, pAMPK/AMPK protein expression increased even more (Figure 7A). These results suggest that during lung branching morphogenesis, AMPK is activated by RA signaling under stimulation or inhibition conditions. Moreover, the increased activation of AMPK under RA signaling inhibition might be due to a compensatory mechanism used to re-establish tissue homeostasis due to the cystic-like phenotype.

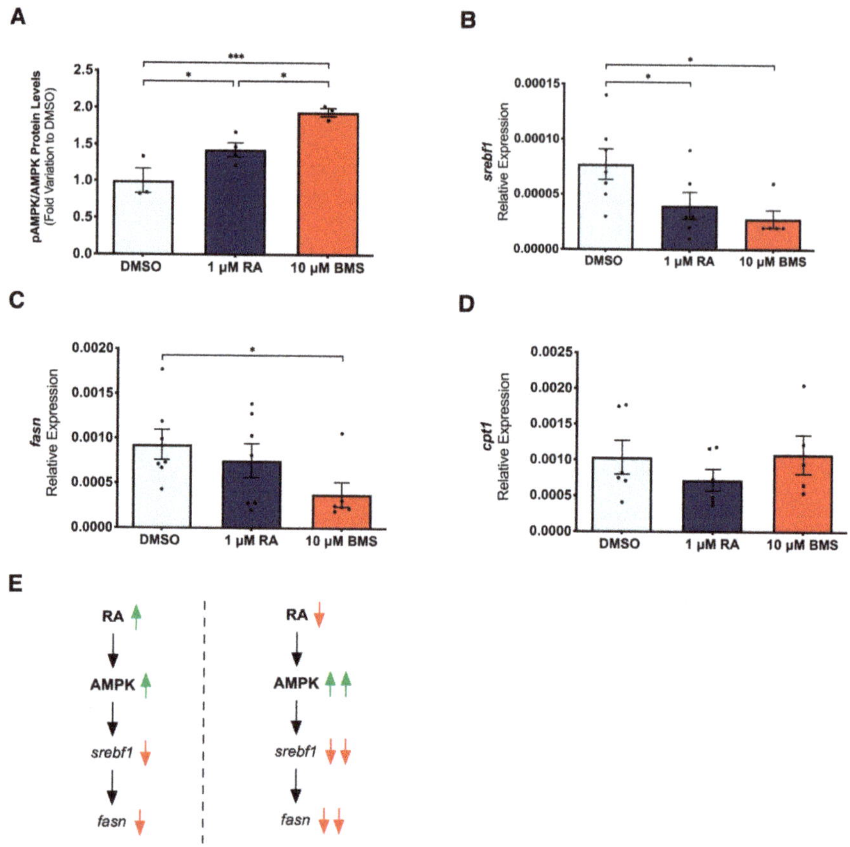

Figure 7. Impact of RA signaling modulation on AMPK pathway and lipid metabolism in early lung branching morphogenesis. D2 lungs were analyzed for (**A**) pAMPK/AMPK protein expression levels ($n \geq 3$/condition), after 48 h of lung explant culture (DMSO, 1 μM of RA, or 10 μM of BMS). pAMPK and AMPK immunoblots blots and total protein are presented in Figure S5. Results are represented in fold variation to DMSO; mRNA relative expression levels of (**B**) *srebf1*, (**C**) *fasn*, and (**D**) *cpt1* ($n \geq 7$/condition), after 48 h of lung explant culture (DMSO, 1 μM of RA, or 10 μM of BMS). Results are expressed as mean ± SEM. One-Way ANOVA and Fisher's LSD tests were performed. Significantly different results are indicated as * $p < 0.05$; *** $p < 0.001$. (**E**) Proposed molecular mechanism involving RA and AMPK signaling pathways, and respective impact on lipid synthesis machinery during lung branching morphogenesis. The symbols refer to increase (↑), decrease (↓), bigger increase (↑↑), or bigger decrease (↓↓).

Lipogenesis is, in part, regulated by sterol regulatory element-binding protein 1 (SREBP1). Our results showed that regulatory element-binding transcription factor 1 (*srebf1*) is present in the embryonic lung, with the expression levels decreasing from DMSO to RA-exposed and BMS-treated conditions (Figure 7B). Fatty acid synthase (FAS) is a rate-limiting enzyme of fatty acid synthesis; it converts acetyl-CoA and malonyl-CoA into palmitate [39]. In the chicken lung, fatty acid synthase (*fasn*) expression decreased from the DMSO to BMS-treated lungs (Figure 7C). Together, our data suggest a potential mechanism during early lung branching, in which RA signaling modulation activates the AMPK pathway, which in turn downregulates *srebf1*. The downregulation of *srebf1* decreases the expression of the downstream target *fasn* and, consequently, fatty acid synthesis is reduced.

Fatty acid β-oxidation is the process by which fatty acids are oxidized to acetyl-CoA to produce energy. Carnitine palmitoyltransferase I (CPT1) catalyzes the rate-limiting step of long-chain fatty acid oxidation; it promotes the translocation of fatty acids from the cytosol to the mitochondrial matrix [39]. In the embryonic chicken lung, *cpt1* expression levels remained unaltered between the three experimental conditions (Figure 7D). These data indicate that RA signaling does not modulate fatty acid oxidation during embryonic lung branching.

3. Discussion

Growing evidence suggests that signaling–metabolism interactions are essential during animal development [28,30,31]. In this study, we asked whether RA signaling, a well-known modulator of lung organogenesis, could regulate the energy metabolism of lung development during branching morphogenesis.

Our work revealed that RA signaling inhibition, upon BMS treatment, affects lung development by decreasing lung branching morphogenesis and inducing a cystic-like phenotype. This phenotype is comparable to rat overexpression of *fgf10*, which induces cystic malformations similar to human CPAM [4]. Likewise, mouse DICER mutant lungs display an upregulation of mesenchymal *fgf10* accompanied by branching arrest and the formation of large epithelial pouches [3]. An inhibition of FGF10 signaling is also associated with cystic appearances in early pulmonary branching [11]. During lung branching morphogenesis, a gradient of RA is produced from the pleura region to the periepithelial mesenchyme that surrounds the distal region of the growing bud [18]. RA availability regulates *fgf10* expression levels in the mesenchymal compartment surrounding the distal bud tips of the developing lung [11,13,50,51]. Our data indicate that cystic-like structures result from defective RA signaling impairing lung branching. Conversely, and as previously demonstrated, RA signaling stimulation increases lung branching morphogenesis with proper morphology [13,32,33].

After establishing the RA/BMS experimental model, we explored the metabolic alterations induced by RA signaling modulation on lung branching morphogenesis. We took advantage of ex vivo lung explant culture to study the whole organ and precisely assess metabolic and associated molecular alterations in a controlled environment [20,52,53]. Our work revealed that RA signaling stimulation is associated with less glucose consumption than BMS-treated lungs despite increasing lung branching. Thus, RA signaling stimulation promotes a better use of glucose or alternative metabolic substrate utilization. Additionally, other glycolysis-related products were studied, revealing that RA signaling stimulation increases pyruvate production and decreases lactate production in the embryonic lung. Alanine production follows the same tendency as glucose consumption. These findings point to an increase in the amount of glucose directed into pyruvate rather than into alanine or lactate under RA signaling stimulation. Moreover, this also suggests an upregulation of the glycolytic pathway under RA stimulation conditions and without variations in *pfk1* at the transcript level. However, PFK-1 can be modulated by many metabolites, including fructose-2,6-bisphosphate, ATP, AMP, and hormones, potentially suggesting an indirect regulation by RA at the protein/enzyme activity level [39].

Both *ldha* and *ldhb* are downstream targets of the RA signaling cascade. At the protein level, we have noticed the same tendency for LDHA and LDHB but without pronounced differences, which can be explained since we used whole lung tissues rather than isolated tissues from specific regions in the Western blot. Such findings show that RA signaling modulates LDH reaction, namely at active branching sites.

In the ^{1}H-NMR spectra, no changes were observed in acetate production, and citrate was not detected. However, there was a significant elevation in succinate levels, which could have had significant effects on various cellular events. This increase in succinate suggests that it may serve as a metabolic signaling molecule, triggering important signaling pathways within the cells [41,42].

In this report, we also showed that proliferation regions are associated with active branching sites and expand substantially upon RA signaling stimulation since this condition increases lung branching. On the contrary, the BMS-treated lungs displayed decreased branching and cystic morphology and were associated with less proliferation overall. Notably, the high proliferation sites match *ldhb* spatial localization upon RA signaling stimulation.

In order to test if RA signaling stimulation promotes pyruvate production to supply mitochondria, we assessed mitochondrial respiration. Curiously, RA signaling stimulation presented an overall respiration profile similar to the control. However, BMS-induced inhibition modulated mitochondria. BMS treatment promoted mitochondrial function, inducing a more oxidative metabolism without influencing mitochondrial biogenesis.

During lung branching morphogenesis, the AMPK pathway is activated by RA signaling under stimulation and inhibition conditions. While AMPK stimulation could be beneficial to a certain extent, in the case of RA signaling stimulation, an overactivation of AMPK by RA signaling inhibition might be related to a compensatory mechanism used to re-establish tissue homeostasis. Similarly, RA inhibits neointimal hyperplasia and suppresses vascular smooth muscle cell proliferation and migration through AMPK signaling activation [54]. Moreover, RA activates the AMPK signaling pathway and sensitizes hepatocellular carcinoma cells to apoptosis induced by sorafenib [55]. RA also activates AMPK in skeletal muscle cells [56]. Our analysis revealed a molecular mechanism in which RA signaling modulation activates the AMPK pathway, downregulating the downstream target *srebf1*. Consequently, the downregulation of *srebf1* decreases the expression of *fasn*, and fatty acid synthesis might be reduced (Fig. 7E). Likewise, in human liver cells, AMPK activation is associated with SREBP-1c inhibition [57]. SREBP influences genes that control epithelial development, proliferation, and cell death, and plays a role in controlling the lung lipid transcriptional network [58]. Moreover, RA treatment inhibits lipid biosynthesis in mice livers [45]. RA signaling does not affect lung branching fatty acid oxidation, which is in contrast to what occurs in other systems [45–47].

In conclusion, this study describes the metabolic changes produced by RA signaling modulation on lung branching morphogenesis. RA signaling inhibition decreases lung branching and induces a cystic-like phenotype that is accompanied by an increase in mitochondrial function. Since lung branching relies on glycolytic lactate-based metabolism [20], the opposite metabolic profile, the OXPHOS-based BMS-induced metabolism, might explain the observed disease phenotype. On the other hand, RA signaling stimulation increased lung branching while maintaining proper morphology. RA signaling stimulation required less glucose consumption and produced less lactate. Nonetheless, RA stimulation displayed a similar OCR mitochondrial profile as the control lungs. Such data support the extracellular accumulation of pyruvate and succinate, which are not used to fuel OXPHOS. Still, pyruvate and succinate might exert additional metabolic or signaling roles (Figure 8). We hypothesize that RA signaling stimulation might promote optimal growth conditions, while RA signaling inhibition promotes a less efficient metabolism for branching morphogenesis. Moreover, an RA-AMPK-dependent molecular mechanism seems to regulate lipid synthesis in the embryonic lung.

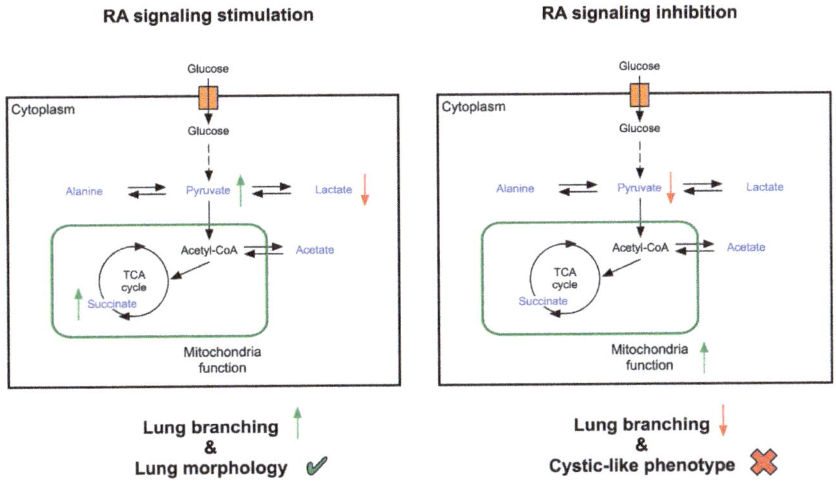

Figure 8. Schematic representation of the metabolic profile of early lung branching morphogenesis upon RA signaling stimulation vs. inhibition. Left panel: Upon RA signaling stimulation there is a metabolic remodeling compared to control lungs. There is a decrease in lactate production and an extracellular accumulation of pyruvate and succinate, which are not used to fuel mitochondria but might exert additional metabolic or signaling roles. In this condition, lung branching morphogenesis is increased and lung tissues display proper lung morphology. Right panel: RA signaling inhibition decreases pyruvate production and increases mitochondrial function, resulting in a more OXPHOS-based metabolic profile. In this condition, lung branching morphogenesis is decreased and lung tissues display a cystic-like phenotype. The symbols refer to increase (↑), decrease (↓), proper lung morphology (✔), or impaired lung morphology (X).

4. Materials and Methods

4.1. Ethical Statement

Under the European Parliament Directive 2010/63/EU of 22 September 2010 and the Portuguese Directive 113/2013 of 7 August 2013 on the protection of animals used for scientific purposes, no ethical approval was required to carry out this work, which was performed at the early stages of chicken embryonic development.

4.2. Tissue Collection

Fertilized chicken eggs, *Gallus gallus*, were incubated between 4.5 and 5.5 days (Embryonic day 4.5–5.5) in an incubator with a 49% humidified atmosphere at 37 °C (Termaks KB400, Fjärås, Sweden). Stage b2 lungs (two secondary buds formed per bronchus) were micro-dissected using a stereomicroscope (Olympus SZX16, Tokyo, Japan) [11]. The dissected lung tissues were processed for ex vivo lung explant culture.

4.3. Ex Vivo Lung Explant Culture

Lungs were dissected in PBS and placed on 8 µm nucleopore polycarbonate membranes (Whatman, Marlborough, MA, USA). The lung explants were cultured in 200 µL of medium 199 containing 5.5 mM glucose (Sigma, St Louis, MI, USA),supplemented with 1% (*v*/*v*) L-glutamine (Invitrogen, Waltham, MA, USA), 0.25 mg/mL of ascorbic acid (Sigma), 5% (*v*/*v*) heat-inactivated fetal calf serum (Invitrogen), 10% (*v*/*v*) chick serum (Invitrogen), and 1% (*v*/*v*) penicillin 5000 IU/mL plus streptomycin 5000 IU/mL (Invitrogen). The lung explants were exposed to increasing doses of BMS (BMS493, Sigma): 0.1 µM, 1 µM, or 10 µM; or to a different experimental setting with 1 µM of RA (Sigma) or 10 µM of BMS. DMSO 0.1% was used as the control. The lung explants were incubated for 48 h at 37 °C with 5% CO_2 (Heraeus HeraCell CO_2 incubator, Hanau, Germany). At 24 h, the culture medium was replaced by a fresh supplemented medium. The lung explants were photographed at 0 h (D0), 24 h (D1), and 48 h (D2) (Olympus U-LH100HG coupled to Olympus SZX16) and then morphometrically analyzed (AxioVision, Carl Zeiss Microscopy, Oberkochen, Germany). Media samples were collected at D0, D1, and D2 for ^{1}H-NMR spectroscopy. D2 tissues were collected for RNA, DNA, and protein extraction. D2 explants were also collected for in situ hybridization, proliferation assay, and seahorse analysis.

4.4. RNA Probes

Total RNA was extracted from D2 lung explants using the TripleXtractor directRNA kit (Grisp, Porto, Portugal). Subsequently, 1 µg of RNA was treated with DNase I (Thermo Fisher Scientific, Waltham, MA, USA) and reverse transcribed into cDNA using the GRS cDNA Synthesis kit (Grisp). *rarβ* [59], *ldha*, and *ldhb* [20] RNA probes were produced as previously described. Antisense digoxigenin-labeled RNA probes were produced using T3 (*rarβ*) or SP6 (*ldha* and *ldhb*) RNA polymerase, according to the manufacturer's instructions (Roche, Mannheim, Germany).

4.5. Whole-Mount In Situ Hybridization

After explant culture, the lungs were fixed in PBS solution with 4% formaldehyde and 2 mM EGTA, pH 7.5, at 4 °C overnight. Afterward, the lung explants were dehydrated in a methanol series and stored at −20 °C. The tissues were rehydrated in a methanol/PBT series and processed for whole-mount in situ hybridization ($n \geq 4$ per gene/condition) [60]. The tissues were permeabilized with proteinase K solution (PBT with 0.05% proteinase K) (Roche). Next, the tissues were incubated in a post-fixing solution (PBT with 10% formaldehyde and 0.4% glutaraldehyde). Subsequently, the tissues were incubated with hybridization solution containing 50% formamide; 6.5% SSC; 1% EDTA, 0.5 M, pH 9.8; 0.5% CHAPS; 0.25% t-RNA; 0.2% heparin; and 0.2% Tween 20; at 70 °C. Then, the tissues were incubated overnight with specific RNA probes in the hybridization solution at 70 °C. The next day, washes were performed with preheated hybridization solution, hybridization solution with MABT (50:50) (5.8% $C_4H_4O_4$, 4.4% NaCl, 7% NaOH, 1% Tween 20, pH 7.5), and MABT. Next, the tissues were treated with blocking solutions [MABT with 20% blocking reagent (Roche); MABT with 20% blocking reagent plus 20% goat serum (Invitrogen)]. Then, the lungs were incubated in MABT, 20% goat serum, 20% blocking reagent, and anti-digoxigenin antibody (1:2000) (Roche) solution overnight. We committed day 3 to performing MABT solution washes. On the last day, the tissues were washed in NTMT solution (0.1 M Tris-HCl, 0.1 M NaCl, 50 mM $MgCl_2$, 1% Tween 20) and then incubated in a developing solution (NTMT with BCIP and NBT) (Roche), protected from light, at 37 °C. The reaction was stopped at the same time for each group of lungs/probes. Lastly, the lung explants were photographed (Olympus U-LH100HG coupled to Olympus SZX16).

4.6. ^{1}H-NMR Spectroscopy

Explant culture media samples (200 µL) were collected at D0, D1, and D2 and analyzed by using ^{1}H-NMR spectroscopy ($n \geq 8$/condition) according to [61]. Spectra were accessed at 25 °C by using a Bruker Avance 600 MHz spectrometer with a 5 mm QXI probe and z-gradient (Bruker Biospin, Ettlingen, Germany). Solvent-suppressed ^{1}H-NMR spectra were acquired with 6 kHz spectral width, 14 s inter-pulse, 3 s water pre-saturation, 45-degree pulse angle, 3.5 s acquisition time, and 128 scans (minimum). Sodium fumarate 10 mM (singlet, at 6.50 ppm) was used as an internal reference. The following metabolites were detected and quantified: H1-α glucose (doublet, 5.22 ppm), pyruvate (singlet, 2.35 ppm), alanine (doublet, 1.46 ppm), lactate (doublet, 1.33 ppm), acetate (singlet, 1.9 ppm), and succinate (singlet, 2.39 ppm). The relative areas of ^{1}H-NMR resonances were quantified by using the NUTSpro^{TM} NMR spectral analysis program (Acorn NMR, Livermore, CA, USA). D0 media samples were used as the reference/control. Metabolite consumption or production was calculated using the mathematical formula $|(D1-D0) + (D2-D0)|$ [20] and normalized to the total amount of protein.

4.7. Quantitative PCR

Total RNA and cDNA were obtained from stage b2 lungs as previously described. A qPCR method was performed as described in [20]. Specific exon–exon spanning primers were designed for the amplification of the targets (*pfk1, g6pd, pgd, tfam, srebf1, fasn,* and *cpt1*) and housekeeping transcripts (*18s* and *actin-β*) (Table S1). Primers were optimized for annealing temperature and PCR cycles by using NZY Taq 2x Green Master Mix (NZYTech, Lisboa, Portugal). Afterward, primers were optimized for the efficiency range. Each qPCR was performed in duplicate using 1 μL of cDNA and the SYBR method with the NZY qPCR Green Master Mix (2x) (NZYTech). The *18s* and *actin-β* housekeeping genes were used to normalize the mRNA expression levels. Data on the relative expression levels were calculated using the mathematical model $2^{(-\Delta Ct)}$ [62].

4.8. Western Blot

Pooled samples of 10 lungs/pool of D2 lung explants were processed for Western blot as described in [12]. Protein was extracted and quantified according to [63]. Then, 40 μg of protein was loaded onto 10% acrylamide minigels and electrophoresed at 100 V in a Mini-PROTEAN Tetra Cell (Bio-Rad, Hercules, CA, USA). Blotting was performed using low-fluorescence PVDF membranes (Bio-Rad) and a Trans-Blot Turbo Transfer System (Bio-Rad). After that, membranes were incubated with AzureRed Fluorescent Total Protein Stain (Azure Biosystems, Dublin, CA, USA) according to the manufacturer's instructions. Immunoblots were probed with primary antibodies for LDHA (1:40,000; #3582, Cell Signaling, Danvers, MA, USA), LDHB (1:10,000; #ab240482, Abcam, Cambridge, UK), AMPKα (1:2000; #2532, Cell Signaling), and Phospho-AMPKα (Thr172) (1:2000; #2531, Cell Signaling). Subsequently, the blots were incubated with anti-rabbit IgG HRP-linked secondary antibody (1:5000; #7074, Cell Signaling) or anti-goat secondary IgG (H+L) HRP cross-adsorbed antibody (1:5000; #R-21459, Invitrogen). The membranes were developed with Clarity or Clarity Max Western ECL substrate (Bio-Rad). To capture the chemiluminescent signal, a Sapphire Biomolecular Imager (Azure Biosystems) was used. Western blot quantifications were performed using AzureSpot Analysis Software (Version 2.2.167) (Azure Biosystems) and normalized to the total protein. Two or more independent experiments were performed per pool of tissue ($n \geq 3$/condition).

4.9. Proliferation Assay and Confocal Microscopy

A proliferation assay ($n \geq 4$/condition) was performed as described in [20]. After 48 h of lung explant culture, half of the explant's media was replaced by fresh media containing EdU (150 μM final concentration). Explants were incubated with the EdU solution for 90 min at 37°C with 5% CO_2 (Heraeus HeraCell CO_2 incubator). Then, the tissues were fixed in PBS with 3.7% formaldehyde. Afterward, the tissues were washed in PBS with 3% BSA and permeabilized in PBS with 0.5% Triton X-100 for 90 min. The issues were washed and processed for the Click-iT Plus EdU reaction according to the manufacturer's instructions (Click-iT™ Plus EdU Cell Proliferation Kit for Imaging, Invitrogen). Alexa Fluor 488 was used to detect EdU, and the nuclei were counterstained with Hoechst 33342 (1:2000). Image acquisition was performed by using an Olympus LPS Confocal FV3000 microscope (Olympus).

4.10. Seahorse Analysis

After 48 h of in vitro lung explant culture, D2 lungs were processed for real-time measurement of oxygen consumption rate (OCR) ($n \geq 13$/condition) using a Seahorse XFe24 (Agilent, Santa Clara, CA, USA). The seahorse Mito Stress Test was performed according to the manufacturer's instructions (Agilent, USA). On the previous day, the Seahorse sensor cartridge (Agilent) was hydrated overnight in Seahorse XF calibrant (Agilent) in a non-CO_2 incubator with humidity at 37 °C. On the protocol day, the Seahorse assay medium was freshly prepared with medium 199 without phenol red (Sigma) and supplemented with 1 mM pyruvate (Sigma) and 1% (v/v) L-glutamine (Invitrogen), pH

7.3, and warmed at 37 °C until use. Drugs were freshly prepared in the Seahorse assay medium and sequentially loaded into the sensor cartridge injection ports with oligomycin 15 μM (Sigma), FFCP 20 μM (Sigma), and rotenone/antimycinn A 8 μM (Sigma) as the final concentrations. After the explant culture, D2 lungs were washed in PBS and placed on the Seahorse assay media. Islet capture microplates (Agilent) were prepared with 500 μL of the Seahorse assay medium; then, D2 tissues were placed on the wells' inner depression, and the islet capture screens (Agilent) were placed into the wells. After preparation, both the islet capture microplates (tissues) and the Seahorse sensor cartridge (drugs) were pre-warmed and calibrated at 37 °C. The Seahorse protocol cycles (5;5;5;8) and moment of injections are represented in Figure S3. Each OCR measurement was performed with 3 min of mixing, 2 min of waiting, and 3 min of measuring. In the end, tissues were collected, washed in PBS several times, and processed for protein extraction and quantification. OCR data (pmol/min) were normalized to the total amount of protein.

4.11. mtDNA Copy Number

A qPCR method was performed to evaluate the mtDNA copy number as described by [64], with minor modifications. Total DNA was extracted from D2 lungs using a GRS Genomic DNA kit (Grisp). DNA integrity was assessed by using electrophoresis in a 0.6% agarose gel. Specific primers were designed for mitochondrial NADH dehydrogenase 1 (*nd1*) and Nuclear Angiotensin II receptor type 1 (*agrt1*), and the primers were produced according to [65] (Table S1). The primers were optimized for annealing temperature and PCR cycles by using NZY Taq 2×Green Master Mix (NZYTech). Then, the primers were optimized for the efficiency range. qPCR was performed in duplicate ($n \geq 8$/condition), using 20 ng of DNA per reaction, and the SYBR method was performed by utilizing the NZY qPCR Green Master Mix (2x) (NZYTech). Ct value differences between *nd1* and *agrt1* were used to quantify the mtDNA copy number using the mathematical model 2^(-ΔCt) [62].

4.12. Statistical Analysis

Statistical analysis was performed using GraphPad Prism 8 (GraphPad Software, Boston, MA, USA). The normality of distribution was tested using the Kolmogorov–Smirnov test. One-Way ANOVA was performed and followed by a post hoc Fisher's Least Significant Difference (LSD) test for multiple comparisons. All the data are presented as mean ± standard error of the mean (SEM) with a statistical significance level of 5% ($p < 0.05$).

5. Conclusions

RA signaling modulation induces metabolic alterations at the transcript, protein, and metabolite levels. This report unveils novel insights into the signaling–metabolism interaction during embryonic lung development and highlights the importance of metabolic regulation in this phase. Furthermore, our data may contribute to understanding the etiology of congenital lung disorders, namely cystic-related disorders. This is a new and unexplored topic, and several questions requiring additional mechanistic understanding have been raised. Still, this study lays the groundwork for future studies in this emerging field.

Supplementary Materials: The following supporting information can be downloaded at https://www.mdpi.com/article/10.3390/ijms25095054/s1, Figure S1: Dose-dependent effect of BMS treatment on RA signaling pathway and lung branching morphogenesis; Figure S2: LDHA and LDHB full-length blots and total protein; Figure S3: Seahorse OCR profile; Figure S4: Mitochondrial biogenesis in lung branching morphogenesis; Figure S5: AMPK and pAMPK full-length blots and total protein; Table S1: Primers and qPCR conditions.

Author Contributions: Conceptualization, H.F.-S., M.G.A., P.F.O., C.C.F.H. and R.S.M.; Formal analysis, H.F.-S., M.G.A., P.F.O., C.C.F.H. and R.S.M. Funding Acquisition, J.C.-P., C.C.F.H. and R.S.M.; Investigation, H.F.-S., M.R.G. and R.S.M.; Methodology, H.F.-S., M.G.A., P.F.O., C.C.F.H. and R.S.M.; Project Administration, J.C.-P., C.C.F.H. and R.S.M.; Resources, J.C.-P., M.G.A., P.F.O., C.C.F.H. and R.S.M.; Supervision, C.C.F.H. and R.S.M.; Validation, H.F.-S., M.G.A., P.F.O., C.C.F.H.

and R.S.M.; Visualization, H.F.-S., C.C.F.H. and R.S.M.; Writing—Original Draft, H.F.-S. and R.S.M.; Writing—Review and Editing, H.F.-S., C.C.F.H. and R.S.M. All authors have read and agreed to the published version of the manuscript.

Funding: This work has been funded by National funds, through the Foundation for Science and Technology (FCT)–project UIDB/50026/2020 (DOI 10.54499/UIDB/50026/2020), UIDP/50026/2020 (DOI 10.54499/UIDP/50026/2020) and LA/P/0050/2020 (DOI 10.54499/LA/P/0050/2020); by ICVS Scientific Microscopy Platform, member of the national infrastructure PPBI–Portuguese Platform of Bioimaging (PPBI-POCI-01-0145-FEDER-022122); and by the projects NORTE-01-0145-FEDER-000013 and NORTE-01-0145-FEDER-000023, supported by Norte Portugal Regional Operational Programme (NORTE 2020), under the PORTUGAL 2020 Partnership Agreement, through the European Regional Development Fund (ERDF). This work was also supported by the European Research Council (ERC) under the European Union's Horizon 2020 research and innovation programme (H2020-ERC-2017-STG-GA 759853-StemCellHabitat). Hugo Fernandes-Silva was supported by a doctoral fellowship (PD/BD/137655/2018) from FCT as part of the Inter-University Doctoral Programme in Ageing and Chronic Disease (PhDOC).

Institutional Review Board Statement: Not applicable.

Informed Consent Statement: Not applicable.

Data Availability Statement: Data are contained within the article.

Acknowledgments: The authors would like to thank Goreti Pinto for the confocal microscopy support.

Conflicts of Interest: The authors declare no conflicts of interest.

References

1. Caldeira, I.; Fernandes-Silva, H.; Machado-Costa, D.; Correia-Pinto, J.; Moura, R.S. Developmental Pathways Underlying Lung Development and Congenital Lung Disorders. *Cells* **2021**, *10*, 2987. [CrossRef] [PubMed]
2. Schittny, J.C. Development of the lung. *Cell Tissue Res.* **2017**, *367*, 427–444. [CrossRef] [PubMed]
3. Harris, K.S.; Zhang, Z.; McManus, M.T.; Harfe, B.D.; Sun, X. Dicer function is essential for lung epithelium morphogenesis. *Proc. Natl. Acad. Sci. USA* **2006**, *103*, 2208–2213. [CrossRef] [PubMed]
4. Gonzaga, S.; Henriques-Coelho, T.; Davey, M.; Zoltick, P.W.; Leite-Moreira, A.F.; Correia-Pinto, J.; Flake, A.W. Cystic adenomatoid malformations are induced by localized FGF10 overexpression in fetal rat lung. *Am. J. Respir. Cell Mol. Biol.* **2008**, *39*, 346–355. [CrossRef] [PubMed]
5. Maina, J.N. A systematic study of the development of the airway (bronchial) system of the avian lung from days 3 to 26 of embryogenesis: A transmission electron microscopic study on the domestic fowl, Gallus gallus variant domesticus. *Tissue Cell* **2003**, *35*, 375–391. [CrossRef] [PubMed]
6. Kim, H.Y.; Varner, V.D.; Nelson, C.M. Apical constriction initiates new bud formation during monopodial branching of the embryonic chicken lung. *Development* **2013**, *140*, 3146–3155. [CrossRef] [PubMed]
7. Maina, J.N. Comparative molecular developmental aspects of the mammalian- and the avian lungs, and the insectan tracheal system by branching morphogenesis: Recent advances and future directions. *Front. Zool.* **2012**, *9*, 16. [CrossRef]
8. Metzger, R.J.; Klein, O.D.; Martin, G.R.; Krasnow, M.A. The branching programme of mouse lung development. *Nature* **2008**, *453*, 745–750. [CrossRef] [PubMed]
9. Moura, R.S.; Correia-Pinto, J. Molecular aspects of avian lung development. In *The Biology of the Avian Respiratory System*; Maina, J.N., Ed.; Springer: Cham, Switzerland, 2017; pp. 129–146.
10. Moura, R.S.; Silva-Goncalves, C.; Vaz-Cunha, P.; Correia-Pinto, J. Expression analysis of Shh signaling members in early stages of chick lung development. *Histochem. Cell Biol.* **2016**, *146*, 457–466. [CrossRef]
11. Moura, R.S.; Coutinho-Borges, J.P.; Pacheco, A.P.; Damota, P.O.; Correia-Pinto, J. FGF signaling pathway in the developing chick lung: Expression and inhibition studies. *PLoS ONE* **2011**, *6*, e17660. [CrossRef]
12. Moura, R.S.; Carvalho-Correia, E.; daMota, P.; Correia-Pinto, J. Canonical Wnt signaling activity in early stages of chick lung development. *PLoS ONE* **2014**, *9*, e112388. [CrossRef] [PubMed]
13. Fernandes-Silva, H.; Vaz-Cunha, P.; Barbosa, V.B.; Silva-Goncalves, C.; Correia-Pinto, J.; Moura, R.S. Retinoic acid regulates avian lung branching through a molecular network. *Cell. Mol. Life Sci.* **2017**, *74*, 4599–4619. [CrossRef] [PubMed]
14. Cunningham, T.J.; Duester, G. Mechanisms of retinoic acid signalling and its roles in organ and limb development. *Nat. Rev. Mol. Cell Biol.* **2015**, *16*, 110–123. [CrossRef] [PubMed]
15. Morrisey, E.E.; Hogan, B.L. Preparing for the first breath: Genetic and cellular mechanisms in lung development. *Dev. Cell* **2010**, *18*, 8–23. [CrossRef] [PubMed]
16. Marquez, H.A.; Cardoso, W.V. Vitamin A-retinoid signaling in pulmonary development and disease. *Mol. Cell. Pediatr.* **2016**, *3*, 28. [CrossRef] [PubMed]
17. Ghyselinck, N.B.; Duester, G. Retinoic acid signaling pathways. *Development* **2019**, *146*, dev167502. [CrossRef] [PubMed]

18. Fernandes-Silva, H.; Araujo-Silva, H.; Correia-Pinto, J.; Moura, R.S. Retinoic Acid: A Key Regulator of Lung Development. *Biomolecules* **2020**, *10*, 152. [CrossRef] [PubMed]
19. Marquez, H.A.; Chen, F. Retinoic Acid Signaling and Development of the Respiratory System. *Subcell. Biochem.* **2020**, *95*, 151–174. [CrossRef] [PubMed]
20. Fernandes-Silva, H.; Alves, M.G.; Araujo-Silva, H.; Silva, A.M.; Correia-Pinto, J.; Oliveira, P.F.; Moura, R.S. Lung branching morphogenesis is accompanied by temporal metabolic changes towards a glycolytic preference. *Cell Biosci.* **2021**, *11*, 134. [CrossRef]
21. Yeager, H., Jr.; Massaro, D. Glucose metabolism and glycoprotein synthesis by lung slices. *J. Appl. Physiol.* **1972**, *32*, 477–482. [CrossRef]
22. Fisher, A.B. Normal and pathologic biochemistry of the lung. *Environ. Health Perspect.* **1976**, *16*, 3–9. [CrossRef] [PubMed]
23. Fisher, A.B. Intermediary metabolism of the lung. *Environ. Health Perspect.* **1984**, *55*, 149–158. [CrossRef] [PubMed]
24. Krejci, A.; Tennessen, J.M. Metabolism in time and space—Exploring the frontier of developmental biology. *Development* **2017**, *144*, 3193–3198. [CrossRef] [PubMed]
25. Miyazawa, H.; Aulehla, A. Revisiting the role of metabolism during development. *Development* **2018**, *145*, dev131110. [CrossRef] [PubMed]
26. Teleman, A.A. Metabolism meets development at Wiston House. *Development* **2016**, *143*, 3045–3049. [CrossRef]
27. Cable, J.; Pourquie, O.; Wellen, K.E.; Finley, L.W.S.; Aulehla, A.; Gould, A.P.; Teleman, A.; Tu, W.B.; Garrett, W.S.; Miguel-Aliaga, I.; et al. Metabolic decisions in development and disease—A Keystone Symposia report. *Ann. N. Y. Acad. Sci.* **2021**, *1506*, 55–73. [CrossRef] [PubMed]
28. Slaninova, V.; Krafcikova, M.; Perez-Gomez, R.; Steffal, P.; Trantirek, L.; Bray, S.J.; Krejci, A. Notch stimulates growth by direct regulation of genes involved in the control of glycolysis and the tricarboxylic acid cycle. *Open Biol.* **2016**, *6*, 150155. [CrossRef] [PubMed]
29. Nellas, I.; Iyer, K.V.; Iglesias-Artola, J.M.; Pippel, M.; Nadler, A.; Eaton, S.; Dye, N.A. Hedgehog signaling can enhance glycolytic ATP production in the Drosophila wing disc. *EMBO Rep.* **2022**, *23*, e54025. [CrossRef]
30. Kuwabara, S.; Yamaki, M.; Yu, H.; Itoh, M. Notch signaling regulates the expression of glycolysis-related genes in a context-dependent manner during embryonic development. *Biochem. Biophys. Res. Commun.* **2018**, *503*, 803–808. [CrossRef]
31. Oginuma, M.; Moncuquet, P.; Xiong, F.; Karoly, E.; Chal, J.; Guevorkian, K.; Pourquie, O. A Gradient of Glycolytic Activity Coordinates FGF and Wnt Signaling during Elongation of the Body Axis in Amniote Embryos. *Dev. Cell* **2017**, *40*, 342–353. [CrossRef]
32. Pereira-Terra, P.; Moura, R.S.; Nogueira-Silva, C.; Correia-Pinto, J. Neuroendocrine factors regulate retinoic acid receptors in normal and hypoplastic lung development. *J. Physiol.* **2015**, *593*, 3301–3311. [CrossRef] [PubMed]
33. Schuger, L.; Varani, J.; Mitra, R., Jr.; Gilbride, K. Retinoic acid stimulates mouse lung development by a mechanism involving epithelial-mesenchymal interaction and regulation of epidermal growth factor receptors. *Dev. Biol.* **1993**, *159*, 462–473. [CrossRef] [PubMed]
34. Germain, P.; Gaudon, C.; Pogenberg, V.; Sanglier, S.; Van Dorsselaer, A.; Royer, C.A.; Lazar, M.A.; Bourguet, W.; Gronemeyer, H. Differential action on coregulator interaction defines inverse retinoid agonists and neutral antagonists. *Chem. Biol.* **2009**, *16*, 479–489. [CrossRef] [PubMed]
35. Takayama, M.; Miyatake, K.; Nishida, E. Identification and characterization of retinoic acid-responsive genes in mouse kidney development. *Genes Cells* **2014**, *19*, 637–649. [CrossRef] [PubMed]
36. Nadendla, E.; Teyssier, C.; Delfosse, V.; Vivat, V.; Krishnasamy, G.; Gronemeyer, H.; Bourguet, W.; Germain, P. An Unexpected Mode of Binding Defines BMS948 as a Full Retinoic Acid Receptor beta (RARbeta, NR1B2) Selective Agonist. *PLoS ONE* **2015**, *10*, e0123195. [CrossRef] [PubMed]
37. Elgamal, R.M.; Bell, G.I.; Krause, S.C.T.; Hess, D.A. BMS 493 Modulates Retinoic Acid-Induced Differentiation During Expansion of Human Hematopoietic Progenitor Cells for Islet Regeneration. *Stem Cells Dev.* **2018**, *27*, 1062–1075. [CrossRef] [PubMed]
38. Mendelsohn, C.; Lohnes, D.; Decimo, D.; Lufkin, T.; LeMeur, M.; Chambon, P.; Mark, M. Function of the retinoic acid receptors (RARs) during development (II). Multiple abnormalities at various stages of organogenesis in RAR double mutants. *Development* **1994**, *120*, 2749–2771. [CrossRef] [PubMed]
39. Nelson, D.; Cox, M. *Lehninger Principles of Biochemistry*, 5th ed.; Worth Publishers: New York, NY, USA, 2008.
40. Vander Heiden, M.G.; Cantley, L.C.; Thompson, C.B. Understanding the Warburg effect: The metabolic requirements of cell proliferation. *Science* **2009**, *324*, 1029–1033. [CrossRef] [PubMed]
41. Murphy, M.P.; O'Neill, L.A.J. Krebs Cycle Reimagined: The Emerging Roles of Succinate and Itaconate as Signal Transducers. *Cell* **2018**, *174*, 780–784. [CrossRef]
42. Guo, Y.; Cho, S.W.; Saxena, D.; Li, X. Multifaceted Actions of Succinate as a Signaling Transmitter Vary with Its Cellular Locations. *Endocrinol. Metab.* **2020**, *35*, 36–43. [CrossRef]
43. Ventura-Clapier, R.; Garnier, A.; Veksler, V. Transcriptional control of mitochondrial biogenesis: The central role of PGC-1alpha. *Cardiovasc. Res.* **2008**, *79*, 208–217. [CrossRef] [PubMed]
44. Gureev, A.P.; Shaforostova, E.A.; Popov, V.N. Regulation of Mitochondrial Biogenesis as a Way for Active Longevity: Interaction Between the Nrf2 and PGC-1alpha Signaling Pathways. *Front. Genet.* **2019**, *10*, 435. [CrossRef] [PubMed]

45. Amengual, J.; Ribot, J.; Bonet, M.L.; Palou, A. Retinoic acid treatment enhances lipid oxidation and inhibits lipid biosynthesis capacities in the liver of mice. *Cell. Physiol. Biochem.* **2010**, *25*, 657–666. [CrossRef] [PubMed]
46. Amengual, J.; Ribot, J.; Bonet, M.L.; Palou, A. Retinoic acid treatment increases lipid oxidation capacity in skeletal muscle of mice. *Obesity* **2008**, *16*, 585–591. [CrossRef] [PubMed]
47. Amengual, J.; Garcia-Carrizo, F.J.; Arreguin, A.; Musinovic, H.; Granados, N.; Palou, A.; Bonet, M.L.; Ribot, J. Retinoic Acid Increases Fatty Acid Oxidation and Irisin Expression in Skeletal Muscle Cells and Impacts Irisin In Vivo. *Cell. Physiol. Biochem.* **2018**, *46*, 187–202. [CrossRef] [PubMed]
48. Garcia, D.; Shaw, R.J. AMPK: Mechanisms of Cellular Energy Sensing and Restoration of Metabolic Balance. *Mol. Cell* **2017**, *66*, 789–800. [CrossRef] [PubMed]
49. Herzig, S.; Shaw, R.J. AMPK: Guardian of metabolism and mitochondrial homeostasis. *Nat. Rev. Mol. Cell Biol.* **2018**, *19*, 121–135. [CrossRef] [PubMed]
50. Malpel, S.; Mendelsohn, C.; Cardoso, W.V. Regulation of retinoic acid signaling during lung morphogenesis. *Development* **2000**, *127*, 3057–3067. [CrossRef] [PubMed]
51. Bellusci, S.; Grindley, J.; Emoto, H.; Itoh, N.; Hogan, B.L. Fibroblast growth factor 10 (FGF10) and branching morphogenesis in the embryonic mouse lung. *Development* **1997**, *124*, 4867–4878. [CrossRef]
52. Carraro, G.; del Moral, P.M.; Warburton, D. Mouse embryonic lung culture, a system to evaluate the molecular mechanisms of branching. *J. Vis. Exp.* **2010**, *40*, e2035. [CrossRef]
53. Yeganeh, B.; Bilodeau, C.; Post, M. Explant Culture for Studying Lung Development. *Methods Mol. Biol.* **2018**, *1752*, 81–90. [CrossRef] [PubMed]
54. Zhang, J.; Deng, B.; Jiang, X.; Cai, M.; Liu, N.; Zhang, S.; Tan, Y.; Huang, G.; Jin, W.; Liu, B.; et al. All-Trans-Retinoic Acid Suppresses Neointimal Hyperplasia and Inhibits Vascular Smooth Muscle Cell Proliferation and Migration via Activation of AMPK Signaling Pathway. *Front. Pharmacol.* **2019**, *10*, 485. [CrossRef] [PubMed]
55. Ishijima, N.; Kanki, K.; Shimizu, H.; Shiota, G. Activation of AMP-activated protein kinase by retinoic acid sensitizes hepatocellular carcinoma cells to apoptosis induced by sorafenib. *Cancer Sci.* **2015**, *106*, 567–575. [CrossRef] [PubMed]
56. Lee, Y.M.; Lee, J.O.; Jung, J.H.; Kim, J.H.; Park, S.H.; Park, J.M.; Kim, E.K.; Suh, P.G.; Kim, H.S. Retinoic acid leads to cytoskeletal rearrangement through AMPK-Rac1 and stimulates glucose uptake through AMPK-p38 MAPK in skeletal muscle cells. *J. Biol. Chem.* **2008**, *283*, 33969–33974. [CrossRef] [PubMed]
57. Li, Y.; Xu, S.; Mihaylova, M.M.; Zheng, B.; Hou, X.; Jiang, B.; Park, O.; Luo, Z.; Lefai, E.; Shyy, J.Y.; et al. AMPK phosphorylates and inhibits SREBP activity to attenuate hepatic steatosis and atherosclerosis in diet-induced insulin-resistant mice. *Cell Metab.* **2011**, *13*, 376–388. [CrossRef] [PubMed]
58. Xu, Y.; Zhang, M.; Wang, Y.; Kadambi, P.; Dave, V.; Lu, L.J.; Whitsett, J.A. A systems approach to mapping transcriptional networks controlling surfactant homeostasis. *BMC Genom.* **2010**, *11*, 451. [CrossRef] [PubMed]
59. Michaille, J.J.; Kanzler, B.; Blanchet, S.; Garnier, J.M.; Dhouailly, D. Characterization of cDNAs encoding two chick retinoic acid receptor alpha isoforms and distribution of retinoic acid receptor alpha, beta and gamma transcripts during chick skin development. *Int. J. Dev. Biol.* **1995**, *39*, 587–596. [PubMed]
60. Moura, R.S. Retinoic Acid as a Modulator of Proximal-Distal Patterning and Branching Morphogenesis of the Avian Lung. In *Retinoid and Rexinoid Signaling*; Springer: New York, NY, USA, 2019; pp. 209–224. [CrossRef]
61. Alves, M.G.; Oliveira, P.F.; Martins, F.O.; Oliveira, P.J.; Carvalho, R.A. Gender-dependent metabolic remodeling during heart preservation in cardioplegic celsior and histidine buffer solution. *J. Cardiovasc. Pharmacol.* **2012**, *60*, 227–233. [CrossRef] [PubMed]
62. Livak, K.J.; Schmittgen, T.D. Analysis of relative gene expression data using real-time quantitative PCR and the 2(-Delta Delta C(T)) Method. *Methods* **2001**, *25*, 402–408. [CrossRef]
63. Kling, D.E.; Lorenzo, H.K.; Trbovich, A.M.; Kinane, T.B.; Donahoe, P.K.; Schnitzer, J.J. MEK-1/2 inhibition reduces branching morphogenesis and causes mesenchymal cell apoptosis in fetal rat lungs. *Am. J. Physiol. Lung Cell. Mol. Physiol.* **2002**, *282*, L370–L378. [CrossRef]
64. Martins, A.D.; Monteiro, M.P.; Silva, B.M.; Barros, A.; Sousa, M.; Carvalho, R.A.; Oliveira, P.F.; Alves, M.G. Metabolic dynamics of human Sertoli cells are differentially modulated by physiological and pharmacological concentrations of GLP-1. *Toxicol. Appl. Pharmacol.* **2019**, *362*, 1–8. [CrossRef] [PubMed]
65. Liu, R.; Jin, L.; Long, K.; Tang, Q.; Ma, J.; Wang, X.; Zhu, L.; Jiang, A.; Tang, G.; Jiang, Y.; et al. Analysis of mitochondrial DNA sequence and copy number variation across five high-altitude species and their low-altitude relatives. *Mitochondrial DNA B Resour.* **2018**, *3*, 847–851. [CrossRef] [PubMed]

International Journal of
Molecular Sciences

MDPI

Article

Perinatal Use of Citrulline Rescues Hypertension in Adult Male Offspring Born to Pregnant Uremic Rats

You-Lin Tain [1,2,3], Chih-Yao Hou [4], Guo-Ping Chang-Chien [5,6,7], Sufan Lin [5,6,7] and Chien-Ning Hsu [8,9,*]

1 Department of Pediatrics, Kaohsiung Chang Gung Memorial Hospital, Kaohsiung 833, Taiwan; tainyl@cgmh.org.tw
2 Institute for Translational Research in Biomedicine, Kaohsiung Chang Gung Memorial Hospital, Kaohsiung 833, Taiwan
3 College of Medicine, Chang Gung University, Taoyuan 330, Taiwan
4 Department of Seafood Science, National Kaohsiung University of Science and Technology, Kaohsiung 811, Taiwan; chihyaohou@webmail.nkmu.edu.tw
5 Institute of Environmental Toxin and Emerging-Contaminant, Cheng Shiu University, Kaohsiung 833, Taiwan; guoping@csu.edu.tw (G.-P.C.-C.); linsufan2003@gmail.com (S.L.)
6 Center for Environmental Toxin and Emerging-Contaminant Research, Cheng Shiu University, Kaohsiung 833, Taiwan
7 Super Micro Mass Research and Technology Center, Cheng Shiu University, Kaohsiung 833, Taiwan
8 School of Pharmacy, Kaohsiung Medical University, Kaohsiung 807, Taiwan
9 Department of Pharmacy, Kaohsiung Chang Gung Memorial Hospital, Kaohsiung 833, Taiwan
* Correspondence: cnhsu@cgmh.org.tw; Tel.: +886-975-368-975

Abstract: The growing recognition of the association between maternal chronic kidney disease (CKD) and fetal programming highlights the increased vulnerability of hypertension in offspring. Potential mechanisms involve oxidative stress, dysbiosis in gut microbiota, and activation of the renin–angiotensin system (RAS). Our prior investigation showed that the administration of adenine to pregnant rats resulted in the development of CKD, ultimately causing hypertension in their adult offspring. Citrulline, known for enhancing nitric oxide (NO) production and possessing antioxidant and antihypertensive properties, was explored for its potential to reverse high blood pressure (BP) in offspring born to CKD dams. Male rat offspring, both from normal and adenine-induced CKD models, were randomly assigned to four groups (8 animals each): (1) control, (2) CKD, (3) citrulline-treated control rats, and (4) citrulline-treated CKD rats. Citrulline supplementation successfully reversed elevated BP in male progeny born to uremic mothers. The protective effects of perinatal citrulline supplementation were linked to an enhanced NO pathway, decreased expression of renal (pro)renin receptor, and changes in gut microbiota composition. Citrulline supplementation led to a reduction in the abundance of *Monoglobus* and *Streptococcus* genera and an increase in *Agothobacterium Butyriciproducens*. Citrulline's ability to influence taxa associated with hypertension may be linked to its protective effects against maternal CKD-induced offspring hypertension. In conclusion, perinatal citrulline treatment increased NO availability and mitigated elevated BP in rat offspring from uremic mother rats.

Keywords: chronic kidney disease; citrulline; nitric oxide; developmental origins of health and disease (DOHaD); asymmetric dimethylarginine; gut microbiota; hypertension

1. Introduction

Hypertension affects one in three adults worldwide, and its roots may begin in early life [1,2]. Identifying and addressing hypertension early on could be a cost-effective strategy to globally reduce its burden. The connection between early-life environmental influences and later-life diseases is known as the "developmental origins of health and disease (DOHaD)" [3]. Adverse maternal conditions during gestation can impact fetal programming, potentially resulting in offspring hypertension [4]. Previous research demonstrated that pregnant rats fed adenine exhibited reduced kidney function along with glomerular and tubulointerstitial damage, hypertension, and increased uremic toxins [5]. These characteristics closely mirror the complex nature of human CKD. Maternal adenine-induced CKD affects fetal programming, leading to offspring hypertension. This hypertension is associated with deficient nitric oxide (NO) signaling, abnormal activation of the renin–angiotensin system (RAS), and alterations in the composition of gut microbiota [5].

Dietary antioxidants play a significant role in the treatment and prevention of various human diseases [6]. Citrulline, a non-essential amino acid initially identified in watermelon, has emerged as a potential antioxidant supplement that contributes to the improvement of cardiometabolic health [7,8]. Evidence suggests that citrulline derived from watermelon serves as an antioxidant by supporting the synthesis of NO along with arginine [9]. Orally ingested citrulline is absorbed by enterocytes and efficiently transported to the kidneys, where it is transformed to arginine. Citrulline supplementation offers pharmacokinetic advantages over arginine by bypassing hepatic first-pass metabolism, leading to increased NO production [10]. As NO is a well-known vasodilator, oral supplementation of arginine or citrulline is believed to effectively reduce blood pressure (BP) [11].

Evidence suggests that early-life oxidative stress increases the future risk of hypertension [12]. Conversely, the perinatal use of dietary antioxidants has been shown to protect adult offspring from hypertension in various animal models [13]. Previously, we found that supplementing citrulline during pregnancy in rats with NO deficiency improved offspring hypertension by enhancing NO production [14]. Nevertheless, the impact of perinatal citrulline supplementation on offspring hypertension programmed by maternal uremia is still unknown.

The objective of this study was to investigate the protective role of citrulline in hypertension with developmental origins. In vivo, this was accomplished by administering perinatal citrulline supplementation in a maternal uremia rat model to delve into potential underlying mechanisms, including NO, the RAS, and gut microbiota.

2. Results

2.1. Body Weight and BP

Figure 1 depicts that at 12 weeks of age, there were no discernible differences in body weight (Figure 1A) and the kidney weight-to-body weight ratio (Figure 1B) among the four groups. Additionally, the plasma creatinine levels, serving as an indicator of kidney function in the CKD group, were comparable to those in the remaining groups (Figure 1C). BP measurements were conducted longitudinally from weeks 3 to 12 (Figure 1D). Notably, during weeks 8–12, maternal CKD resulted in an increase in offspring's systolic BP, which was reversed by perinatal citrulline treatment (Figure 1D).

2.2. NO Pathway

Table 1 presents the findings regarding NO parameters in the plasma, including citrulline, arginine, asymmetric dimethylarginine (ADMA), and symmetric dimethylarginine (SDMA). Following perinatal citrulline supplementation, the citrulline concentration exhibited a significant increase in the NC group as opposed to the CKD and CKDC groups. Plasma concentrations of arginine were notably higher in the NC and CKDC groups when compared to the CKD group. The levels of ADMA and SDMA were elevated due to maternal CKD, and citrulline supplementation mitigated the increase in ADMA in the CKDC group. A noteworthy reduction in the arginine-to-ADMA ratio was observed in the CKD group, which was a trend that was counteracted by citrulline treatment (Table 1).

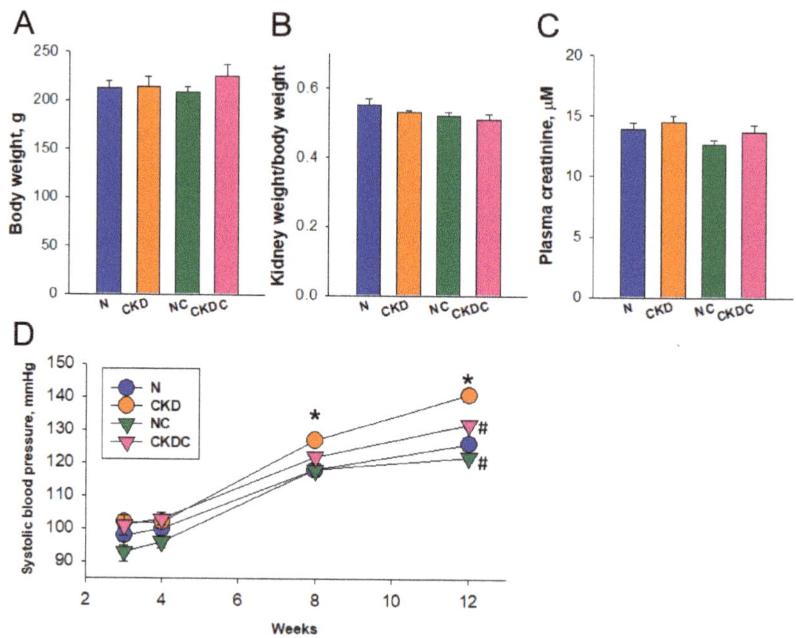

Figure 1. (**A**) Offspring body weight, (**B**) ratio of kidney weight-to-body weight, and (**C**) plasma creatinine concentration at 12 weeks of age. (**D**) Systolic blood pressure in offspring from 3 to 12 weeks of age with a sample size of n = 8 per group. * $p < 0.05$ vs. N; # $p < 0.05$ vs. CKD.

Table 1. NO parameters in the plasma of 12-week-old offspring.

Groups	N	CKD	NC	CKDC
Citrulline, μM	65.6 ± 3.1	63.4 ± 1.9	74.1 ± 3 #	63.3 ± 1.8 †
Arginine, μM	173.6 ± 15.3	154.3 ± 3.5	191.7 ± 7.6 #	172.4 ± 3.7 #†
ADMA, μM	2.04 ± 0.09	2.65 ± 0.09 *	1.81 ± 0.19 #	2.07 ± 0.04 #†
SDMA, μM	1.5 ± 0.08	2.16 ± 0.12 *	1.7 ± 0.16 #	1.91 ± 0.09
Ratio of arginine-to-ADMA	84.6 ± 5.3	57.7 ± 3.2 *	111.1 ± 7.5 *#	83.6 ± 2.9 #†

N = 8/group. * $p < 0.05$ vs. N; # $p < 0.05$ vs. CKD; † $p < 0.05$ vs. NC.

We next analyzed protein levels of dimethylarginine dimethylaminohydrolase-1 and -2 (DDAH1 and DDAH2; ADMA-metabolizing enzymes), endothelial NOS (eNOS), and neuronal NOS (nNOS) by Western blot. Their expression in the offspring's kidneys is illustrated in Figure 2. Maternal rats with CKD led to a decrease in renal protein levels of eNOS and nNOS, which is an effect that was prevented by maternal citrulline treatment (Figure 2B,C). Figure 2D,E show that the renal expression of ADMA-metabolizing enzymes DDAH-1 and -2 was comparable across the four experimental groups.

The findings suggest that maternal CKD hinders the NO pathway by reducing eNOS and nNOS protein levels, the ratio of arginine to ADMA, and by elevating ADMA and SDMA concentrations. Maternal citrulline supplementation appears to counteract these effects, restoring NO availability by increasing arginine, the ratio of arginine to ADMA, eNOS, and nNOS, and decreasing ADMA.

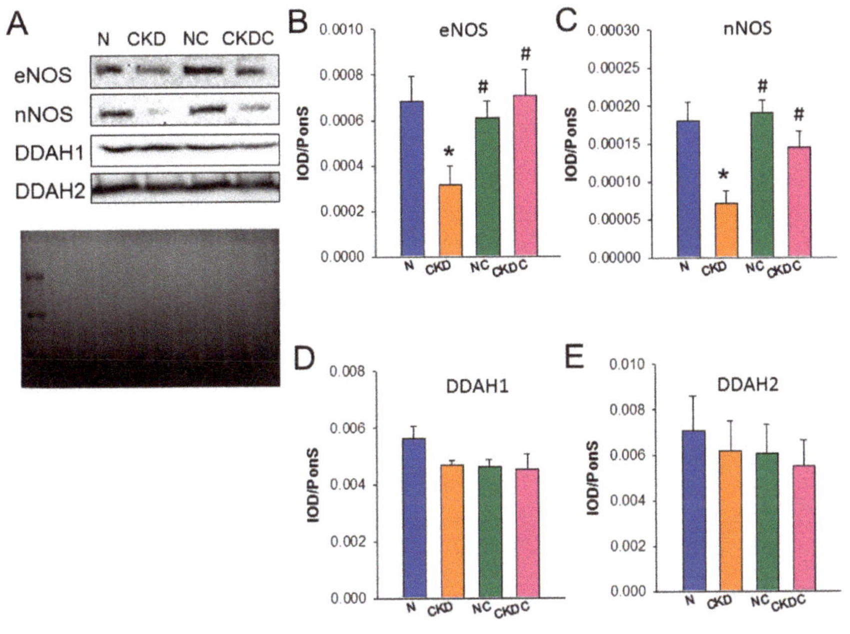

Figure 2. The results of Western blot analyses for (**A**) endothelial nitric oxide synthase (eNOS), neuronal NOS (nNOS), dimethylarginine dimethylaminohydrolase-1 (DDAH1), and -2 (DDAH2) in the offspring's kidneys, with Ponceau S staining employed as a loading control. The relative abundance of (**B**) eNOS, (**C**) nNOS, (**D**) DDAH1, and (**E**) DDAH2 was quantified and presented. * $p < 0.05$ vs. N; # $p < 0.05$ vs. CKD.

2.3. RAS

Quantitative real-time polymerase chain reaction (qPCR) was employed to analyze various components of the RAS system. The components assessed included (pro)renin receptor (PRR), renin, angiotensin-converting enzyme (ACE), angiotensinogen (AGT), and angiotensin II type 1 receptor (AT1R). Renal mRNA content of renin, PRR, AGT, ACE, and AT1R did not differ between the N and CKD groups (Figure 3). Among the four groups, CKDC rats exhibited the lowest renal expression of PRR.

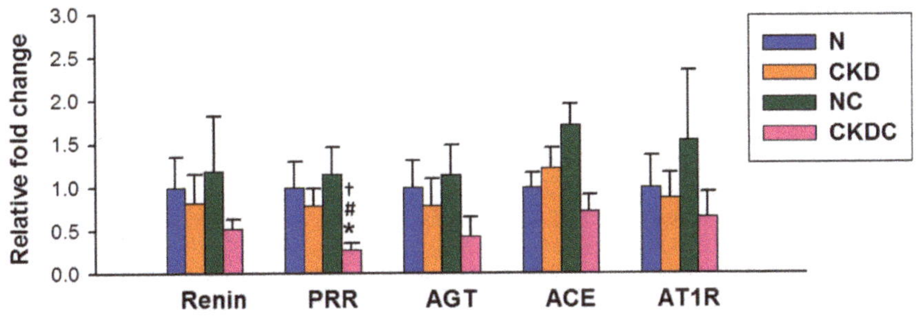

Figure 3. Renal mRNA expression of RAS components. * $p < 0.05$ vs. N; # $p < 0.05$ vs. CKD; † $p < 0.05$ vs. NC.

2.4. Gut Microbiota Composition

Alpha diversity, representing the species richness and evenness within each sample, was assessed using Pielou's evenness and the Shannon index. Both alpha-diversity indices exhibited no notable differences among the four groups (Figure 4A,B).

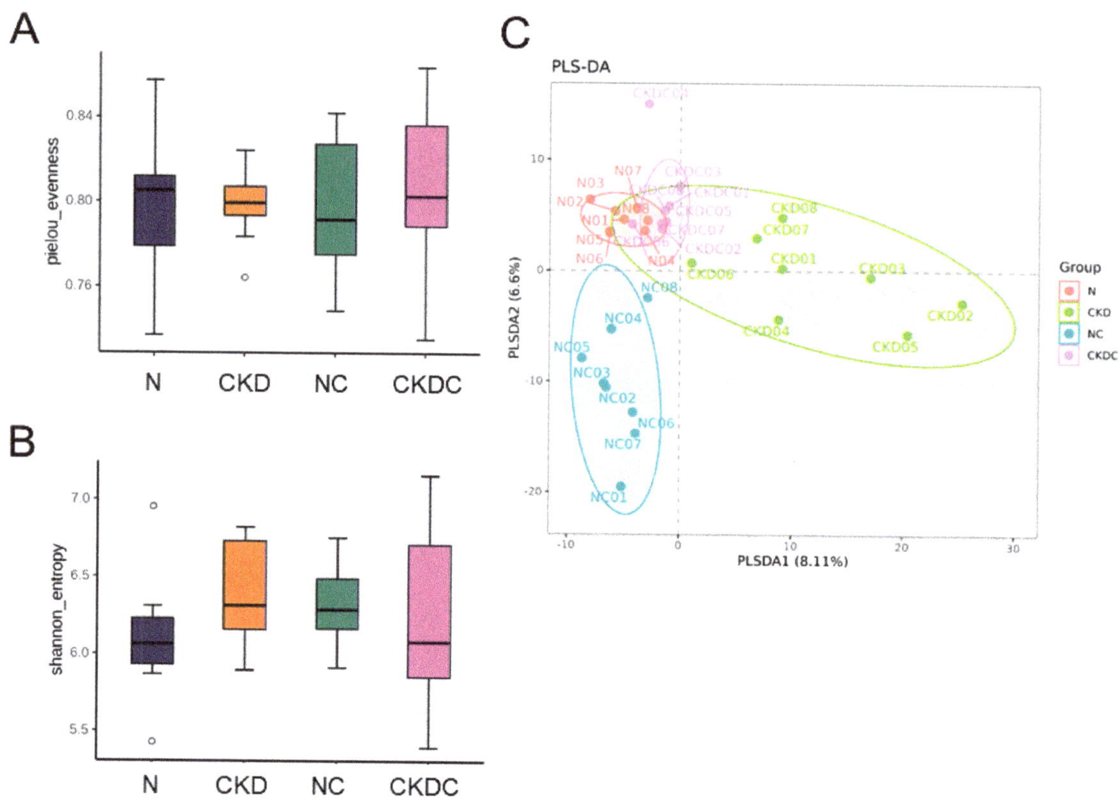

Figure 4. (**A**) Pielou's evenness and (**B**) Shannon index, illustrating alpha diversity across the four groups. (**C**) The Partial Least Squares Discriminant Analysis (PLS-DA) plots depict the clustering of fecal microbiota from the four groups. Each dot represents an individual's microbiota, with color indicating the respective group.

The Partial Least Squares Discriminant Analysis (PLS-DA) revealed distinct clustering of gut samples in the N group compared to the other groups. This indicated differences in gut microbiota between the N group and CKD group ($p = 0.001$ by ANOSIM), between the N group and NC group ($p = 0.02$ by ANOSIM), and between the N group and CKDC group ($p = 0.016$ by ANOSIM) (Figure 4C). However, when comparing the CKD group with the CKDC group, the observed differences did not reach statistical significance ($p = 0.069$ by ANOSIM).

To further assess the distinctions in gut microflora among the four groups, linear discriminant analysis effect size (LEfSe) analysis was conducted (Figure 5). CKD offspring rats exhibited a noteworthy rise in the abundance of genera *Turicibacter*, *Alistipes*, and *Neglectibacter*. Citrulline treatment led to an increased level of genera *Murimonas*, *Faecalimonas*, *Sinanaerobacter*, and *Breznakia* in the NC group. Additionally, Figure 5 highlighted that the genus *Peptococcus* was overrepresented in the CKDC group. Among these, certain taxa were found to be correlated with hypertension, including *Turicibacter*, *Alistipes*, *Faecalimonas*, and *Peptococcus* [14,15].

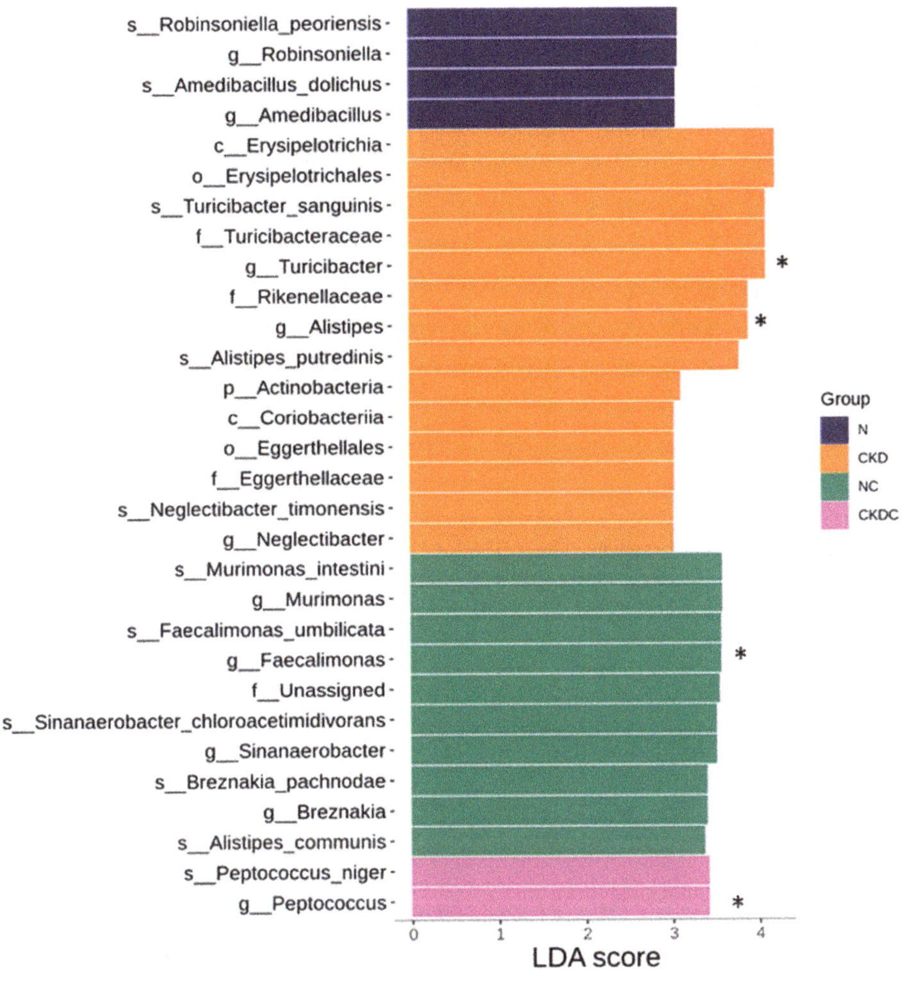

Figure 5. Linear discriminant analysis effect size (LEfSe) illustration. Different colors denote groups of microbes that have significant effects in different groups with linear discriminant analysis (LDA) > 3. * indicates taxa that are linked to hypertension.

To test further whether certain microorganisms are involved in the protective role of citrulline against maternal CKD-primed offspring hypertension, we examined different compositions and abundance between the CKD group and CKDC group. At the genus level, compared with the CKD group, the proportions of *Monoglobus* and *Streptococcus* were lower in the CKDC group (Figure 6A,B). Species-based comparison showed the abundance of *Agothobacterium Butyriciproducens* was amplified by maternal citrulline supplementation in the CKDC rats in comparison to the CKD rats (Figure 6C).

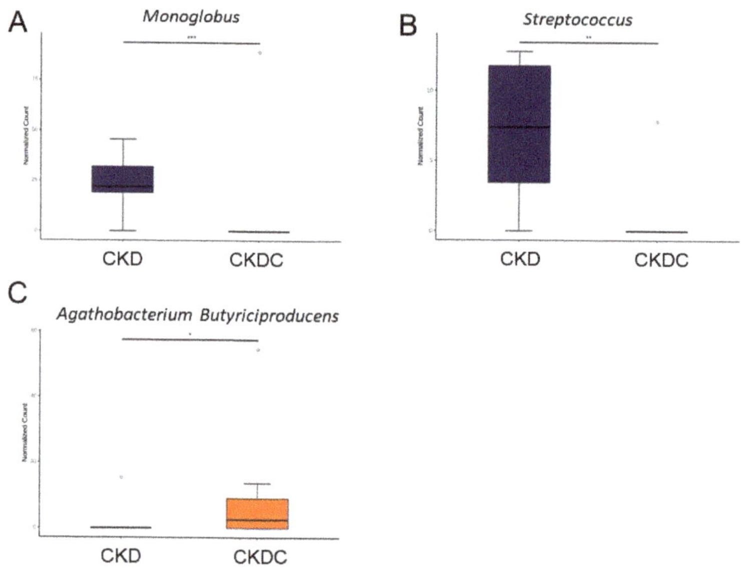

Figure 6. Relative abundance of (**A**) *Monoglobus* (genus), (**B**) *Streptococcus* (genus), and (**C**) *Agathobacterium Butyriciproducens* (species) between the CKD and CKDC group. * $p < 0.05$; ** $p < 0.01$; *** $p < 0.005$.

3. Discussion

Early-life oxidative stress serves as a critical mechanism in the developmental programming of hypertension, and antioxidant therapy emerges as a potential preventive strategy [12,13]. Our study presents the first evidence that offspring hypertension induced by maternal CKD can be prevented through perinatal citrulline supplementation. Key findings include: (1) maternal CKD induces offspring hypertension, which is a condition prevented by perinatal citrulline treatment; (2) hypertension in offspring primed by maternal CKD is linked to an inhibited NO pathway characterized by reduced eNOS and nNOS protein levels, a diminished ratio of arginine to ADMA, and increased ADMA and SDMA concentrations; (3) maternal citrulline treatment safeguards adult offspring from hypertension by restoring NO, decreasing renal PRR expression, and influencing gut microbiota; (4) the protective action of citrulline aligns with a decreased abundance of the genera *Monoglobus* and *Streptococcus* and an increase in *Agathobacterium Butyriciproducens*.

Previous research suggests that citrulline, functioning as an antioxidant, modulates NO and prevents oxidative stress-induced cardiovascular disease [8–10]. Our study extends the application of citrulline during gestation and lactation to mitigate offspring hypertension associated with maternal CKD. In alignment with previous studies involving models of maternal NO deficiency [16] and prenatal dexamethasone exposure [17], our data suggests that maternal citrulline supplementation enhances NO availability, averting offspring hypertension.

Our study's results demonstrate that maternal CKD diminishes eNOS and nNOS protein levels in offspring kidneys, reduces the ratio of arginine to ADMA, and increases ADMA and SDMA, thus limiting NO production. Maternal citrulline therapy effectively reverses the inhibitory effects on NOS protein abundance and restores the balance between arginine and ADMA to enhance the NO pathway. Considering the dysregulated ADMA/NO pathway as a mediator of oxidative stress in hypertension [13], the beneficial action of citrulline may be linked to its ability to improve NO availability.

An activated classic RAS axis is known to increase BP through increased oxidative stress [18]. While the imbalance between NO and RAS is closely linked to hypertension pathophysiology [19], little is known about whether citrulline treatment can modulate the RAS to control BP. Activating PRR promotes vasoconstriction [20], and maternal citrulline supplementation reduces PRR, favoring lower BP. Although not statistically significant, our results suggest that citrulline treatment tends to reduce most classic RAS components. Future studies may explore whether citrulline's protective effect against offspring hypertension correlates with RAS blockade.

Another potential protective mechanism of citrulline against maternal uremia-programmed hypertension may be associated with alterations in gut microbiota. The redox status influences gut health [21], and dietary antioxidants may benefit health by modulating gut microbiota [22]. Citrulline has been demonstrated to contribute to the maintenance of both the integrity of the intestinal barrier and the balance of microbiota [23]. However, there are no reports on the impacts of citrulline on gut microbiota in hypertension.

The PLS-DA analysis in our study did not reveal distinct clustering patterns between the CKD and CKDC groups, suggesting that citrulline supplementation may have a limited role in shaping offspring's gut microbiota compared to maternal CKD. Nevertheless, citrulline still contributes to the low relative abundance of *Monoglobus* and *Streptococcus* and a high proportion of *Agathobacterium Butyriciproducens* in the gut microbiota of the CKDC group. The genus *Monoglobus* has been associated with hypertension [24], and *Streptococcus* spp., opportunistic pathogenic taxa, are often found in hypertensive gut microbiomes [25]. Our findings indicate that the protective effects of citrulline against hypertension in the offspring of uremic dams may be associated with its capacity to impact taxa associated with hypertension. *Agathobaculum butyriciproducens*, a butyrate-producing probiotic, has shown beneficial effects on cognitive deficits and Alzheimer's disease pathologies [26]. Butyrate, a short-chain fatty acid (SCFA), can regulate BP through the activation of its receptors [27]. Previously, we observed that butyrate supplementation throughout gestation and lactation prevented offspring hypertension programmed by maternal CKD [28]. Citrulline likely has the potential to enhance SCFA-producing probiotics and, consequently, reduce BP. A previous study identified *Peptococcus* as bacteria depleted in subjects with metabolic syndrome [29]. Based on our LEfSe analysis, citrulline supplementation, which enhances *Peptococcus* abundance, may be attributed to its beneficial action in preventing hypertension.

Several limitations of the present study need acknowledgment. Firstly, we did not analyze gut microbiota and derived metabolites in offspring rats at different developmental stages and their dams. Gut microbial alterations in adult offspring rats may be attributed to postnatal plasticity rather than primary programmed processes responding to early-life environmental cues. Secondly, while we understand that the mechanisms mentioned may not entirely cover the antioxidant actions of citrulline against maternal CKD-programmed hypertension, a comprehensive examination of the complete mechanisms involved would aid in the development of novel antioxidant preventive therapies. Finally, our data, although useful in demonstrating that citrulline treatment has beneficial effects on male rat offspring's BP, is limited to experimentation in this model. The beneficial effects of citrulline supplementation were associated with the restoration of the NO pathway and modifications in gut microbiota, yet additional underlying mechanisms remain to be fully elucidated. While dietary antioxidants present a promising strategy for oxidative-stress-induced hypertension, conclusive results in humans are still pending [30,31]. Further investigations are required in other models of programmed hypertension and in humans before clinical translation.

4. Materials and Methods

4.1. Animals

We procured female Sprague–Dawley (SD) rats aged 6–8 weeks from BioLASCO Taiwan Co., Ltd. (Taipei, Taiwan) and housed them in our AAALAC-accredited animal facility. All procedures adhered to the regulations set by the Institutional Animal Care and Use Committee (IACUC) at our hospital with the permit number 2022091601.

Figure 7 illustrates the experimental protocol. Eight-week-old female rats (n = 12) were randomly divided into two groups. The rats were assigned to either a normal diet or a 0.5% adenine diet for a duration of 3 weeks, as previously described [5]. Individual females were paired overnight with a proven fertile male until the identification of a copulatory plug. Pregnant rats were then randomized into four groups: rats receiving a normal diet (N), adenine-treated rats (CKD), control rats receiving citrulline supplementation (0.1% citrulline in drinking water) throughout gestation and lactation (NC), and CKD rats receiving citrulline supplementation (CKDC). The citrulline dosage followed a previously established protocol [16]. Litter sizes at birth were reduced to eight pups. Given the higher prevalence of hypertension in males compared to females [32], only male offspring from each litter were selected for subsequent experiments.

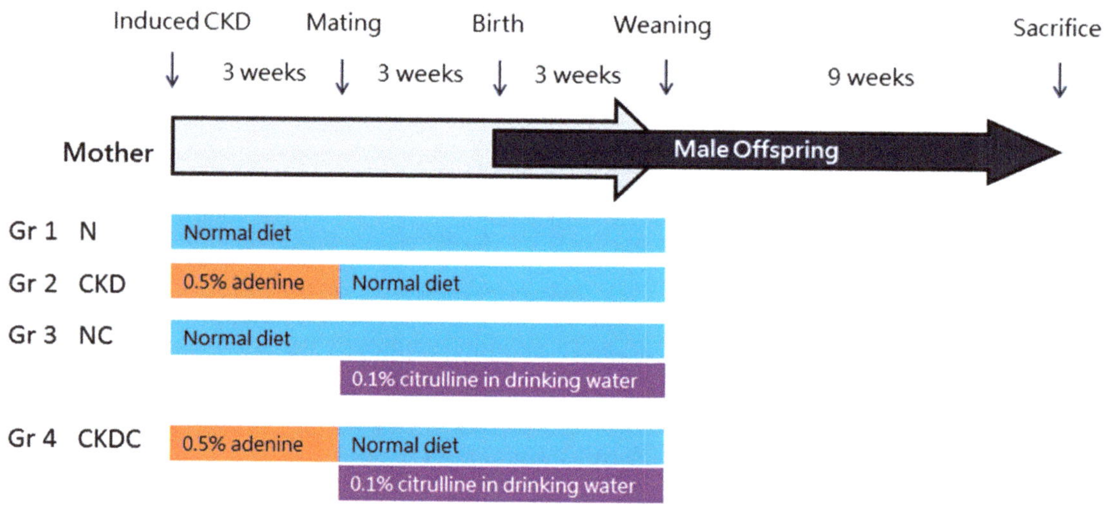

Figure 7. Experimental protocol used in the current study. N = dams received a normal diet; CKD = dams treated with an adenine diet; NC = dams received a normal diet and received citrulline supplementation; CKDC = adenine-treated dams received citrulline supplementation.

Rat offspring were assigned to four groups (8 animals each): N, CKD, NC, and CKDC. To acclimate the rats, we utilized the CODA BP system (a tail-cuff method, Kent Scientific Corporation, Torrington, CT, USA) for BP measurements every four weeks. At 12 weeks of age, blood draws and sacrifices were performed to assess kidney weights. Kidney tissues were snap-frozen and stored accordingly. Prior to sacrifice, fecal samples from each offspring were stored in a −80 °C freezer. The creatinine concentrations in rat offspring blood were determined using high-performance liquid chromatography (HPLC, HP Agilent 1100, Agilent Technologies Inc., Santa Clara, CA, USA).

4.2. NO Parameters

The HPLC method was employed to analyze plasma levels of NO-related parameters. The measurements were made on an Agilent 1100 HPLC (Agilent Technologies Inc.) by using O-phthalaldehyde/3-mercaptopropionic acid (OPA/3-MPA) as a derivatization agent with fluorescence detection. The ratio of arginine to ADMA was calculated, which provides information on NO availability [33].

4.3. Western Blot

Equal amounts of 200 μg of kidney cortical proteins were loaded per lane onto a polyacrylamide gel and subjected to electrophoresis. Following separation, the proteins were transferred onto nitrocellulose membranes. Membranes underwent treatment with a

0.1% Ponceau S solution (Sigma-Aldrich, St. Louis, MO, USA) for 10 min on a shaker, which was followed by rinsing with distilled water to eliminate background staining. Ponceau S staining served for total protein normalization.

The transferred proteins were then probed using specific antibodies, including a mouse eNOS antibody (1:250; BD610297BD, Biosciences, San Jose, CA, USA), a mouse nNOS antibody (1:200; SC-5302, Santa Cruz, CA, USA), a mouse DDAH1 antibody (1:500; SC-271337, Santa Cruz), or a rabbit DDAH2 antibody (1:2000; Ab184166, Abcam, Cambridge, UK). Subsequent to washing, the blots were incubated with the corresponding secondary antibody conjugated to horseradish peroxidase. Immunopositive bands were scanned using an imaging densitometer (Quantity One, Bio-Rad, Hercules, CA, USA) to quantify integrated optical density (IOD). Protein abundance was expressed as IOD normalized by Ponceau S stain (PonS, representing the total protein loaded). Complete blots and corresponding images of Ponceau S staining can be found in Supplementary Material.

4.4. Analysis of RAS Components Using qPCR

Total RNA was isolated from renal cortical tissues for qPCR analysis on a thermal cycler (iCycler, Bio-Rad, Hercules, CA, USA) in duplicate. The internal control utilized in this study was the 18S ribosomal RNA (R18S). PCR primers for both RAS components and R18S are detailed in Table 2. Relative quantification was determined through the comparative $2^{-\Delta\Delta CT}$ method.

Table 2. Primers for qPCR.

Gene	Accession No	Sense	Antisense
AGT	XM_032887807.1	5 gcccaggtcgcgatgat 3	5 tgtacaagatgctgagtgaggcaa 3
Renin	J02941.1	5 aacattaccagggcaactttcact 3	5 accccttcatggtgatctg 3
ACE	U03734.1	5 caccggcaaggtctgctt 3	5 cttggcatagtttcgtgaggaa 3
PRR	AB188298.1	5 gaggcagtgaccctcaacat 3	5 ccctcctcacacaacaaggt 3
AT1R	NM_030985.4	5 gctgggcaacgagtttgtct 3	5 cagtccttcagctggatcttca 3
R18S	X01117	5 gccgcggtaattccagctcca 3	5 cccgcccgctcccaagatc 3

4.5. 16S rRNA Sequencing

Microbial community DNA was extracted from fecal samples and subsequently underwent 16S rRNA sequencing at Biotools Co., Ltd. (New Taipei City, Taiwan) [5]. The amplification of the V1–V9 region of the 16S rRNA gene with barcoded primers was prepared for a multiplexed SMRTbell library (PacBio, Menlo Park, CA, USA) and the sequencing procedure. To construct a phylogenetic tree, QIIME2 phylogeny fast tree utilized a set of sequences representing the amplicon sequence variants (ASVs) [34,35]. Alpha diversity analysis, evaluating microbiota richness and evenness within a single sample, utilized the Shannon index and Pielou's evenness. Beta diversity analysis relied on ANOSIM and PLS-DA. Significantly differential taxa were identified using LEfSe analysis with an LDA score exceeding 3.

4.6. Statistics

The data are expressed as means ± standard error of the mean (SEM). Group distinctions were evaluated utilizing either one-way ANOVA or two-way ANOVA, depending on the context. Subsequent to the ANOVA, Tukey post hoc analysis was conducted to elucidate differences between specific groups. A significance level of $p < 0.05$ was employed to determine statistical significance. All statistical analyses were carried out using SPSS 17.0 software (SPSS, Inc., Chicago, IL, USA).

5. Conclusions

This study represents one of the initial observations highlighting the potential of citrulline supplementation during gestation and lactation to prevent offspring hypertension complicated by maternal uremia. Given the reversible nature of offspring hypertension through citrulline, a deeper understanding of its extent and the involved mechanisms could contribute to the development of optimal antioxidants as preventive therapies, thereby mitigating the health burden imposed by elevated BP on future generations.

Supplementary Materials: The following supporting information can be downloaded at https://www.mdpi.com/article/10.3390/ijms25031612/s1.

Author Contributions: Conceptualization, Y.-L.T. and C.-N.H.; Formal analysis, Y.-L.T., S.L. and C.-N.H.; Data curation, Y.-L.T., C.-Y.H. and C.-N.H.; Investigation, C.-Y.H., G.-P.C.-C. and S.L.; Methodology, C.-Y.H., G.-P.C.-C. and S.L.; Writing—original draft, Y.-L.T. and C.-N.H.; Writing—review and editing, Y.-L.T. and C.-N.H.; Funding acquisition, Y.-L.T. All authors have read and agreed to the published version of the manuscript.

Funding: This work was funded by CMRPG8N0171 from the Kaohsiung Chang Gung Memorial Hospital, Taiwan.

Institutional Review Board Statement: All animal studies were approved by the Institutional Animal Ethics Committee (IACUC) of Chang Gung Memorial Hospital (Permit # 2022091601).

Informed Consent Statement: Not applicable.

Data Availability Statement: Data are contained within the article.

Acknowledgments: We would like to thank the Institute of Environmental Toxin and Emerging-Contaminant, the Center for Environmental Toxin and Emerging Contaminant Research, and the Super Micro Mass Research and Technology Center, Cheng Shiu University, Kaohsiung, for technical support. We also appreciate the support from the Center for Laboratory Animals, Kaohsiung Chang Gung Memorial Hospital.

Conflicts of Interest: The authors declare no conflicts of interest.

References

1. Bromfield, S.; Muntner, P. High blood pressure: The leading global burden of disease risk factor and the need for worldwide prevention programs. *Curr. Hypertens. Rep.* **2013**, *15*, 134–136. [CrossRef]
2. Paauw, N.D.; van Rijn, B.B.; Lely, A.T.; Joles, J.A. Pregnancy as a critical window for blood pressure regulation in mother and child: Programming and reprogramming. *Acta Physiol.* **2017**, *219*, 241–259. [CrossRef]
3. Suzuki, K. The developing world of DOHaD. *J. Dev. Orig. Health Dis.* **2018**, *9*, 266–269. [CrossRef]
4. Paixão, A.D.; Alexander, B.T. How the kidney is impacted by the perinatal maternal environment to develop hypertension. *Biol. Reprod.* **2013**, *89*, 144. [CrossRef] [PubMed]
5. Hsu, C.N.; Yang, H.W.; Hou, C.Y.; Chang-Chien, G.P.; Lin, S.; Tain, Y.L. Maternal adenine-induced chronic kidney disease programs hypertension in adult male rat offspring: Implications of nitric oxide and gut microbiome derived metabolites. *Int. J. Mol. Sci.* **2020**, *21*, 7237. [CrossRef] [PubMed]
6. Gordon, M.H. Significance of dietary antioxidants for health. *Int. J. Mol. Sci.* **2012**, *13*, 173–179. [CrossRef] [PubMed]
7. Aguayo, E.; Martínez-Sánchez, A.; Fernández-Lobato, B.; Alacid, F. L-Citrulline: A Non-Essential Amino Acid with Important Roles in Human Health. *Appl. Sci.* **2021**, *11*, 3293. [CrossRef]
8. Burton-Freeman, B.; Freeman, M.; Zhang, X.; Sandhu, A.; Edirisinghe, I. Watermelon and L-Citrulline in Cardio-Metabolic Health: Review of the Evidence 2000–2020. *Curr. Atheroscler. Rep.* **2021**, *23*, 81. [CrossRef]
9. Allerton, T.D.; Proctor, D.N.; Stephens, J.M.; Dugas, T.R.; Spielmann, G.; Irving, B.A. L-Citrulline Supplementation: Impact on Cardiometabolic Health. *Nutrients* **2018**, *10*, 921. [CrossRef] [PubMed]
10. Cynober, L.; Moinard, C.; De Bandt, J.P. The 2009 ESPEN Sir David Cuthbertson. Citrulline: A new major signaling molecule or just another player in the pharmaconutrition game? *Clin. Nutr.* **2010**, *29*, 545–551. [CrossRef] [PubMed]
11. Khalaf, D.; Krüger, M.; Wehland, M.; Infanger, M.; Grimm, D. The Effects of Oral l-Arginine and l-Citrulline Supplementation on Blood Pressure. *Nutrients* **2019**, *11*, 1679. [CrossRef] [PubMed]
12. Thompson, L.P.; Al-Hasan, Y. Impact of oxidative stress in fetal programming. *J. Pregnancy* **2012**, *2012*, 582748. [CrossRef] [PubMed]
13. Tain, Y.L.; Hsu, C.N. Oxidative Stress-Induced Hypertension of Developmental Origins: Preventive Aspects of Antioxidant Therapy. *Antioxidants* **2022**, *11*, 511. [CrossRef]
14. Yang, T.; Santisteban, M.M.; Rodriguez, V.; Li, E.; Ahmari, N.; Carvajal, J.M.; Zadeh, M.; Gong, M.; Qi, Y.; Zubcevic, J.; et al. Gut dysbiosis is linked to hypertension. *Hypertension* **2015**, *65*, 1331–1340. [CrossRef] [PubMed]
15. Calderón-Pérez, L.; Gosalbes, M.J.; Yuste, S.; Valls, R.M.; Pedret, A.; Llauradó, E.; Jimenez-Hernandez, N.; Artacho, A.; Pla-Pagà, L.; Companys, J.; et al. Gut metagenomic and short chain fatty acids signature in hypertension: A cross-sectional study. *Sci. Rep.* **2020**, *10*, 6436. [CrossRef] [PubMed]

16. Tain, Y.L.; Huang, L.T.; Lee, C.T.; Chan, J.Y.; Hsu, C.N. Maternal citrulline supplementation prevents prenatal N(G)-nitro-L-arginine-methyl ester (L-NAME)-induced programmed hypertension in rats. *Biol. Reprod.* **2015**, *92*, 7. [CrossRef] [PubMed]
17. Tain, Y.L.; Sheen, J.M.; Chen, C.C.; Yu, H.R.; Tiao, M.M.; Kuo, H.C.; Huang, L.T. Maternal citrulline supplementation prevents prenatal dexamethasone-induced programmed hypertension. *Free Radic. Res.* **2014**, *48*, 580–586. [CrossRef]
18. Fanelli, C.; Zatz, R. Linking oxidative stress, the renin-angiotensin system, and hypertension. *Hypertension* **2011**, *57*, 373–374. [CrossRef]
19. Kopkan, L.; Cervenka, L. Renal interactions of renin-angiotensin system, nitric oxide and superoxide anion: Implications in the pathophysiology of salt-sensitivity and hypertension. *Physiol. Res.* **2009**, *58*, S55–S68. [CrossRef]
20. Song, R.; Yosypiv, I.V. (Pro)renin Receptor in Kidney Development and Disease. *Int. J. Nephrol.* **2011**, *2011*, 247048. [CrossRef]
21. Campbell, E.L.; Colgan, S.P. Control and dysregulation of redox signalling in the gastrointestinal tract. *Nat. Rev. Gastroenterol. Hepatol.* **2019**, *16*, 106–120. [CrossRef]
22. Wan, M.L.Y.; Co, V.A.; El-Nezami, H. Dietary polyphenol impact on gut health and microbiota. *Crit. Rev. Food Sci. Nutr.* **2021**, *61*, 690–711. [CrossRef]
23. Uyanga, V.A.; Amevor, F.K.; Liu, M.; Cui, Z.; Zhao, X.; Lin, H. Potential Implications of Citrulline and Quercetin on Gut Functioning of Monogastric Animals and Humans: A Comprehensive Review. *Nutrients* **2021**, *13*, 3782. [CrossRef] [PubMed]
24. Li, J.; Ma, G.; Xie, J.; Xu, K.; Lai, H.; Li, Y.; He, Y.; Yu, H.; Liao, X.; Wang, X.; et al. Differential Gut Microbiota, Dietary Intakes in Constipation Patients with or without Hypertension. *Mol. Nutr. Food Res.* **2023**, *67*, e2300208. [CrossRef] [PubMed]
25. Yan, Q.; Gu, Y.; Li, X.; Yang, W.; Jia, L.; Chen, C.; Han, X.; Huang, Y.; Zhao, L.; Li, P.; et al. Alterations of the Gut Microbiome in Hypertension. *Front. Cell Infect. Microbiol.* **2017**, *7*, 381. [CrossRef]
26. Go, J.; Chang, D.H.; Ryu, Y.K.; Park, H.Y.; Lee, I.B.; Noh, J.R.; Hwang, D.Y.; Kim, B.C.; Kim, K.S.; Lee, C.H. Human gut microbiota Agathobaculum butyriciproducens improves cognitive impairment in LPS-induced and APP/PS1 mouse models of Alzheimer's disease. *Nutr. Res.* **2021**, *86*, 96–108. [CrossRef] [PubMed]
27. Pluznick, J.L. Microbial short-chain fatty acids and blood pressure regulation. *Curr. Hypertens. Rep.* **2017**, *19*, 25. [CrossRef] [PubMed]
28. Hsu, C.N.; Yu, H.R.; Lin, I.C.; Tiao, M.M.; Huang, L.T.; Hou, C.Y.; Chang-Chien, G.P.; Lin, S.; Tain, Y.L. Sodium butyrate modulates blood pressure and gut microbiota in maternal tryptophan-free diet-induced hypertension rat offspring. *J. Nutr. Biochem.* **2022**, *108*, 109090. [CrossRef]
29. Si, J.; Lee, C.; Ko, G. Oral Microbiota: Microbial Biomarkers of Metabolic Syndrome Independent of Host Genetic Factors. *Front. Cell Infect. Microbiol.* **2017**, *7*, 516. [CrossRef]
30. Sinha, N.; Dabla, P.K. Oxidative stress and antioxidants in hypertension-a current review. *Curr. Hypertens. Rev.* **2015**, *11*, 132–142. [CrossRef]
31. Griendling, K.K.; Camargo, L.L.; Rios, F.J.; Alves-Lopes, R.; Montezano, A.C.; Touyz, R.M. Oxidative Stress and Hypertension. *Circ. Res.* **2021**, *128*, 993–1020. [CrossRef] [PubMed]
32. Bernstein, S.R.; Kelleher, C.; Khalil, R.A. Gender-based research underscores sex differences in biological processes, clinical disorders and pharmacological interventions. *Biochem. Pharmacol.* **2023**, *215*, 115737. [CrossRef]
33. Bode-Böger, S.M.; Scalera, F.; Ignarro, L.J. The L-arginine paradox: Importance of the L-arginine/asymmetrical dimethylarginine ratio. *Pharmacol. Ther.* **2007**, *114*, 295–306. [CrossRef]
34. Bolyen, E.; Rideout, J.R.; Dillon, M.R.; Bokulich, N.A.; Abnet, C.C.; Al-Ghalith, G.A.; Alexander, H.; Alm, E.J.; Arumugam, M.; Asnicar, F.; et al. Reproducible, interactive, scalable and extensible microbiome data science using QIIME 2. *Nat. Biotechnol.* **2019**, *37*, 852–857. [CrossRef] [PubMed]
35. Price, M.N.; Dehal, P.S.; Arkin, A.P. FastTree 2—Approximately maximum-likelihood trees for large alignments. *PLoS ONE* **2010**, *5*, e9490. [CrossRef] [PubMed]

International Journal of
Molecular Sciences

MDPI

Article

From Local to Systemic: The Journey of Tick Bite Biomarkers in Australian Patients

Wenna Lee [1,2,3,4,*], Amanda D. Barbosa [1,5,6], Amy Huey-Yi Lee [7], Andrew Currie [3,8], David Martino [4,9], John Stenos [10], Michelle Long [10], Miles Beaman [11], Nathan T. Harvey [4,12], Nina Kresoje [4,9], Patrycja Skut [2], Peter J. Irwin [5], Prasad Kumarasinghe [13,14,15], Roy A. Hall [16], Rym Ben-Othman [17], Stephen Graves [10], Tobias R. Kollmann [2,18,†] and Charlotte L. Oskam [1,3,*,†]

1 Centre for Biosecurity and One Health, Harry Butler Institute, Murdoch University, Murdoch, WA 6150, Australia
2 The Kids Research Institute Australia, Nedlands, WA 6009, Australia; tkollm@mac.com (T.R.K.)
3 School of Medical, Molecular, and Forensic Sciences, College of Environmental and Life Sciences, Murdoch University, Murdoch, WA 6150, Australia
4 UWA Medical School, University of Western Australia, Crawley, WA 6009, Australia
5 School of Veterinary Medicine, College of Environmental and Life Sciences, Murdoch University, Murdoch, WA 6150, Australia
6 CAPES Foundation, Ministry of Education of Brazil, Brasilia-DF 70040-020, Brazil
7 Department of Molecular Biology and Biochemistry, Simon Fraser University, Burnaby, BC V5A 1S6, Canada
8 Personalised Medicine Centre, Health Futures Institute, Murdoch University, Murdoch, WA 6150, Australia
9 Wal-yan Respiratory Research Centre, The Kids Research Institute Australia, Nedlands, WA 6009, Australia
10 Australian Rickettsial Reference Laboratory, Barwon Biomedical Research, University Hospital Geelong, Barwon Health, Geelong, VIC 3220, Australia
11 Faculty of Health and Medical Sciences, Pathology & Laboratory Medicine, University of Western Australia, Crawley, WA 6009, Australia
12 PathWest Laboratory Medicine, Department of Anatomical Pathology, QEII Medical Centre, Nedlands, WA 6009, Australia
13 School of Medicine, University of Western Australia, Crawley, WA 6009, Australia
14 College of Science, Health, Education and Engineering, Murdoch University, Murdoch, WA 6150, Australia
15 Western Dermatology, Hollywood Medical Centre, Nedlands, WA 6009, Australia
16 Australian Infectious Diseases Research Centre, School of Chemistry and Molecular Biosciences, The University of Queensland, St Lucia, QLD 4072, Australia
17 RAN BioLinks Ltd., 10212 Yonge Street, 202, Richmond Hill, ON L4C 3B6, Canada
18 Department of Microbiology & Immunology, Faculty of Medicine, Dalhousie University, Halifax, NS B3H 4R2, Canada
* Correspondence: wenna.lee@thekids.org.au (W.L.); c.oskam@murdoch.edu.au (C.L.O.)
† These authors contributed equally to this work as senior authors.

Abstract: Tick bites and tick-related diseases are on the rise. Diagnostic tests that identify well-characterised tick-borne pathogens (TBPs) possess limited capacity to address the causation of symptoms associated with poorly characterised tick-related illnesses, such as debilitating symptom complexes attributed to ticks (DSCATT) in Australia. Identification of local signals in tick-bitten skin that can be detected systemically in blood would have both clinical (diagnostic or prognostic) and research (mechanistic insight) utility, as a blood sample is more readily obtainable than tissue biopsies. We hypothesised that blood samples may reveal signals which reflect relevant local (tissue) events and that the time course of these signals may align with local pathophysiology. As a first step towards testing this hypothesis, we compared molecular signatures in skin biopsies taken from the tick-bite location of human participants, as published in our previous study, together with peripheral blood signatures obtained concurrently. This approach captures differentially expressed molecules across multiple omics datasets derived from peripheral blood (including cellular and cell-free transcriptomics, proteomics, metabolomics, and DNA methylation), and skin biopsies (spatial transcriptomics). Our original data revealed that extracellular matrix organisation and platelet degranulation pathways were upregulated in the skin within

72 h of a tick bite. The same signals appeared in blood, where they then remained elevated for three months, displaying longitudinally consistent alterations of biological functions. Despite the limited sample size, these data represent proof-of-concept that molecular events in the skin following a tick bite can be detectable systemically. This underscores the potential value of blood samples, akin to a liquid biopsy, to capture biomarkers reflecting local tissue processes.

Keywords: emerging diseases; MULTI-OMICS; systems biology; tick-borne diseases

1. Introduction

The escalating prevalence and public health threat posed by tick bites calls for innovative approaches that would enhance our understanding of disease processes [1]. The need for suitable biomarkers is underscored by the difficulty of diagnosing tick-borne diseases (TBDs) since patients may exhibit a wide range of symptoms yet show 'normal' test results using current diagnostic methods [2]. Starting from the assumption that TBDs ultimately arise from the tick bite site, we hypothesised that samples such as blood may allow the detection of biological signals that reflect the pathophysiological events originating from the tissue at that site. Such liquid biopsy methods have revolutionised disease detection and monitoring in prenatal testing and cancer diagnostics [3–6]. Importantly, interrogating and analysing molecular biomarkers present in a bodily fluid such as blood have allowed for the identification of biomarkers as promising monitoring tools, offering minimally invasive access to diseased tissue and near real-time insights into disease status and progression [3–6]. We reasoned that applying liquid biopsy methodologies could potentially lead not only to improved understanding and earlier diagnosis of TBDs but also to improved treatment outcomes [1,5].

We designed a multi-omics study to identify both single omics that capture the most expressed differences as well as an integrative method that enables different omics to complement each other to enhance the robustness of our findings. To assess the feasibility of a liquid biopsy approach to capture disease progression for TBDs, tick-bitten skin biopsies (solid biopsies) and their contralateral controls were collected concurrently with peripheral blood samples (liquid biopsies) at time points of; (i) enrolment, (ii) one week, and (iii) three months after tick bite. Multi-omics analysis of blood samples was performed at all three time points, encompassing cellular and cell-free transcriptomics, proteomics, metabolomics, and epigenetics (DNA methylation) to ascertain the most informative omics capable of capturing disease progression. The multi-omics data generated in this study were compared with the spatial transcriptomics data we reported previously (Lee. et al., 2024) [7]. Notably, two of these—cell-free transcriptomics and proteomics—hold particular potential as liquid biopsy modalities as they are known to be capable of detecting perturbations from within cancerous tissue [8]. Specifically, four genes—LRIG3, WIF1, LIN7B and RRM2—associated with processes such as cell growth regulation, signalling pathways, and immune responses [9] were identified as differentially expressed in both skin and blood samples.

2. Results

Six participants were recruited in Perth, Western Australia, in accordance with the inclusion criteria previously described [7]. Participant details (Table S1) (post quality assurance and control, processing, and normalisation) (Figure S1) are described in Lee et al., 2024. All participants had been bitten by ticks identified as *Amblyomma triguttatum* (the ornate kangaroo tick), with instars recorded (1 × larva, 3 × nymphs, and 2 × adults (1 × male and 1 × female)). This native Australian tick is well known to feed on humans opportunistically [10].

2.1. Local (Solid Biopsy) Signatures with Spatial Transcriptomics

We have previously published the analysis of the spatial (tissue) transcriptomic data [7]. We here include these data but amended to support integration with the systemic blood-based (liquid biopsy) data sets.

There were 1380 differentially expressed genes (DEGs) identified (Figure 1a); 1018 were downregulated and 362 were upregulated when the transcripts in all tick-bitten skin sections were compared to those obtained from contralateral control skin sections. All DEGs and adjusted *p*-values are listed in Table S1. The most significant downregulated spatial DEGs were ELOVL5 (which translates to a membrane protein localised to the endoplasmic reticulum associated with elongation of the long-chain polyunsaturated fatty acids) with an adjusted *p*-value of 4.82×10^{-10} and TOMMs (a component of the mitochondrial outer membrane translocase complex) with an adjusted *p*-value of 2.02×10^{-9}. Among the downregulated spatial DEGs, olfactory receptor activity (GO ID:0004984) was the most prominent molecular function, with the detection of chemical stimulus involved in sensory perception of smell (GO ID:0050911) as the most significant of the biological processes and intermediate filament cytoskeleton (GO ID:0045111) as the most significant cellular compartment these DEGs were localised to (Figure 1b). All GO results and the adjusted *p*-values are listed in Table S2. The pathway analysis with the identified downregulated spatial DEGs identified four significant pathways. "Expression and translocation of olfactory receptors" and "olfactory signalling pathway", both comprising 70 DEGs, were identified as the top two most significant pathways derived from the downregulated DEGs, with adjusted *p*-values of 1.03×10^{-17} and 1.75×10^{-17}, respectively (Figure 1c). All pathways identified and the adjusted *p*-values are listed in Table S3.

For spatially upregulated DEGs, the most significant molecular function was associated with extracellular matrix structural constituent (GO ID:0005201) with an adjusted *p*-value of 6.96×10^{-10}, while cadherin binding (GO ID:0045296) with an adjusted *p*-value of 7.80×10^{-8} had the highest gene count of 27 (Figure 1d). The most prominent biological process implicated was epidermis development (GO ID:0008544), and the collagen-containing extracellular matrix (GO ID:0062023) was the most significant cellular compartment where these DEGs were localised (Figure 1d). The most significantly upregulated spatial DEGs were CXCL14 (a cytokine gene involved in inflammatory and immune processes) and GSN (coding for an actin-binding protein with multiple isoforms) with adjusted *p*-values of 5.69×10^{-9} and 6.46×10^{-9}, respectively. An overexpression analysis (ORA) was carried out with the upregulated spatial DEGs and identified 67 significant pathways, listed in Table S3. The most significant pathways were "extracellular matrix organisation" comprising 29 DEGs and "ECM proteoglycans" comprising 15 DEGs, both with adjusted *p*-values of 1.49×10^{-7} (Figure 1e).

2.2. Peripheral Blood Cell-Free Transcriptomics

In the comparison between T1 and T0, we identified 1442 DEGs in the cf compartment of peripheral blood, 909 upregulated and 533 downregulated (Figure 2a). The most significant upregulated DEGs identified were PPP1R3C (associated with carbohydrate and glycogen metabolism) and COLEC11 (associated with immunological processes), with adjusted *p*-values of 2.59×10^{-24} and 2.11×10^{-19}, respectively. All cfDEGs, their fold change, and the adjusted *p*-values are listed in Table S1. The most downregulated DEGs at T1 were identified as POSTN (associated with heparin-binding molecular function in cellular adhesion biological processes) and TNFRSF17 (a cell-membrane receptor associated with adaptive immunological processes), with adjusted p-values of 2.30×10^{-16} and 1.60×10^{-14}, respectively. While 1105 pathways at T1 were identified (based on p-value), such as 'GPCR ligand binding' comprising 45 DEGs (p-value of 2.92×10^{-4}), none of the pathways retained significance after adjustment of the p-value.

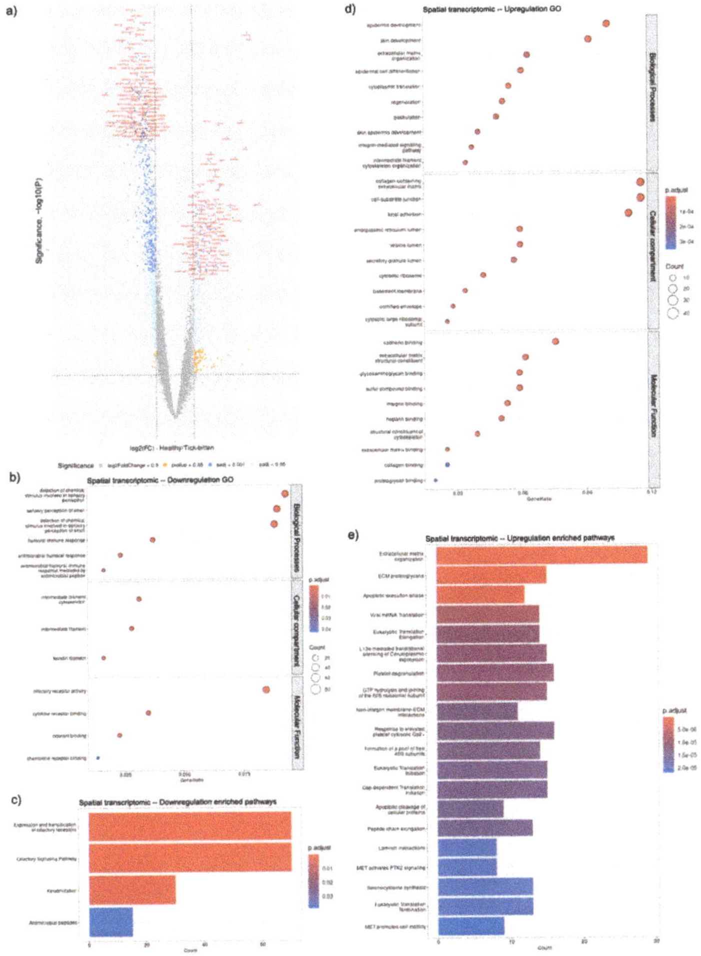

Figure 1. Differential expressed gene analysis comparing transcripts of tick-bitten skin to contralateral controls. (**a**) Volcano plot of differentially expressed genes (DEGs), identified with p-value < 0.05 and fold change of 1.5 in either direction. Grey: Not significant or under a log2(FC) < 0.5; Yellow: p-value < 0.05; Cyan: False discovery rate adjusted p-value < 0.05; Blue: False discovery rate adjusted p-value < 0.001; (**b**) Dotplot of downregulated spatial DEGs in terms of significance in cellular compartment, molecular function, and biological processes. The size of the dots represents the number of genes within the enriched pathways while the dot colour represents the enrichment scores based on adjusted p-values; (**c**) Enriched pathway analysis of downregulated spatial DEGs in terms of significance based on adjusted p-value represented with the bar colour; (**d**) Dotplot of upregulated spatial DEGs in terms of significance in the cellular compartment, molecular function, and biological processes. The size of the dots represents the number of genes within the enriched pathways while the dot colour represents the enrichment scores based on adjusted p-values; (**e**) Enriched pathway analysis of upregulated spatial DEGs in terms of significance based on adjusted p-value, represented with the bar colour.

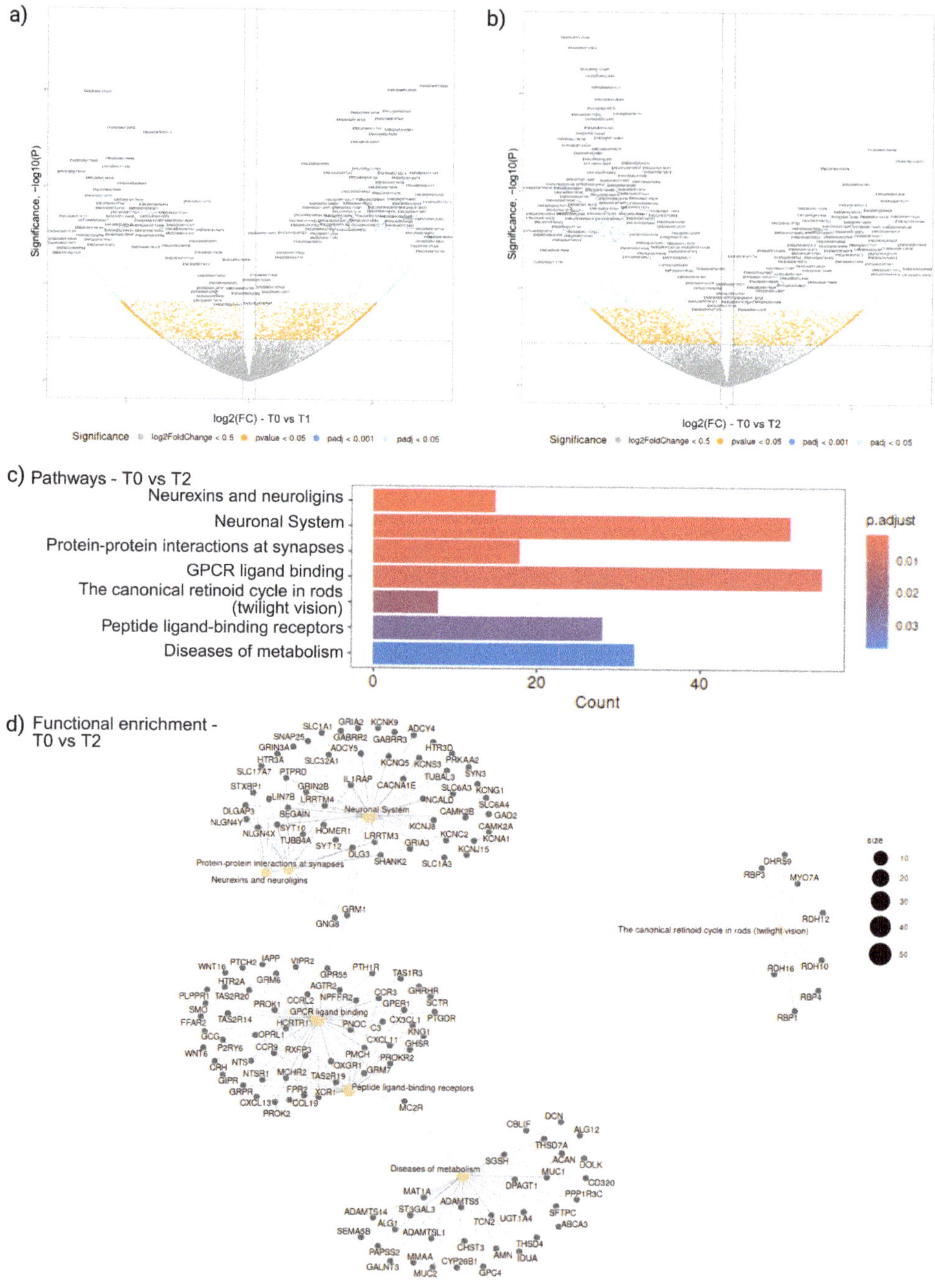

Figure 2. Differential gene expression analysis of cell-free transcriptomics. Volcano plot of DEGs at T1 (**a**) and T2 (**b**), when compared to intake sample at T0, identified with p-value < 0.05 and fold change of 1.5 in either direction. Grey: Not significant or under a log2(FC) < 0.5; Yellow: p-value < 0.05; Cyan: False discovery rate adjusted p-value < 0.05; Blue: False discovery rate adjusted p-value < 0.001; (**c**) Bar plot of the significant pathways in when T2 was compared to T0, ranked according to significance; (**d**) Functional enrichment visualisation of pathways found when T2 was compared to T0.

When T2 and T0 were compared, 1676 cfDEGs were identified, 493 upregulated and 1183 downregulated (Figure 2b). The most significant upregulated DEGs at T2 were AGMAT (a hydrolase for arginine metabolism) and ZNF391 (a DNA-binding zinc finger transcription factor), with adjusted p-values of 1.31×10^{-19} and 2.01×10^{-14}, respectively. The most significant downregulated T2 DEGs identified were SMARCA1 (associated with transcription regulation) and BRSK1 (a kinase associated with DNA damage and cell cycle), with adjusted p-values of 7.69×10^{-32} and 7.57×10^{-27}, respectively. Additionally, seven distinct pathways at T2 were identified with adjusted p-values below 0.05 (Figure 2c). Notably, 'Neurexins and neuroligins', comprising 15 DEGs, and 'Neuronal System', consisting of 51 DEGs, were among the pathways showing significant alterations, with adjusted p-values of 2.02×10^{-3} and 3.98×10^{-3}, respectively (Figure 2c). From the functional enrichment plot (Figure 2d), the three most significant pathways identified on Figure 2c, can be observed clustering together.

When the upregulated and downregulated DEGs within the cell-free RNA (cfRNA) dataset were analysed separately, only the 1183 downregulated cfRNA at T2 yielded significant findings, in particular the 13 pathways, including "Neuronal System" with an adjusted p-value of 4.71×10^{-3}, comprising 41 DEGs, identified as most significant (Table S3). Most downregulated T2 cfRNA DEGs were located in the collagen-containing extracellular matrix (GO ID:0062023), while the most significant molecular activity was symporter activity (GO ID:0015293). Biological processes most downregulated at T2 included sensory perception of light stimulus (GO ID:0050953) and visual perception (GO ID:0007601) (Figure S2).

2.3. Peripheral Blood Cellular Transcriptomics

The comparison of T1 and T2 time points with T0 yielded few DEGs (Figure 3a,b). When T1 was compared to T0, three downregulated DEGs were identified (Figure 3a). All cellular DEGs, their fold change, and the adjusted p-values are listed in Table S1. With an adjusted p-value of 1.87×10^{-6}, heparin-binding, LAMC2, associated with cell adhesion, was identified as the most significant downregulated DEG. RSAD2 and IFIT3, both involved in processes such as anti-viral defence, innate immunity, and general immunity, with adjusted p-values of 1.21×10^{-2} and 4.15×10^{-3}, respectively, were also identified as downregulated when at T1 when compared to T0. Enriched pathway analysis identified 14 significant pathways, with 'Interferon alpha/beta signalling' and 'Interferon Signalling' as the two most significant with adjusted p-values of 1.91×10^{-3} and 7.37×10^{-3}, respectively, and associated with RSAD2 and IFIT3 (Figure 3c.) While these two most significant pathways were linked to only each other, dense connections were observed in the other 12 pathways identified (Figure 3d). When T2 was compared to T0, there were no DEGs (Figure 3b), nor pathways identified.

2.4. Peripheral Blood Plasma Proteomics

The comparison of T1 and T2 time points with T0 yielded an abundance of differentially expressed proteins (DEPs) (Figure 4a,b). When T1 was compared to T0, 79 DEPs were identified, 33 upregulated and 46 downregulated (Figure 4a). The most significantly upregulated T1 DEPs were RL12 (RNA-binding ribonucleoprotein) and TIMP-1 (a secreted growth factor and enzyme inhibitor), with adjusted p-values of 1.39×10^{-3} and 3.85×10^{-3}, respectively. The most downregulated DEPs were CX3CL-1 (a cytokine associated with cell inflammatory responses) and ITA5 (an integrin receptor), with adjusted p-values of 8.23×10^{-5} and 1.84×10^{-3}, respectively, both pathways associated with cell adhesion and host-virus interactions. Pathway analysis of the T1 DEPs identified three pathways, with 'Platelet degranulation' and 'Response to elevated platelet cytosolic Ca2+' as the most significant, both with adjusted p-values of 2.51×10^{-3} and comprising six DEPs (Figure 4c).

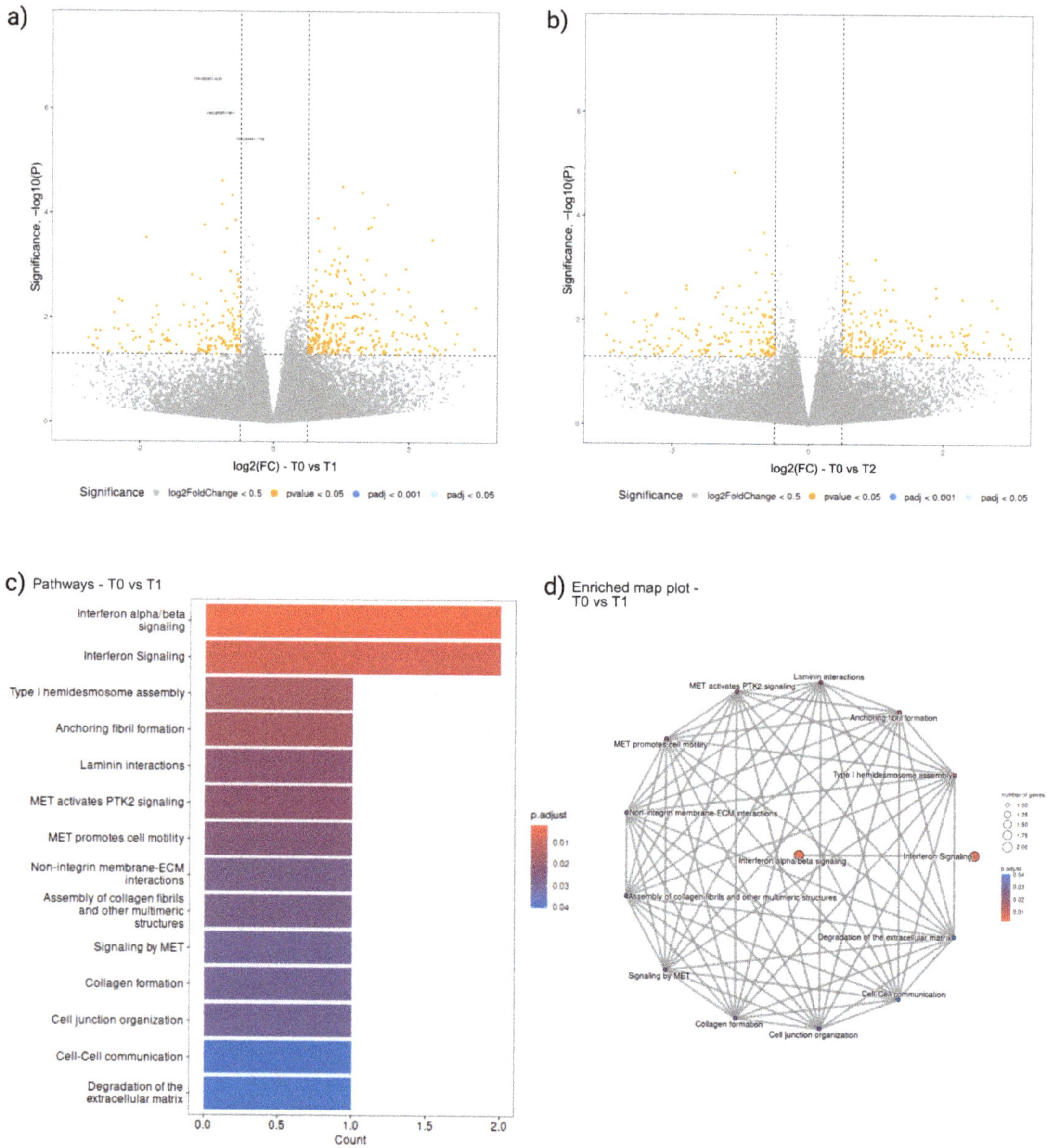

Figure 3. Differential gene expression analysis of cellular transcriptomics. Volcano plot of DEGs at T1 (**a**) and T2 (**b**), when compared to intake sample at T0, identified with p-value < 0.05 and fold change of 1.5 in either direction. Grey: Not significant or under a log2(FC) < 0.5; Yellow: p-value < 0.05; Cyan: False discovery rate adjusted p-value < 0.05; Blue: False discovery rate adjusted p-value < 0.001; (**c**) Bar plot of the significant pathways in when T1 was compared to T0, ranked according to significance; (**d**) Enrichment map plot of the DEGs identified in T1 when compared to intake at T0.

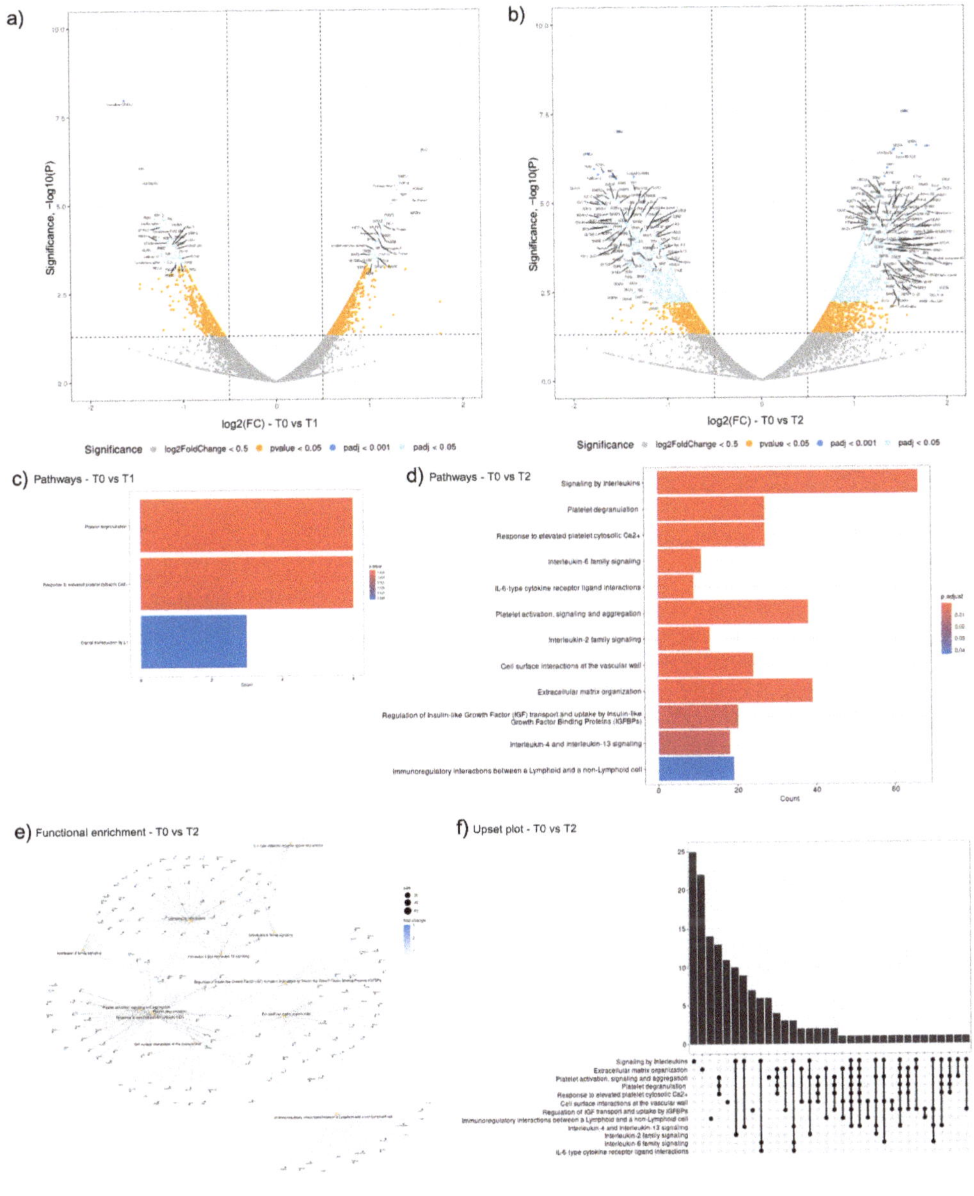

Figure 4. Differential expression analysis of proteomics samples. Volcano plot of DEPs at T1 (**a**) and T2 (**b**), when compared to intake sample at T0, identified with p-value < 0.05 and fold change of 1.5 in either direction. Grey: Not significant or under a log2(FC) < 0.5; Yellow: p-value < 0.05; Cyan: False discovery rate adjusted p-value < 0.05; Blue: False discovery rate adjusted p-value < 0.001; (**c**) Bar plot of the significant pathways in when T1 was compared to T0, ranked according to significance; (**d**) Bar plot of the significant pathways in when T2 was compared to T0, ranked according to significance; (**e**) Functional enrichment visualisation of pathways found when T2 was compared to T0; (**f**) Upset plot showing the number of DEPs in the intersection in the pathways identified when T2 was compared to T0. Horizontal rows represent the pathways identified and vertical bars represent the number of DEPs in each intersection.

Among the 46 downregulated T1 DEPs identified, cytokine receptor binding (GO ID:0005126) was the most prominent molecular function, comprising six DEPs and an adjusted p-value of 5.22×10^{-3} (Figure S3a). There were no significant pathways identified for T1 downregulated DEPs (Table S3). For 33 upregulated T1 DEPs (Figure S3b), the most significant molecular function was associated with endopeptidase inhibitor activity (GO ID:0004866), the most significant biological process affected was the positive regulation of mononuclear cell migration (GO ID:0071677) and the platelet alpha granule (GO ID:0031091) was the most significant cellular compartment these DEPs were localised at (Supplemental Figure S2b). ORA pathway analysis was carried out with the upregulated T1 DEPs and identified eight significant pathways (Figure S3c), listed in Table S3. The most significant pathways were "Platelet degranulation" and "Response to elevated platelet cytosolic Ca2+" both comprising six DEPs, both with adjusted p-values of 5.50×10^{-6} (Figure S3c).

When T2 was compared to T0, 1031 DEPs were identified, 636 upregulated and 395 downregulated (Figure 4b). The most significantly upregulated T2 DEPs were SMAC (associated with the apoptosis process) and NRBF2 (associated with the autophagy and transcription regulation processes), with adjusted p-values of 2.30×10^{-4} and 4.40×10^{-4}, respectively. The most significantly downregulated T2 DEPs were identified as LRFN2 (a leucine-rich protein domain associated with postsynaptic cellular membranes) and DPYL4 (a dihydropyrimidinase-related protein) with adjusted p-values of 3.78×10^{-4} and 4.40×10^{-4}, respectively. Pathway analysis of T2 DEPs identified 12 significant pathways, with 'Signalling by Interleukins' (comprising 66 DEPs) and 'Platelet degranulation' (comprising 27 DEPs) as the most significant with adjusted p-values of 9.11×10^{-8} and 5.61×10^{-6}, respectively (Figure 4d). From the functional enrichment plot (Figure 4e) at T2, it can be observed that the five significant interleukin signalling pathways shared multiple common DEPs. Apart from a handful of DEPs, the three haemostasis-related pathways shared common DEPs among one another that were different to the interferon signalling ones (Figure 4e,f).

In the 395 downregulated T2 DEPs, cytokine receptor binding (GO ID:0005126) was again identified as the most significant molecular function with an adjusted p-value of 2.29×10^{-6} and comprising 23 genes (Figure S4a). The external side of the plasma membrane (GO ID:0009897), comprising 28 genes, was identified as the most significant cellular compartment and the most significant biological process was leukocyte proliferation (GO ID:0070661), with a gene count of 26 (Figure S4a). The most significantly downregulated T2 DEP pathways were "Regulation of Complement cascade" with an adjusted p-value of 1.35×10^{-2} and "Complement cascade" with an adjusted p-value of 1.88×10^{-2}, both comprised five DEPs (Figure S4b).

The 636 upregulated T2 DEPs were localised most significantly in platelet alpha granules (GO ID:0031091) with an adjusted p-value of 1.21×10^{-10} and platelet alpha granule lumen (GO ID:0031093) with an adjusted p-value of 1.88×10^{-10}, comprising 21 and 18 DEPs, respectively (Supplemental Figure S3c). The most significant molecular activity identified was growth factor activity (GO ID:0008083) and four significant biological processes associated with peptidyl tyrosine (GO ID:0050731, 0018108, 0018212, 0050730) were identified (Table S2 and Figure S4c). "Platelet degranulation" and "Response to elevated platelet cytosolic Ca2+", the same two pathways identified at T1, were even more significantly upregulated at T2, both with 24 DEPs and adjusted p-values of 1.24×10^{-7} and 1.42×10^{-7}, respectively (Figure S4d).

2.5. Integration

Data Integration Analysis for Biomarker discovery using the Latent cOmponent (DIABLO) identified markers that exhibited the highest correlation across our multi-omics datasets (Figure S5) and then singled out a minimal number of markers with the greatest predictive value (Table S4). These selected markers were then used to construct molecular signatures for distinguishing between different time points. Fine-tuning the performance of the final DIABLO model revealed that employing two components (Figure S6) achieved the lowest classification error rate, taking into account prediction distance, and enabled the selection of variables across the datasets derived from disparate types of biological measurements (Supplemental Material). Component one comprised 15 cfRNAs, 10 cellular RNAs, 20 proteins, six metabolites and 10 epigenetic methylation sites, while component two comprised 10 cfRNAs, five cellular RNAs, 10 proteins, nine metabolites and 20 epigenetics methylation sites (Table S4). Within the six participants, component one from all omics datasets is highly correlated (Figure S7), with the lowest correlation between mRNA and proteomics at 0.81 and the highest correlation between mRNA and metabolomics at 0.95. For component two, the highest correlation was between cfRNA and epigenetics at 0.97, while the lowest was between proteomics and metabolomics at 0.7 (Figure S8).

Visualising correlations, both within and between omics datasets using a cut-off set at 0.85, the circos plot displays a blend of positive and negative correlations among the selected features across components one and two (Figure 5a). Notably, these correlations span different omics datasets, with the majority of these associations being observed with the proteomics variables. The pseudogene ENSG00000203414, exhibiting a variance value of −0.89 (Table S4), displayed a negative correlation with three selected proteins: seq.3506.49 (lymphotoxin LTA/LTB), seq. 3199.54 (Kallikrein-12), and seq. 2871.73 (DNA repair protein RAD51) (Figure 5a). Using a pair-wise similarity matrix, a relevance network was constructed to offer insights into the interactions and connections between the selected features across the different omics datasets using a predefined cut-off threshold of 0.75. The features selected for the biomarker panel constituted five interconnected clusters on the network (Figure 5b). The Group 4 cluster, visualised as a green cluster in Figure 5b, contained the most features, comprising 32 variables, namely one mRNA, 19 proteins, four metabolites, three cfRNAs, and five methylation sites. Pathway analysis of the Group 4 cluster identified three pathways associated with the TCA cycle, namely 'pyruvate metabolism', 'pyruvate metabolism and the citric acid (TCA) cycle' and 'the citric acid (TCA) cycle and respiratory electron transport' in order of significance based on adjusted p values.

The loading plots reveal which variables contribute more for each omic type in each component, highlighting distinctive patterns (Figure 6a,b). The most significant molecular variances across all omics datasets when comparing the three time points (T0, T1, and T2) reveal that molecules at T2, featuring negative coefficients, contribute the most discriminative features in component one, effectively segregating T2 from the other time points (Figure 6a). In component two, discriminant features are evident across datasets, with molecular variances at T1 displaying negative coefficients, while the molecular variances at T0 exhibit positive coefficients, effectively distinguishing T0 from T1 (Figure 6b). For a comprehensive list of the molecular features identified by DIABLO and featured in both the loading plots and clustered image maps see Table S4.

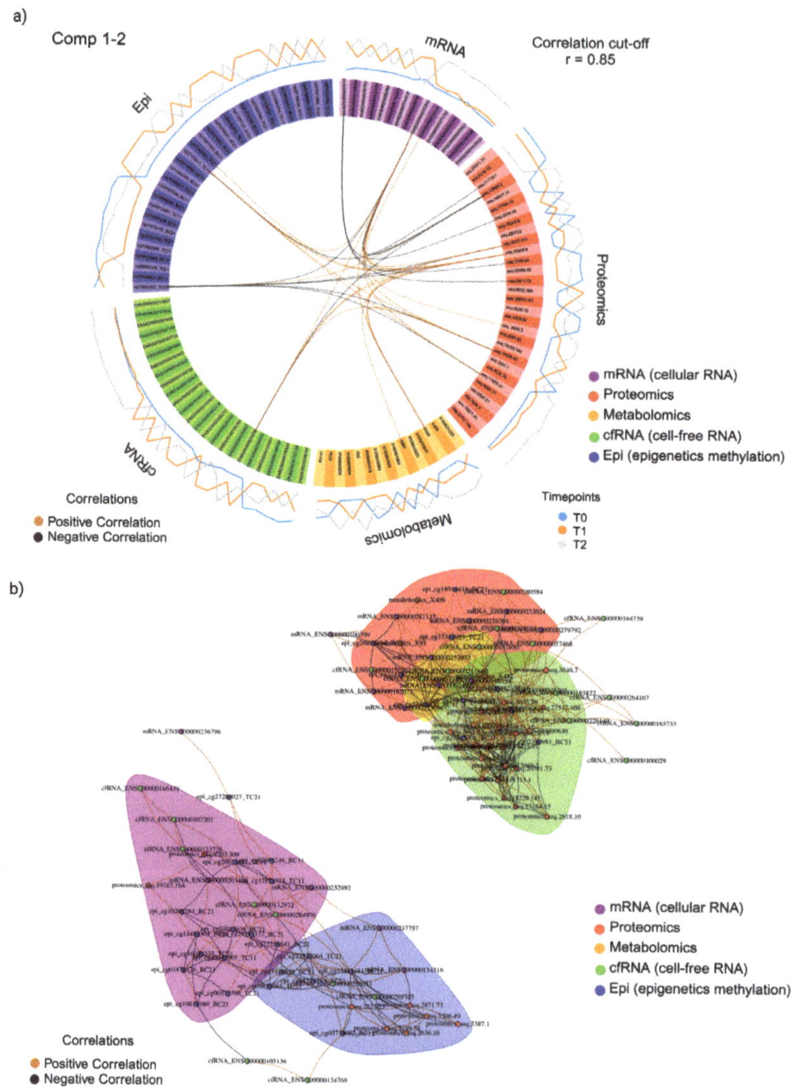

Figure 5. A multi-omics expression signature has been established by examining the correlations among selected features across multi-omics datasets over two components. (**a**) A DIABLO circos plot was generated using a correlation cut-off set above 0.85. The plot encompasses multiple omics datasets, which are distinguished by colour: cellular mRNA (purple), proteomics (red), metabolomics (yellow), cfRNA (green), and epigenetics (blue), organized in respective quadrants. Within the plot, the internal lines connecting the selected variables are indicative of positive (brown) or negative (black) correlations. The external lines represent the expression activity of the corresponding variable at different time points, where T0 is denoted in blue, T1 in orange, and T2 in grey; (**b**) A relevance network has been constructed to illustrate the associations among selected features on components 1 and 2. This network is based on a pairwise similarity matrix derived from multiple omics datasets, with each dataset node represented by colour: cellular mRNA (purple), proteomics (red), metabolomics (yellow), cfRNA (green), and epigenetics (blue). Within the network, the colour of the edges serves as a visual indicator of the relationships between the nodes. Positive correlations are represented by orange edges, while negative correlations are denoted by black edges.

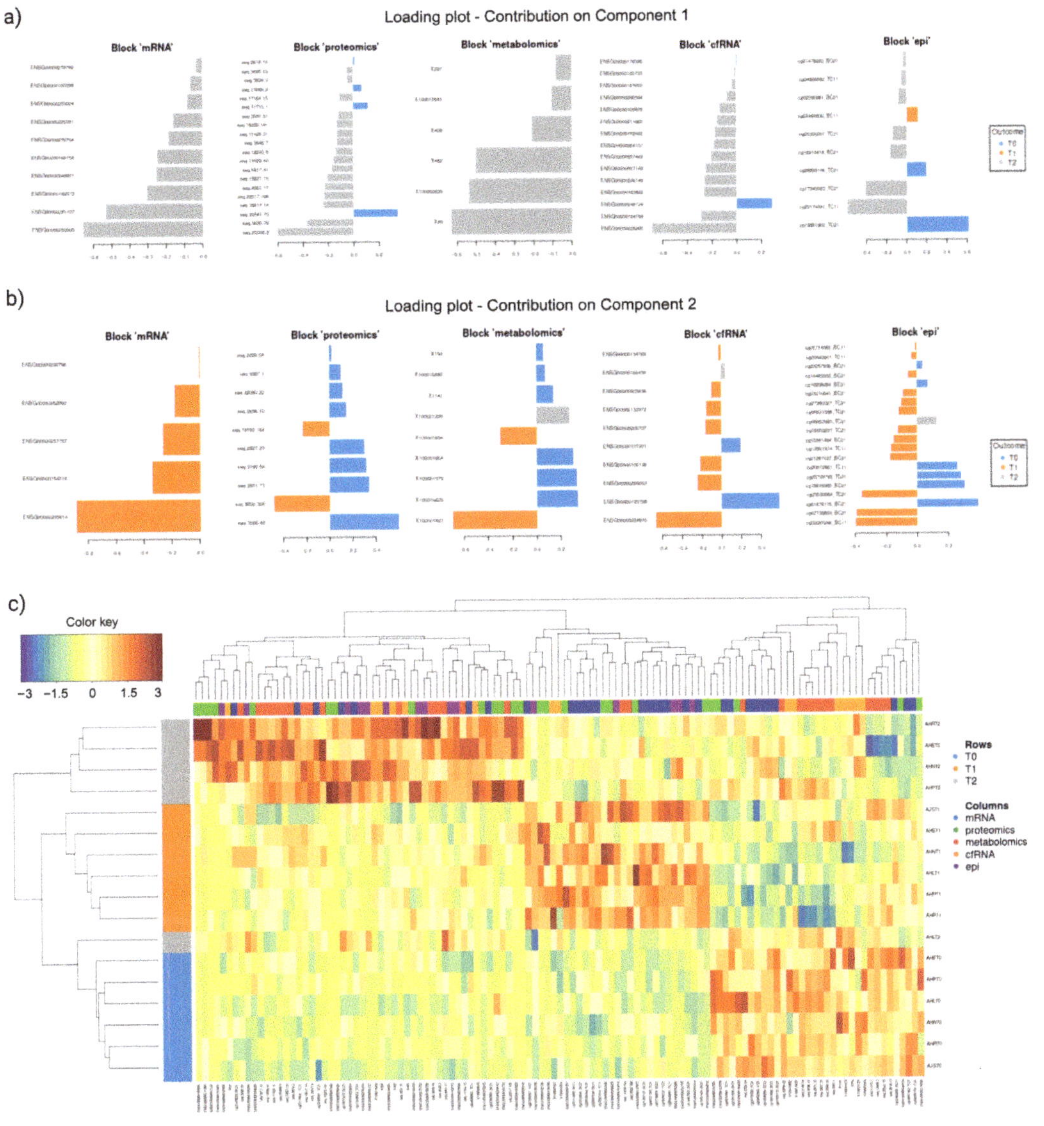

Figure 6. (**a**) Loading plot of DIABLO selected discriminant features on component one; (**b**) Loading plot of DIABLO selected discriminant features on component two. For a and b, from left to right, the loading plot shows the most variable molecular features (y-axis) across time points in mRNA, proteomics, metabolomics, cfRNA, and epigenetics datasets, with the length of the bar representing the contribution weight; (**c**) Hierarchically clustered image map for DIABLO-selected variables, where the rows represent the time points, T0 (blue), T1 (orange), and T2 (grey), and the columns represent the dataset the selected feature belongs to cellular mRNA (blue), proteomics (green), metabolomics (red), cfRNA (orange), and epigenetics (purple).

Hierarchical clustering employing the Euclidean method was conducted for component one features, revealing distinct molecular patterns associated with each timepoint across datasets (Figure 6c). Within these patterns, a consistent set of discriminant features was observed, reflecting a similar expression profile across time points. In the clustered image heatmap (Figure 6c), all samples, except for participant AHL at T2 (AHLT2), exhibited clustering based on their respective time points. This clustering was indicative of the presence of overexpression signatures in the majority of T2 samples (Figure 6c). Participant AHL at T2 failed to exhibit the same overexpression signatures observed in the other T2 samples. Instead, this particular sample loosely clustered with other T0 samples (Figure 6c). It is of note that serological analysis of participant AHL indicated seroconversion to *Coxiella burnetii*, the causative agent of Q-fever, at T2 (Table S1).

Pathway analysis of the DIABLO selected features in component one identified eight pathways, with four TCA cycle associated pathways identified as the most significant—namely 'pyruvate metabolism and citric acid (TCA) cycle' (comprising 10 features), 'citric acid (TCA) cycle and respiratory electron transport' (comprising 10 features), 'pyruvate metabolism' (comprising six features), and 'citric acid cycle (TCA cycle)' (comprising four features), with adjusted p-values of 1.37×10^{-12}, 9.79×10^{-8}, 9.79×10^{-8}, and 5.73×10^{-5}, respectively. (Figure 7a). Component two of DIABLO selected features identified the same four TCA cycle-associated pathways as the most significant in the same order (Figure 7b), comprising 10, 11, 6, and 4 features, with adjusted p-values of 3.13×10^{-13}, 1.08×10^{-9}, 4.57×10^{-8}, and 3.70×10^{-5}, respectively (Table S4).

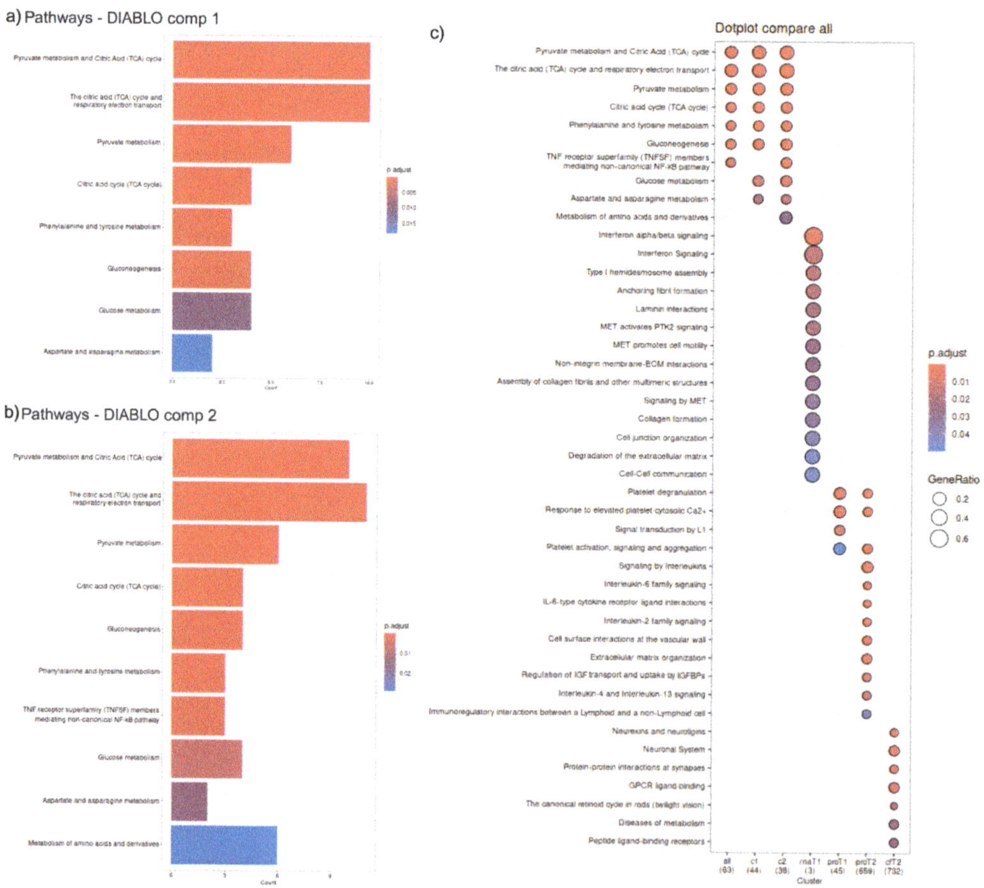

Figure 7. Pathway analysis of DIABLO selected features. (**a**) DIABLO component 1 pathways, ranked according to significance; (**b**) DIABLO component 2 pathways, ranked according to significance; (**c**) Comparison of DIABLO pathways with pathways identified in single omics analysis.

The pathways derived from DIABLO selected features were compared to the pathways identified with single omics analysis (Figure 7c), revealing different profiles of biological pathways. DIABLO component one, made up of the most predictive features of T2, comprised 44 selected features. In comparison, 659 proteins and 732 cfRNA were discriminative of T2 in single omics. DIABLO component two, represents the most distinctive features that can discriminate between T1 and T2, comprising 38 features. In contrast, three cellular RNAs and 45 proteins were discriminative of T1.

2.6. Correlational Analysis of Local (Solid Biopsy) and Systemic (Liquid Biopsy) Signals

The data from local events (skin tissue) [7] and peripheral blood samples were compared and contrasted to investigate the intersections between local and systemic signals. Out of the 1380 DEGs identified in the skin, 1177 (85%) occurred uniquely in the skin (Figure 8). Similarly, cfRNA T1(645 DEGs), cfRNAT2 (824 DEGs), proteomics T1(21 DEGs), and proteomics T2 (861 DEGs) sample sets had DEG/DEPs that were not common in the other (tissue) data set (Figure 8). However, local (tissue) samples shared several common DE molecules with systemic (liquid) samples (blue columns), namely proteomics T2 (52 DEGs), cfRNA T1(47 DEGs), and cfRNA T2 (47 DEGs).

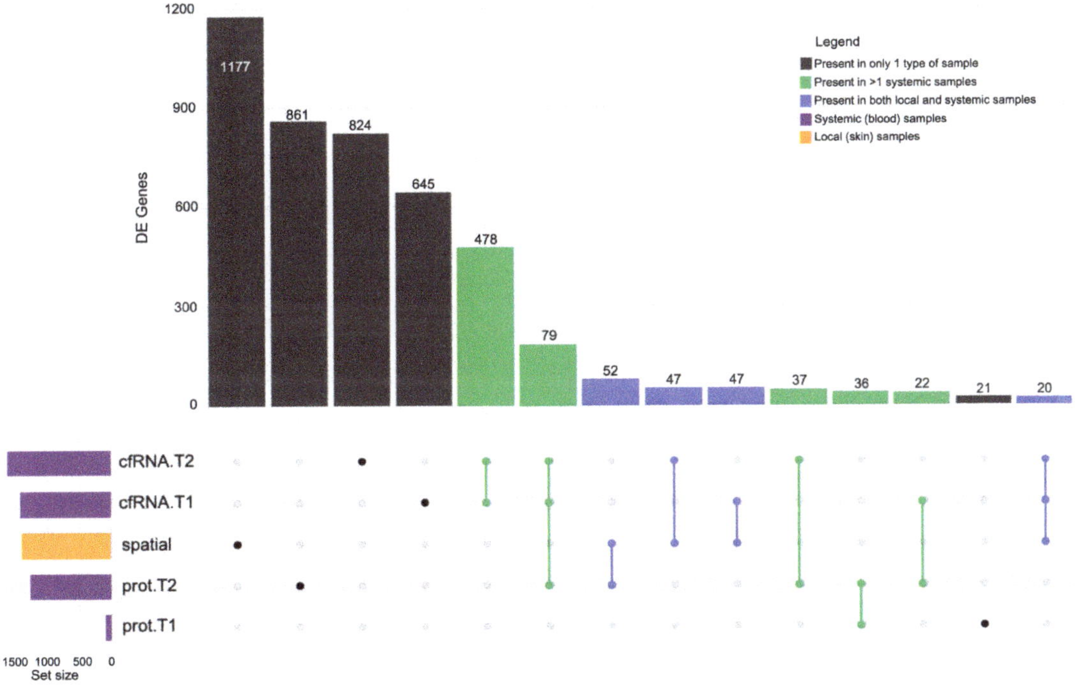

Figure 8. Upset plot comparing the DEGs from local skin (horizontal orange bar) samples set to systemic blood (horizontal purple bars) sample sets, comprising differentially expressed cell-free RNA and proteomics derived from peripheral blood drawn at three time points (T0, T1, and T2). DEGs present only in one type of sample were presented in grey, in more than one systemic sample type in green and in both local and systemic sample types in blue.

The extracellular matrix (ECM) organisation (R-HSA-1474244) pathway experienced an upregulation in the tick-bitten skin on intake at T0 when compared to the contralateral control, comprising 29 DEGs (Figure 1e). Four proteins at T1 and 27 proteins at T2, that were part of the ECM organisation pathway, were upregulated in comparison to

T0. Interestingly, 31 downregulated cfRNA DEGs, that were part of the ECM organisation pathway, were observed at T2. The other consistently identified pathways across time points were "Platelet degranulation" (R-HSA-114608) and "Response to elevated platelet cytosolic Ca2+" (R-HSA-76005) which were observed to comprise 16 upregulated DEGs locally in the skin (Figure 1e), followed by 6 DEPs at T1 (Figure S3c), and 24 DEPs at T2 (Figure S4d) when compared to T0.

Enrichment analysis carried out with GO on the DEGs showed that the heparin-binding (GO ID:0008201) molecular function exhibited an initial upregulation locally (Figure 1d) with 18 DEGs and concurrently within the first-week post-intake, displaying an upregulation in proteomics at T1 (Figure S3b) with a further increase of three DEGs. However, its expression was subsequently downregulated in both T2 samples when compared to T0, namely proteomics (Figure S3a) with 11 DEPs downregulated and cfRNA (Figure S2) with 18 DEGs. Furthermore, 31 DEGs were downregulated in the skin (Figure 1b) which was part of the cytokine receptor binding (GO ID:0005126) molecular function, which experienced a further downregulation of 6 proteins at T1 (Figure S3a). For the cytokine receptor binding function at T2, 23 proteins were downregulated (Figure S4a) while a set of 26 proteins from the same molecular function were upregulated (Figure S4c).

Gene ontology also pinpointed the cellular compartment, collagen-containing extracellular matrix (GO ID:0062023), across data sets with 41 upregulated DEGs from the skin identified (Figure 1d), followed by an additional upregulation of 5 plasma proteins at T1. Localised at the collagen-containing extracellular matrix at T2, were 32 upregulated plasma proteins (Figure S4c) and 48 downregulated cfDEGs (Figure S2). Biological processes observed across sample types include upregulated processes such as negative regulation of peptidase activity (GO ID:0010466), negative regulation of endopeptidase activity (GO ID:0010951), and regulation of peptidase activity (GO ID:0052547) and 15 DEGs, respectively identified in the skin (Table S2). These three biological processes identified went on to experience further upregulation in DEPs at T1 and T2.

3. Discussion

The diverse symptomology experienced by humans following a tick bite may impede the timely diagnosis of TBDs, and this has consequently hindered insight into the early underlying pathophysiological mechanisms [11,12]. This study aimed to address this knowledge gap by correlating local with systemic signals. Spatial transcriptomics of paired skin biopsies [7] enabled the identification of local tick-bite signals on the day of enrolment. As we show here, concurrent (i.e., at the time of skin biopsy) and subsequent, longitudinal sampling of peripheral blood after tick bite allowed systemic signals to be observed and compared to the time of enrolment, at one week, and at three months, offering a glimpse into the trajectory of perturbations of the local biological processes.

Our previous investigation [7] revealed an abundance of local skin signatures that, as expected, exhibited a larger range of DEGs and enriched pathways than systemic blood signatures described in this study. This is consistent with comparable research where tissue biopsies demonstrated superior detection rates of clinically significant perturbations compared to liquid biopsies [13–15]. Nonetheless, liquid biopsies retain their value in scenarios where tissue testing is challenging, such as in cases necessitating sequential sampling [13–15]. Comparisons between tick-bitten and contralateral control skin collected upon enrolment (T0) revealed an upregulation in DEGs that are part of extracellular matrix (ECM) organisation and platelet degranulation pathways [7]. Importantly, as we demonstrated here, these pathways also share a pattern of increased activity in the systemic (blood) proteomics dataset, at one week and three months post tick bite. These signals, in both the local and systemic data, could reflect the active involvement of the platelet degranulation pathway in haemostasis and the ECM organisation pathway in the repair of the cutaneous wound inflicted by tick bites [16].

In the proteomics dataset, the ECM organisation pathway demonstrated a marginal increase in activity from T0 to T2. Conversely, the cfRNA data did not detect any perturbations in the ECM organisation pathway between T0 and T1 but exhibited a marked downregulation of ECM organisation pathway activity at T2, potentially indicating the conclusion of the wound repair process as the participants approached the three-month mark post tick bite. In cases of unresolved tick-associated conditions, it is suggested that these signals may align with dermatological lesions in these patients and may persist [17–19]. Similarly, changes to haemostasis post tick bite, as demonstrated in this study, can also lead to coagulopathies like thrombocytopenia from the over-consumption of coagulation mediators such as platelets [20,21]. Therefore, monitoring of these signals post-treatment could provide prognostic insight into the persistence or resolution of symptoms.

Fluctuating expression patterns of gene ontologies across different types of samples (local and systemic) and over multiple time points were demonstrated in heparin binding and cytokine receptor binding molecular functions. A tick bite not only initiates haemostasis and antimicrobial activities via heparin-binding activities but also introduces pathogen-associated molecular patterns from the pathogens inoculated and the release of damage-associated molecular patterns by injured host cells, both of which are recognised and can bind to host pattern recognition receptors [1,22]. The engagement of these receptors can trigger a cascade of molecular mechanisms within the host, including the activation of inflammasomes and fluctuating cytokine activities [1,23]. These fluctuations in cytokine binding activities often reflect the dynamic and complex interplay of the human response following a tick bite [23,24].

This pilot study suggests that shared signals between local and systemic datasets in tick-bitten individuals exist and, more importantly, can be measured and tracked through serial peripheral blood samples. To further validate the potential of this approach as being capable of detecting local molecular perturbations discernible at a systemic level for TBDs, and to expand this methodology to a broader range of applications, including other vector-borne diseases [5,6], it is now reasonable to design larger follow-up studies around this proof-of-concept platform.

While the experimental design translates into a robust workflow, the sample size in this study is a limitation. Furthermore, power calculations were not undertaken due to the absence of a clinical endpoint or phenotype. Other limitations include the sex ratio of the cohort and the relatively short duration of tick attachment in many cases. The cfRNA investigation conducted here is a relatively new technique [4,25,26]. With strict adherence to transcriptomics analysis guidelines, an abundance of cfRNA DEGs was detected at all follow-up time points when compared to intake at T0. However, downstream analysis of the identified cfRNA DEGs by querying publicly available, curated databases, yielded a paucity of results in terms of enriched pathways or gene ontogeny. This limitation may stem from the recency of gene annotations within these databases and the untargeted list of input genes utilised in this analysis, primarily due to the lack of prior knowledge concerning mRNA associated with TBDs [25–27]. Future iterations of these databases can potentially be employed to analyse historical samples, such as the one generated from this study, to derive additional insights.

Despite these limitations, the results of this study represent the success of a 'proof-of-concept technical pilot' that attests to the feasibility of this liquid biopsy approach. We emphasise that this pilot was not intended for biological interpretations of human tick bite cases. Yet, despite the small sample size, these data demonstrate that identifying local molecular signals common to local tissue and systemic alterations in blood is possible. Conceptually, this study shows that systemic signals, particularly liquid biopsy signals derived from peripheral blood can capture local skin-based tick-associated signals. This provides a novel avenue to detect and investigate tick-associated illnesses that were undetectable with current methods. In the long term, these findings will also need to be compared to the effects of other bites (for example, bites from fleas, spiders, and mosquitoes) in contrast to purely mechanical damage to the skin.

4. Materials and Methods

4.1. Overview

The experimental study design, detailed in Figure 9, included sample processing that adhered to guidelines set by the Murdoch University Human Research Ethics Committee (permit 2019/124) as part of a broader research effort investigating human tick-associated diseases in Australia [28]. Written informed consent was obtained from all participants before enrolment and collection of their samples.

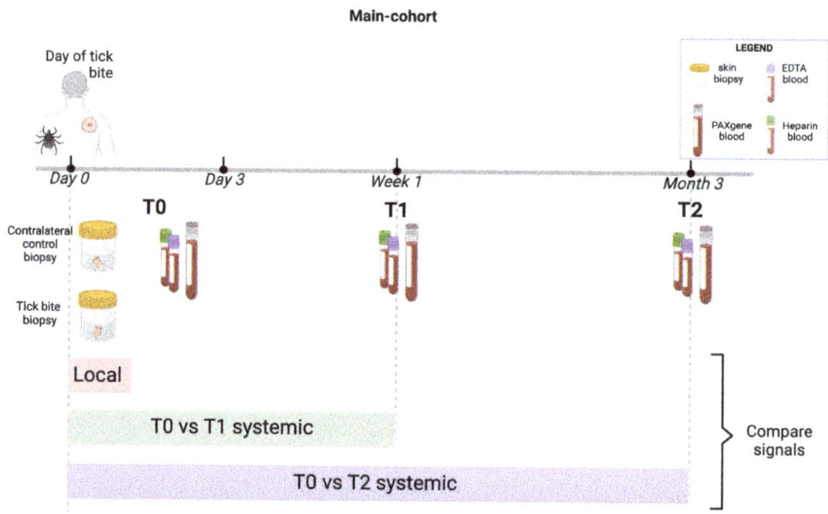

Figure 9. Tick-bitten participant sample collection schedule. On enrolment (T0 = within 72 h (up to Day 3) from tick bite), two skin biopsies and a set of blood samples were collected from study participants. On subsequent follow-up time points (T1 = one-week post enrolment and T2 = three months post enrolment), only venous blood samples were collected. Created with BioRender.com.

4.2. Data Selection Criteria and Extraction

Each participant contributed a set of samples, including the tick responsible for the bite, two skin biopsies (one extracted from the tick bite site, the other from the contralateral (control) site), whole blood samples collected into PAXgene® Blood RNA tubes (PreAnalytiX GmbH; Hombrechtikon, Switzerland), EDTA, and lithium heparin tubes (Figure 9). Inclusion criteria comprised participants (a) who were enrolled and sampled within 72 h of tick bite and (b) were able to provide samples for at least one of the cohort aims (local or systemic studies). Exclusion criteria include (a) children under 18 years of age; (b) current pregnancy; (c) previous diagnosis of chronic fatigue syndrome, fibromyalgia, myalgic encephalomyelitis, Lyme disease (LD) or chronic "LD-like" illness); and patients with coagulopathy or on anticoagulant therapy (except aspirin) [28]. All participants except one (AHP, who declined consent for the contralateral control biopsy) provided the full sets of samples. As such, participant AHP was omitted from the spatial analysis due to a lack of a control sample but was included in the whole blood systemic analysis.

4.3. Participant Details

A set of clinical information pertaining to the selected participants was extracted, encompassing both participant metadata and processing metadata, including quality assurance and control, processing, and normalisation details (Table S1).

4.4. Spatial Transcriptomics

Spatial transcriptomics data for each participant (from Lee et al., 2024) were downloaded for analysis using R v4.3.1 (R Foundation for Statistical Computing, Vienna, Austria) [29]/Bioconductor packages v3.18.0 and NanoString-validated packages (NanoString Technologies, Inc., Seattle, WA, USA) for R, namely GeoMxWorkflows (v1.6.0) [30] and GeoMxTools (v 3.5.0) [31,32] as previously described [7]. DEGs were ascertained by employing a linear mixed-effect model (LMM) with a random slope, where genes with an observed fold change of 1.5 in either direction and adjusted p-value of 0.05 (Benjamini-Hochberg).

4.5. Cell-Free Transcriptomics

Whole blood, collected via venepuncture directly into EDTA Vacutainer tubes (Becton Dickinson, NJ, USA), was centrifuged at 1000× g at room temperature for 15 min to obtain platelet-rich plasma (PRP). The cell-depleted PRP was meticulously collected via an RNase-free micropipette and was centrifuged at 2500× g at room temperature for 15 min to obtain platelet-poor plasma (PPP) and stored at −80 °C until further usage for cfRNA extraction. Haemolysis scoring on the PRP was measured at 414nm using the NanoDrop 2000 (Thermo Fisher Scientific; Waltham, MA, USA) and cfRNA was extracted using the Qiagen miRNeasy Serum/Plasma Advanced Kit (Qiagen, Hilden, Germany) in accordance with manufacturer's instructions (Figure S1). Extracted cfRNA was checked for genomic DNA (gDNA) contamination with SsoAdvanced Universal SYBR Green Supermix (BioRad, Hercules, CA, USA) via qPCR on C1000 TouchTM Thermal Cycler (Bio-Rad Laboratories, Inc., Hercules, CA, USA) and CFX Real-Time PCR Detection Systems (Bio-Rad Laboratories, Inc., Hercules, CA, USA), and RNA quality was assessed via electrophoresis using the Eukaryote Total RNA Pico assay on the 2100 bioanalyzer (Agilent Technologies Inc., Santa Clara, CA, USA). The Illumina RNA Prep with Enrichment (L) Tagmentation kit (Illumina; San Diego, CA, USA) was used to construct a library with index adapters from IDT for Illumina DNA/RNA Index Set A (Illumina; San Diego, CA, USA) and exome panel with enrichment oligos (Illumina; San Diego, CA, USA). High output (50–60 Gb) RNA sequencing of 75 bp paired-end reads was carried out on the Illumina NextSeq 500/550 platform (Illumina; San Diego, CA, USA) at Norgen BioTek Corp (Figure S1). Initial quality control checks were conducted using FastQC v0.12.0 (Babraham Institute, Cambridge, UK) and MultiQC v1.13 (Seqera Labs, Barcelona, Spain) [33]. To facilitate transcript alignment, STAR v2.7.10b (Cold Spring Harbor Laboratory, Cold Spring Harbor, NY, USA) [34] was used to create a reference genome utilising the hg38 human genome (Ensembl GRCh38.86) [35] (Figure S1). Subsequently, the untrimmed FASTQ sequences, including adapters and indexes, were aligned to hg38 transcripts using STAR (Figure S1). The resulting binary alignment map (BAM) files were indexed using Samtools 1.16.1 (Genome Research Ltd., Hinxton, UK), yielding sequence alignment map (SAM) files [36]. Read counts were then computed from the SAM files using HTseq v2.0.2 [37], leading to the generation of raw RNA count files.

Differential gene expression analysis was conducted using the DESeq2 version 1.38.3 [38] package within R v4.3.1 [29], Bioconductor packages v3.18.0 (Figure S1). Genes with a normalised count of under five in less than three samples as well as globin genes (specifically "ENSG00000206172", "ENSG00000188536", and "ENSG00000244734") were bioinformatically excluded [39]. DEGs were defined as genes exhibiting a fold change of at least 1.5 in either direction using the Wald test and an adjusted p-value of 0.05 (Benjamini-Hochberg). Pathway analysis was carried out using ReactomePA (v1.44.0) [40].

4.6. Cellular Transcriptomics

Whole blood samples collected via venepuncture directly into PAXgene blood RNA tubes (PreAnalytiX GmbH; Hombrechtikon, Switzerland) were inverted 10 times immediately after collection and left to incubate at room temperature for at least 2 h before being stored at −80 °C until RNA extraction. Cellular RNA was manually extracted from whole blood collected in PAXgene blood RNA tubes using a PAXgene blood miRNA kit (PreAnalytiX GmbH; Hombrechtikon, Switzerland) following the manufacturer's instructions (Figure S1). Total cellular RNA was quantified on the NanoDrop 2000 (Thermo Fisher Scientific; Waltham, MA, USA) and the RNA quality was assessed via electrophoresis using the LabChip GX nucleic acid analyser (PerkinElmer; Waltham, MA, USA). The pure RNA samples obtained were used to generate libraries using the Illumina stranded mRNA Library Prep (Illumina; San Diego, CA, USA), perform qPCR quantification for further quality control and perform RNA sequencing of 150 bp paired-end reads at a sequencing depth of 50 million on the Illumina NovaSeq 6000 platform (Illumina; San Diego, CA, USA) at Australian Genome Research Facility. The raw FASTQ data were transferred via RSYNC into a Linux virtual machine for pre-processing. FASTQ sequence reads with more than one file were concatenated and initial quality control was performed using FastQC v0.12.0 and MultiQC v1.13 [33]. STAR v2.7.10b was used to generate a reference genome with Ensembl GRCh38.86 [35] and align the untrimmed FASTQ sequences (inclusive of adapters and indexes) to hg38 transcripts. The BAM files generated by STAR were indexed by Samtools 1.16.1 and SAM files [36]. HTseq v2.0.2 [37] was used to perform read counts on the SAM files and generate raw RNA count files. Differential gene expression analysis was performed as described above.

4.7. Plasma Proteomics

Whole blood samples collected via venepuncture directly into lithium heparin Vacutainer® tubes (Becton Dickinson, Franklin Lakes, NJ, USA) were centrifuged at 1000× g at room temperature for 10 min. The plasma collected via a sterile micropipette (while ensuring the red cells remain untouched) was aliquoted and stored at −80 °C until further analysis. The plasma samples were then analysed in one batch on the SomaScan® platform by SomaLogic Inc. (Boulder, CO, USA) (Figure S1). Single-stranded DNA aptamers, known as SOMAmers®, were used to target approximately 7000 proteins in each plasma sample and the measurement of these SOMAmers® was performed using microarray for relative protein quantification. The raw data was normalised against hybridisation controls and pooled calibrator replicates and analysis was carried out using R v4.3.1 [29] with the Differential expression testing with linear mixed models for repeated measures (Dream) [41] using a LMM. DEPs were determined as proteins with an observed fold change of 1.5 in either direction with an adjusted p-value of 0.05 (Benjamini-Hochberg).

4.8. Plasma Metabolomics

Whole blood, collected via venepuncture directly into lithium heparin Vacutainer® tubes (Becton Dickinson, Franklin Lakes, NJ, USA) was centrifuged at 1000× g at room temperature for 10 min. The plasma layer free of blood cells was collected as described above, aliquoted, and stored at −80 °C awaiting further analysis. The plasma samples were then analysed in one batch on the untargeted metabolomics platform at Metabolon Inc. (Durham, NC, USA) (Figure S1). The extraction of metabolites from the plasma samples was carried out via methanol precipitation on the automated MicroLab STAR® system (Hamilton Company, Reno, NV, USA) and divided into five aliquots to be run on four separate methods. The HD4 Waters ACQUITY method of ultra-performance liquid chromatography (UPLC) (Waters Corporation, Milford, MA, USA) was carried out in tandem with a Q-Exactive mass spectrometry (MS/MS) (Thermo Fisher Scientific, Waltham, MA, USA) with a heated electrospray ionization (HESI) coupled to an Orbitrap mass analyser (Thermo Fisher Scientific, Waltham, MA, USA) set at 35,000 mass resolution to identify and provide a relative quantification of approximately 5400 metabolites in all methods. Aliquot one was run on a reverse-phase (RP) UPLC-MS/MS coupled to a positive ion mode HESI optimised for hydrophilic compounds, aliquot two was run on RP/UPLC-MS/MS coupled to a positive ion mode HESI optimised for hydrophobic compounds, aliquot three was run on RP/UPLC-MS/MS coupled to a negative ion mode HESI, aliquot four was run on HILIC/UPLC-MS/MS coupled to a negative ion mode HESI and aliquot five was reserved as a spare. The raw data was batch-normalised, imputed and log-transformed for analysis on R v4.3.1 [29]. Metabolites were considered differentially expressed based on an observed change of 1.5-fold in either direction using a linear model with an adjusted *p*-value of 0.05 (Benjamini-Hochberg).

4.9. Whole Blood Cell DNA Methylation

Whole blood, collected via venepuncture directly into lithium heparin Vacutainer® tubes (Becton Dickinson, Franklin Lakes, NJ, USA), was centrifuged at 1000× *g* at room temperature for 10 min. The plasma was removed using a sterile micropipette, leaving the packed red cell pellet and the white cell buffy coat untouched. The cell pellet was then resuspended and stored at −80 °C until gDNA extraction. The gDNA was extracted using the automated Chemagic™ 360 instrument (Revvity, Waltham, MA, USA) and Chemagic™ DNA Blood 400 Kit H96 (cat# CMG-1901, Revvity, Waltham, Massachusetts, USA), in accordance with manufacturer instructions (Figure S1). The concentrations of the purified gDNA samples were then quantified using the Qubit™ dsDNA High Sensitivity kit and NanoDrop 2000 (both from Thermo Fisher Scientific; Waltham, MA, USA) as part of a quality control step. Methylation assay was conducted using the Illumina MethylationEPIC v2.0 BeadChip kit (Illumina; San Diego, CA, USA) to interrogate the methylation of more than 935,000 CpG sites of the human methylome. This assay was performed by the Australian Genome Research Facility in Melbourne. Minfi v3.17 [42] was used to pre-process the raw .iDAT files, incorporating control probes to assess sample quality. This process included between-array normalisation, evaluation of the P-detection call rate and the removal of probes exhibiting off-target effects. Analysis was carried out using R v4.3.1 [29] with Dream [41] using a LMM. Differentially methylated genes (DMGs) were determined with an observed fold change of 1.5 in either direction with an adjusted *p*-value of 0.05 (Benjamini–Hochberg).

4.10. Data Integration

The multi-variate DIABLO approach, a module in the MixOmics R package [43], was used to provide insight into the progression of tick bites from five highly-dimensional, discrete datasets, namely, cellular transcriptomics, cell-free transcriptomics, proteomics, metabolomics, and whole blood DNA methylation. The data sets were processed in R v4.3.1 [29] to remove zero values and low counts (defined as cumulative methylation under 0.1 and a cumulative count of under 10 across all samples for all other omics). A multinomial generalised linear model was used to perform LASSO regression for feature selection of the most variable features. The final processed datasets were normalised (z-score normalisation), centred and log-transformed to obtain input matrices for integration [44,45].

Sparse projection to latent structures-determinant analysis (sPLS-DA) [44] was used to perform the integration of input matrices and build components with DIABLO. The time points (T0, T1, and T2) were set as response variables, and DIABLO was used to maximise the covariance between the input matrices and response variables to identify the main determinants to differentiate time points. A basic DIABLO model was constructed with a design matrix with a correlation of 0.1 to identify correlated variables across data sets for 10 components. Hyperparametric performance tuning was carried out to ascertain the optimal number of components (based on the classification error rates after 5-fold of 20-*n* repeats) and features (based on 5-fold of 2-*n* repeats cross-validation) to run the final DIABLO model. The performance of the final DIABLO model was assessed with a diagnostic plot for each relevant component and the correlation as well as clustering quality of the selected features from each dataset was assessed with DIABLO (Supplemental Materials). Pathway analysis on the features selected by the DIABLO model was performed by ReactomePA (v1.44.0) [40] and the *p*-values adjusted with the Benjamini–Hochberg method [46].

4.11. Correlation Analysis and Visualisation

All differentially expressed proteins or transcripts were mapped to common gene identifiers using biomaRt (v 2.58.0) [47] and the genome-wide human annotation org.Hs.eg.db (v3.18.0) on Bioconductor [48]. Enrichment analysis of DEGs was carried out with gene ontogeny (GO) [49,50] to categorise subontologies of biological domains using clusterProfiler (v.4.10.0) [51] and enrichplot (v. 1.22.0) [52]. Over-representation analysis (ORA) to identify pathways was carried out using ReactomePA (v1.44.0) [40].

Supplementary Materials: The following supporting information can be downloaded at: https://www.mdpi.com/article/10.3390/ijms26041520/s1.

Author Contributions: Conceptualization, P.J.I., R.B.-O., T.R.K. and C.L.O.; Data curation, W.L.; Formal analysis, W.L., A.H.-Y.L., D.M. and N.K.; Funding acquisition, J.S., M.B., P.J.I., R.A.H., S.G. and C.L.O.; Investigation, W.L., A.D.B., A.H.-Y.L., N.T.H. and P.K.; Methodology, W.L., A.H.-Y.L., D.M., N.T.H., P.J.I., P.K., R.A.H., R.B.-O., S.G., T.R.K. and C.L.O.; Project administration, A.D.B., M.L. and P.J.I.; Supervision, A.D.B., A.H.-Y.L., A.C., M.B., P.J.I., T.R.K. and C.L.O.; Writing—original draft, W.L.; Writing—review & editing, W.L., A.D.B., A.H.-Y.L., A.C., D.M., J.S., M.L., M.B., N.T.H., N.K., P.S., P.J.I., P.K., R.A.H., R.B.-O., S.G., T.R.K. and C.L.O. All authors have read and agreed to the published version of the manuscript.

Funding: This paper is supported by an Australian Government Research Training Program (RTP) Scholarship to W.L. and funding support from NHMRC GNT1169949 to P.J.I., S.G., C.L.O. & M.B.

Institutional Review Board Statement: The Murdoch University Human Research Ethics Committee approved this research under permit number 2019/124.

Informed Consent Statement: Informed consent was obtained from all subjects involved in the study.

Data Availability Statement: All data are publicly available. The transcriptomics data presented in this publication were submitted to the NCBI Gene Expression Omnibus under accession number GSE286962. All other datasets including cfRNA, proteomics, metabolomics, and epigenetics datasets presented in this publication have been archived on FigShare under DOI: https://doi.org/10.6084/m9.figshare.25772382.v1.

Acknowledgments: We acknowledge Amanda J. Cox and Nicholas P. West from the School of Pharmacy and Medical Sciences and Menzies Health Institute at Griffith University for their invaluable guidance and technical expertise in spatial biology.

Conflicts of Interest: Rym Ben Othman's current affiliation is at RAN BioLinks and she has no commercial interest to declare with regards to this work. All the other authors declare no competing interests.

References

1. Lee, W.; Barbosa, A.D.; Irwin, P.J.; Currie, A.; Kollmann, T.R.; Beaman, M.; Lee, A.H.; Oskam, C.L. A Systems Biology Approach to Better Understand Human Tick-Borne Diseases. *Trends Parasitol.* **2022**, *39*, 53–69. [CrossRef] [PubMed]
2. Shor, S.; Green, C.; Szantyr, B.; Phillips, S.; Liegner, K.; Burrascano, J.; Bransfield, R.; Maloney, E.L. Chronic Lyme Disease: An Evidence-Based Definition by the ILADS Working Group. *Antibiotics* **2019**, *8*, 269. [CrossRef] [PubMed]
3. Szilágyi, M.; Pös, O.; Márton, É.; Buglyó, G.; Soltész, B.; Keserű, J.; Penyige, A.; Szemes, T.; Nagy, B. Circulating Cell-Free Nucleic Acids: Main Characteristics and Clinical Application. *Int. J. Mol. Sci.* **2020**, *21*, 6827. [CrossRef] [PubMed]
4. Cabús, L.; Lagarde, J.; Curado, J.; Lizano, E.; Pérez-Boza, J. Current Challenges and Best Practices for Cell-Free Long RNA Biomarker Discovery. *Biomark. Res.* **2022**, *10*, 62. [CrossRef]
5. Alexandrou, G.; Mantikas, K.-T.; Allsopp, R.; Yapeter, C.A.; Jahin, M.; Melnick, T.; Ali, S.; Coombes, R.C.; Toumazou, C.; Shaw, J.A.; et al. The Evolution of Affordable Technologies in Liquid Biopsy Diagnostics: The Key to Clinical Implementation. *Cancers* **2023**, *15*, 5434. [CrossRef]
6. Alix-Panabières, C.; Marchetti, D.; Lang, J.E. Liquid Biopsy: From Concept to Clinical Application. *Sci. Rep.* **2023**, *13*, 21685. [CrossRef]
7. Lee, W.; Ben-Othman, R.; Skut, P.; Lee, A.H.-Y.; Barbosa, A.D.; Beaman, M.; Currie, A.; Harvey, N.T.; Kumarasinghe, P.; Hall, R.A.; et al. Molecular Analysis of Human Tick-Bitten Skin Yields Signatures Associated with Distinct Spatial and Temporal Trajectories—A Proof-of-Concept Study. *Heliyon* **2024**, *10*, e33600. [CrossRef]
8. Chen, Z.; Li, C.; Zhou, Y.; Yao, Y.; Liu, J.; Wu, M.; Su, J. Liquid Biopsies for Cancer: From Bench to Clinic. *MedComm* **2023**, *4*, e329. [CrossRef]
9. Sayers, E.W.; Bolton, E.E.; Brister, J.R.; Canese, K.; Chan, J.; Comeau, D.C.; Connor, R.; Funk, K.; Kelly, C.; Kim, S.; et al. Database Resources of the National Center for Biotechnology Information. *Nucleic Acids Res.* **2021**, *50*, D20–D26. [CrossRef]
10. Barker, S.C.; Barker, D. Ticks in Australia: Endemics; Exotics; Which Ticks Bite Humans? *Microbiol. Aust.* **2018**, *39*, 194–199. [CrossRef]
11. Lantos, P.M.; Rumbaugh, J.; Bockenstedt, L.K.; Falck-Ytter, Y.T.; Aguero-Rosenfeld, M.E.; Auwaerter, P.G.; Baldwin, K.; Bannuru, R.R.; Belani, K.K.; Bowie, W.R.; et al. Clinical Practice Guidelines by the Infectious Diseases Society of America (IDSA), American Academy of Neurology (AAN), and American College of Rheumatology (ACR): 2020 Guidelines for the Prevention, Diagnosis and Treatment of Lyme Disease. *Clin. Infect. Dis.* **2021**, *72*, e1–e48. [CrossRef] [PubMed]
12. Krause, P.J.; Auwaerter, P.G.; Bannuru, R.R.; Branda, J.A.; Falck-Ytter, Y.T.; Lantos, P.M.; Lavergne, V.; Meissner, H.C.; Osani, M.C.; Rips, J.G.; et al. Clinical Practice Guidelines by the Infectious Diseases Society of America (IDSA): 2020 Guideline on Diagnosis and Management of Babesiosis. *Clin. Infect. Dis.* **2021**, *72*, e49–e64. [CrossRef] [PubMed]
13. Chae, Y.K.; Davis, A.A.; Jain, S.; Santa-Maria, C.; Flaum, L.; Beaubier, N.; Platanias, L.C.; Gradishar, W.; Giles, F.J.; Cristofanilli, M. Concordance of Genomic Alterations by Next-Generation Sequencing in Tumor Tissue versus Circulating Tumor DNA in Breast Cancer. *Mol. Cancer Ther.* **2017**, *16*, 1412–1420. [CrossRef] [PubMed]
14. Lin, L.H.; Allison, D.H.R.; Feng, Y.; Jour, G.; Park, K.; Zhou, F.; Moreira, A.L.; Shen, G.; Feng, X.; Sabari, J.; et al. Comparison of Solid Tissue Sequencing and Liquid Biopsy Accuracy in Identification of Clinically Relevant Gene Mutations and Rearrangements in Lung Adenocarcinomas. *Mod. Pathol.* **2021**, *34*, 2168–2174. [CrossRef]
15. Liu, M.C.; MacKay, M.; Kase, M.; Piwowarczyk, A.; Lo, C.; Schaeffer, J.; Finkle, J.D.; Mason, C.E.; Beaubier, N.; Blackwell, K.L.; et al. Longitudinal Shifts of Solid Tumor and Liquid Biopsy Sequencing Concordance in Metastatic Breast Cancer. *JCO Precis. Oncol.* **2022**, e2100321. [CrossRef]
16. Potekaev, N.N.; Borzykh, O.B.; Medvedev, G.V.; Pushkin, D.V.; Petrova, M.M.; Petrov, A.V.; Dmitrenko, D.V.; Karpova, E.I.; Demina, O.M.; Shnayder, N.A. The Role of Extracellular Matrix in Skin Wound Healing. *J. Clin. Med.* **2021**, *10*, 5947. [CrossRef]
17. Moniuszko-Malinowska, A.; Czupryna, P.; Dunaj, J.; Pancewicz, S.; Garkowski, A.; Kondrusik, M.; Grygorczuk, S.; Zajkowska, J. Acrodermatitis Chronica Atrophicans: Various Faces of the Late Form of Lyme Borreliosis. *Postep. Dermatol. Alergol.* **2018**, *35*, 490–494. [CrossRef]
18. Gindl, A.; Schötta, A.-M.; Berent, S.; Markowicz, M.; Stockinger, H.; Thalhammer, F.; Stary, G.; Strobl, J. Persistent Lyme Disease with Cutaneous *Borrelia* Biofilm Formation. *Br. J. Dermatol.* **2022**, *186*, 1041–1043. [CrossRef]
19. Talbot, N.C.; Spillers, N.J.; Luther, P.; Flanagan, C.; Soileau, L.G.; Ahmadzadeh, S.; Viswanath, O.; Varrassi, G.; Shekoohi, S.; Cornett, E.M.; et al. Lyme Disease and Post-Treatment Lyme Disease Syndrome: Current and Developing Treatment Options. *Cureus* **2023**, *15*, e43112. [CrossRef]
20. Mans, B.J.; Neitz, A.W.H. Adaptation of Ticks to a Blood-Feeding Environment: Evolution from a Functional Perspective. *Insect Biochem. Mol. Biol.* **2004**, *34*, 1–17. [CrossRef]
21. Papamanoli, A.; Kaplun, O.; Lobo, Z.; LeMaitre, B.; Romano, C.L.; Campbell, S.R.; Psevdos, G. Tick-Borne Disease–Associated Thrombocytopenia: Epidemiological Description Among United States Veterans and Tick Surveillance Data From Suffolk County, Long Island, New York. *Infect. Dis. Clin. Pract.* **2022**, *30*, e1090. [CrossRef]

22. Lin, Y.-P.; Yu, Y.; Marcinkiewicz, A.L.; Lederman, P.; Hart, T.M.; Zhang, F.; Linhardt, R.J. Non-Anticoagulant Heparin as a Pre-Exposure Prophylaxis Prevents Lyme Disease Infection. *ACS Infect. Dis.* **2020**, *6*, 503–514. [CrossRef] [PubMed]
23. Torina, A.; Villari, S.; Blanda, V.; Vullo, S.; La Manna, M.P.; Shekarkar Azgomi, M.; Di Liberto, D.; de la Fuente, J.; Sireci, G. Innate Immune Response to Tick-Borne Pathogens: Cellular and Molecular Mechanisms Induced in the Hosts. *Int. J. Mol. Sci.* **2020**, *21*, 5437. [CrossRef]
24. Baquer, F.; Jaulhac, B.; Barthel, C.; Paz, M.; Wolfgramm, J.; Müller, A.; Boulanger, N.; Grillon, A. Skin Microbiota Secretomes Modulate Cutaneous Innate Immunity against *Borrelia Burgdorferi* s.s. *Sci. Rep.* **2023**, *13*, 16393. [CrossRef] [PubMed]
25. Larson, M.H.; Pan, W.; Kim, H.J.; Mauntz, R.E.; Stuart, S.M.; Pimentel, M.; Zhou, Y.; Knudsgaard, P.; Demas, V.; Aravanis, A.M.; et al. A Comprehensive Characterization of the Cell-Free Transcriptome Reveals Tissue- and Subtype-Specific Biomarkers for Cancer Detection. *Nat. Commun.* **2021**, *12*, 2357. [CrossRef] [PubMed]
26. Wang, J.; Huang, J.; Hu, Y.; Guo, Q.; Zhang, S.; Tian, J.; Niu, Y.; Ji, L.; Xu, Y.; Tang, P.; et al. Terminal Modifications Independent Cell-Free RNA Sequencing Enables Sensitive Early Cancer Detection and Classification. *Nat. Commun.* **2024**, *15*, 156. [CrossRef]
27. Wadi, L.; Meyer, M.; Weiser, J.; Stein, L.D.; Reimand, J. Impact of Outdated Gene Annotations on Pathway Enrichment Analysis. *Nat. Methods* **2016**, *13*, 705–706. [CrossRef]
28. Barbosa, A.D.; Long, M.; Lee, W.; Austen, J.M.; Cunneen, M.; Ratchford, A.; Burns, B.; Kumarasinghe, P.; Ben-Othman, R.; Kollmann, T.R.; et al. The Troublesome Ticks Research Protocol: Developing a Comprehensive, Multidiscipline Research Plan for Investigating Human Tick-Associated Disease in Australia. *Pathogens* **2022**, *11*, 1290. [CrossRef]
29. R Core Team. *R: A Language and Environment for Statistical Computing*; R Foundation for Statistical Computing: Vienna, Austria, 2020.
30. Reeves, J.; Divakar, P.; Ortogero, N.; Griswold, M.; Yang, Z.; Zimmerman, S.; Vitancol, R.; David, H. *GeoMxWorkflows: GeoMx Digital Spatial Profiler (DSP) Data Analysis Workflows*, R Package version 1.6.0; NanoString Technologies: Seattle, WA, USA, 2023.
31. Aboyoun, P.; Ortogero, N.; Yang, Z. *NanoStringNCTools: NanoString nCounter Tools*; NanoString Technologies: Seattle, WA, USA, 2023.
32. Ortogero, N.; Yang, Z.; Vitancol, R.; Griswold, M.; Henderson, D. *GeomxTools: NanoString GeoMx Tools*; NanoString Technologies: Seattle, WA, USA, 2023.
33. Ewels, P.; Magnusson, M.; Lundin, S.; Käller, M. MultiQC: Summarize Analysis Results for Multiple Tools and Samples in a Single Report. *Bioinformatics* **2016**, *32*, 3047–3048. [CrossRef]
34. Dobin, A.; Davis, C.A.; Schlesinger, F.; Drenkow, J.; Zaleski, C.; Jha, S.; Batut, P.; Chaisson, M.; Gingeras, T.R. STAR: Ultrafast Universal RNA-Seq Aligner. *Bioinformatics* **2013**, *29*, 15–21. [CrossRef]
35. Schneider, V.A.; Graves-Lindsay, T.; Howe, K.; Bouk, N.; Chen, H.-C.; Kitts, P.A.; Murphy, T.D.; Pruitt, K.D.; Thibaud-Nissen, F.; Albracht, D.; et al. Evaluation of GRCh38 and de Novo Haploid Genome Assemblies Demonstrates the Enduring Quality of the Reference Assembly. *Genome Res.* **2017**, *27*, 849–864. [CrossRef] [PubMed]
36. Li, H.; Handsaker, B.; Wysoker, A.; Fennell, T.; Ruan, J.; Homer, N.; Marth, G.; Abecasis, G.; Durbin, R. 1000 Genome Project Data Processing Subgroup The Sequence Alignment/Map Format and SAMtools. *Bioinformatics* **2009**, *25*, 2078–2079. [CrossRef] [PubMed]
37. Putri, G.H.; Anders, S.; Pyl, P.T.; Pimanda, J.E.; Zanini, F. Analysing High-Throughput Sequencing Data in Python with HTSeq 2.0. *Bioinformatics* **2022**, *38*, 2943–2945. [CrossRef] [PubMed]
38. Love, M.I.; Huber, W.; Anders, S. Moderated Estimation of Fold Change and Dispersion for RNA-Seq Data with DESeq2. *Genome Biol.* **2014**, *15*, 550. [CrossRef]
39. Lee, A.H.; Shannon, C.P.; Amenyogbe, N.; Bennike, T.B.; Diray-Arce, J.; Idoko, O.T.; Gill, E.E.; Ben-Othman, R.; Pomat, W.S.; van Haren, S.D.; et al. Dynamic Molecular Changes during the First Week of Human Life Follow a Robust Developmental Trajectory. *Nat. Commun.* **2019**, *10*, 1092. [CrossRef]
40. Yu, G.; Petyuk, V. *ReactomePA: Reactome Pathway Analysis*; Bioconductor: R Foundation for Statistical Computing: Vienna, Austria, 2023.
41. Hoffman, G.E.; Roussos, P. Dream: Powerful Differential Expression Analysis for Repeated Measures Designs. *Bioinformatics* **2021**, *37*, 192–201. [CrossRef]
42. Hansen, K.D.; Aryee, M.; Irizarry, R.A.; Jaffe, A.E.; Maksimovic, J.; Houseman, E.A.; Fortin, J.-P.; Triche, T.; Andrews, S.V.; Hickey, P.F. *Minfi: Analyze Illumina Infinium DNA Methylation Arrays*; Bioconductor: R Foundation for Statistical Computing: Vienna, Austria, 2023.
43. Singh, A.; Shannon, C.P.; Gautier, B.; Rohart, F.; Vacher, M.; Tebbutt, S.J.; Lê Cao, K.-A. DIABLO: An Integrative Approach for Identifying Key Molecular Drivers from Multi-Omics Assays. *Bioinformatics* **2019**, *35*, 3055–3062. [CrossRef]
44. Lê Cao, K.-A.; Boitard, S.; Besse, P. Sparse PLS Discriminant Analysis: Biologically Relevant Feature Selection and Graphical Displays for Multiclass Problems. *BMC Bioinform.* **2011**, *12*, 253. [CrossRef]
45. Qin, S.; Kim, J.; Arafat, D.; Gibson, G. Effect of Normalization on Statistical and Biological Interpretation of Gene Expression Profiles. *Front. Genet.* **2013**, *3*, 160. [CrossRef]

46. Benjamini, Y.; Hochberg, Y. Controlling the False Discovery Rate: A Practical and Powerful Approach to Multiple Testing. *J. R. Stat. Soc. Ser. B* **1995**, *57*, 289–300. [CrossRef]
47. Durinck, S.; Huber, W.; Davis, S.; Pepin, F.; Buffalo, V.S.; Smith, M. *biomaRt: Interface to BioMart Databases (i.e., Ensembl)*; Bioconductor: R Foundation for Statistical Computing: Vienna, Austria, 2023.
48. Carlson, M. *org.Hs.eg.db: Genome Wide Annotation for Human*; Bioconductor: R Foundation for Statistical Computing: Vienna, Austria, 2019.
49. Ashburner, M.; Ball, C.A.; Blake, J.A.; Botstein, D.; Butler, H.; Cherry, J.M.; Davis, A.P.; Dolinski, K.; Dwight, S.S.; Eppig, J.T.; et al. Gene Ontology: Tool for the Unification of Biology. *Nat. Genet.* **2000**, *25*, 25–29. [CrossRef] [PubMed]
50. The Gene Ontology Consortium; Aleksander, S.A.; Balhoff, J.; Carbon, S.; Cherry, J.M.; Drabkin, H.J.; Ebert, D.; Feuermann, M.; Gaudet, P.; Harris, N.L.; et al. The Gene Ontology Knowledgebase in 2023. *Genetics* **2023**, *224*, iyad031. [CrossRef] [PubMed]
51. Wu, T.; Hu, E.; Xu, S.; Chen, M.; Guo, P.; Dai, Z.; Feng, T.; Zhou, L.; Tang, W.; Zhan, L.; et al. clusterProfiler 4.0: A Universal Enrichment Tool for Interpreting Omics Data. *Innovation* **2021**, *2*, 100141. [CrossRef] [PubMed]
52. Yu, G.; Hu, E.; Gao, C.-H. *Enrichplot: Visualization of Functional Enrichment Result*; Bioconductor: R Foundation for Statistical Computing: Vienna, Austria, 2023.

MDPI AG
Grosspeteranlage 5
4052 Basel
Switzerland
Tel.: +41 61 683 77 34

International Journal of Molecular Sciences Editorial Office
E-mail: ijms@mdpi.com
www.mdpi.com/journal/ijms

www.ingramcontent.com/pod-product-compliance
Lightning Source LLC
Chambersburg PA
CBHW061454210926
51850CB00008B/1717
* 9 7 8 3 7 2 5 8 7 3 6 4 7 *